Critical Role of Animal Science Research in Food Security and Sustainability

Committee on Considerations for the Future of Animal Science Research

Science and Technology for Sustainability Program

Policy and Global Affairs

Board on Agriculture and Natural Resources

Division on Earth and Life Studies

THE NATIONAL ACADEMIES PRESS
Washington, D.C.
www.nap.edu

NATIONAL RESEARCH COUNCIL
OF THE NATIONAL ACADEMIES

THE NATIONAL ACADEMIES PRESS 500 Fifth Street, NW Washington, DC 20001

NOTICE: The project that is the subject of this report was approved by the Governing Board of the National Research Council, whose members are drawn from the councils of the National Academy of Sciences, the National Academy of Engineering, and the Institute of Medicine. The members of the committee responsible for the report were chosen for their special competences and with regard for appropriate balance.

This study was supported by a grant from the Association of American Veterinary Medical Colleges, Bill & Melinda Gates Foundation under award number OPP1097068, Innovation Center for U.S. Dairy, National Cattlemen's Beef Association, National Pork Board, Tyson Foods, Inc., the U.S. Department of Agriculture under award number 59-0208-2-169, and the U.S. Poultry & Egg Association. Any opinions, findings, conclusions, or recommendations expressed in this publication are those of the author(s) and do not necessarily reflect the views of the organizations or agencies that provided support for the project.

International Standard Book Number 13: 978-0-309-31644-6
International Standard Book Number 10: 0-309-31644-8
Library of Congress Control Number: 2015935247

Additional copies of this report are available from the National Academies Press, 500 Fifth Street, NW, Keck 360, Washington, DC 20001; (800) 624-6242 or (202) 334-3313; http://www.nap.edu.

Copyright 2015 by the National Academy of Sciences. All rights reserved.

Printed in the United States of America.

THE NATIONAL ACADEMIES
Advisers to the Nation on Science, Engineering, and Medicine

The National Academy of Sciences is a private, nonprofit, self-perpetuating society of distinguished scholars engaged in scientific and engineering research, dedicated to the furtherance of science and technology and to their use for the general welfare. Upon the authority of the charter granted to it by the Congress in 1863, the Academy has a mandate that requires it to advise the federal government on scientific and technical matters. Dr. Ralph J. Cicerone is president of the National Academy of Sciences.

The National Academy of Engineering was established in 1964, under the charter of the National Academy of Sciences, as a parallel organization of outstanding engineers. It is autonomous in its administration and in the selection of its members, sharing with the National Academy of Sciences the responsibility for advising the federal government. The National Academy of Engineering also sponsors engineering programs aimed at meeting national needs, encourages education and research, and recognizes the superior achievements of engineers. Dr. C. D. Mote, Jr., is president of the National Academy of Engineering.

The Institute of Medicine was established in 1970 by the National Academy of Sciences to secure the services of eminent members of appropriate professions in the examination of policy matters pertaining to the health of the public. The Institute acts under the responsibility given to the National Academy of Sciences by its congressional charter to be an adviser to the federal government and, upon its own initiative, to identify issues of medical care, research, and education. Dr. Victor J. Dzau is president of the Institute of Medicine.

The National Research Council was organized by the National Academy of Sciences in 1916 to associate the broad community of science and technology with the Academy's purposes of furthering knowledge and advising the federal government. Functioning in accordance with general policies determined by the Academy, the Council has become the principal operating agency of both the National Academy of Sciences and the National Academy of Engineering in providing services to the government, the public, and the scientific and engineering communities. The Council is administered jointly by both Academies and the Institute of Medicine. Dr. Ralph J. Cicerone and Dr. C. D. Mote, Jr., are chair and vice chair, respectively, of the National Research Council.

www.national-academies.org

Committee on Considerations for the Future of Animal Science Research

Bernard D. Goldstein (Chair) (IOM), Professor Emeritus, Department of Environmental and Occupational Health, University of Pittsburgh Graduate School of Public Health
Louis D'Abramo, William L. Giles Distinguished Professor of Wildlife, Fisheries and Aquaculture, Mississippi State University
Gary F. Hartnell, Senior Fellow, Chemistry Technology, Monsanto Company
Joy Mench, Professor of Animal Science and Director of the Center for Animal Welfare, University of California, Davis
Sara Place, Assistant Professor of Sustainable Beef Cattle Systems, Oklahoma State University
Mo Salman, Professor of Veterinary Epidemiology, Colorado State University, and Jefferson Science Fellow, U.S. Department of State
Dennis H. Treacy, Executive Vice President and Chief Sustainability Officer, Smithfield Foods, Inc.
B. L. Turner II (NAS), Gilbert F. White Professor of Environment and Society, Arizona State University
Gary W. Williams, Professor of Agricultural Economics and Co-Director, Agribusiness, Food, and Consumer Economics Research Center, Texas A&M University
Felicia Wu, John A. Hannah Distinguished Professor of Food Science and Human Nutrition and Agricultural, Food and Resource Economics, Michigan State University

Science and Technology for Sustainability Program Staff

Richard Bissell, Executive Director, Policy and Global Affairs Division
Jerry L. Miller, Director
Jennifer Saunders, Senior Program Officer
Dominic Brose, Program Officer
Emi Kameyama, Program Associate
Brent Heard, Sustainability Fellow

Board on Agriculture and Natural Resources Staff

Robin Schoen, Director
Kara Laney, Program Officer

Preface

By 2050 the world's population is projected to grow by one-third, reaching between 9 billion and 10 billion people. With trade globalization, increased urbanization, and expected growth in global affluence, a substantial increase in per capita meat, dairy, and fish consumption also is anticipated. Sustainably meeting the nutritional needs of this population and its demand for animal products will require a significant research and development (R&D) investment so that the productivity of today can be sufficiently enhanced to meet the much heightened demands of the year 2050. The challenges to be met go beyond research into enhanced animal agricultural productivity. Research will be required into how to anticipate and meet significant changes in the global environment impacting on animal agriculture, how to improve equitable distribution of animal agricultural products today and in the future, and how best to improve engagement and respectful bidirectional communication between those engaged in animal agriculture and the public. Ensuring sustainable agricultural growth will be critical to addressing this global challenge to food security.

According to the Food and Agriculture Organization of the United Nations (FAO), in 2010, the food animal sector contributed 40 percent of the global value of agricultural output and supported the livelihoods and food security of almost a billion people. This sector, one of the fastest growing areas for agricultural development, is driven by income growth and supported by technological and structural change. Several challenges compound this rapid growth, including increasing pressure on the availability of land, water, and energy to sustainably increase animal agricultural productivity, and the potential adverse impacts of global climate change on agricultural productivity. A further challenge is the overuse of medically important antibiotics leading to an increased risk of infectious disease in humans and in animals, particularly with increasing globalization leading to more rapid spread of disease. The FAO notes that the "speed of change has often significantly outpaced the capacity of governments and societies to provide the necessary policy and regulatory framework to ensure an appropriate balance between the provision of private and public goods."[1]

Advances in animal agriculture have been a result of R&D and new technologies, particularly in areas such as food safety, genetics, reproductive efficiencies, nutrition, animal welfare, disease control, biotechnology, and the environment. Although investment in agricultural R&D continues to be highly productive, the field has witnessed a marked decrease in funding in terms of real dollars. Recognizing this gap between the animal agricultural research enterprise and the need to address challenges related to global food security and animal protein demand globally and domestically, an ad hoc committee of experts was convened to prepare a report to identify critical areas of R&D, technologies, and resource needs for research in the field of animal agriculture, both nationally and internationally. Specifically, the committee was asked to assess global demand for products of animal origin in 2050 within the framework of ensuring global food security; evaluate how environmental changes and natural resource constraints may affect the ability to sustainably meet future global demand for animal products in a wide variety of production systems in the United States and internationally; and identify factors that may affect the ability of the United States to sustainably meet demand for animal products, including the need for trained human capital, product safety and quality, and effective communication and adoption of new knowledge, information, and technologies. The committee was also tasked with identifying the needs for human capital development, technology transfer, and information systems for emerging and evolving animal production systems in developing countries, including social science and economic research into understanding the need for, and acceptability of, new knowledge and technologies; and with describing the evolution of sustainable animal production systems relevant to production and production efficiency metrics in the United States and in developing countries.

The task given to the committee was based on three underlying assumptions, which the committee did not reexamine in depth. First, global animal protein consumption will continue to increase based on population growth and increased per capita animal protein consumption. There is a wealth of literature to support this expectation, although assumptions about the various social, economic, and environmental

---

[1] FAO. 2009. The State of Food and Agriculture 2009: Livestock in the Balance. Rome: FAO.

factors that will drive animal food demand in 2050 are, at best, informed conjectures. Second, restricted resources (e.g., water, land, energy, capital) and environmental changes, including climate change, will drive complex agricultural decisions with impacts on research needs. If the natural ecosystem is defined as one that is unaffected by humans, then agriculture is inherently disruptive. Third, current and foreseeable rapid advances in basic biological sciences, as well as social sciences and economics, provide an unparalleled opportunity to maximize the yield of investments in animal science R&D. The following report is a result of the committee's deliberations based on these assumptions.

The committee operated under a fast-track approach which began in March 2014 and concluded with the submission of the revised report in December 2014. This approach constrained deliberations to those areas clearly within the boundaries of the task. This report points to many directions that can be expanded upon and advanced in future deliberations on the subject of animal agriculture research needs. The committee recognizes that it is not unusual for a National Research Council (NRC) committee charged with evaluating research to find that the area has been relatively underfunded. Accordingly, the committee went out of its way to develop analyses that evaluated this contention. These are detailed in Chapter 5.

The committee recognizes that there is literature advocating the reduction of the amount of animal protein in our diets in order to improve our health and well-being and reduce the impact of these agricultural systems on the environment. We do recommend research to better understand the impact of these dietary preferences as well as other social issues, such as animal welfare. The committee does not, however, directly address how these issues might affect the demand for animal agriculture, because it was specifically tasked with identifying research needs for animal agriculture in light of the projected global increase in human consumption of animal agricultural products. The committee does note that as long as animal protein continues to be consumed, there is value to R&D that improves the efficiency of its production.

In the first paragraph of the Preface of its 2010 report, the NRC Committee on Twenty-First Century Systems Agriculture stated:

> Since the National Research Council published the report Alternative Agriculture in 1989, there has been a remarkable emergence of innovations and technological advances that are generating promising changes and

opportunities for sustainable agriculture in the United States. At the same time, the agricultural sector worldwide faces numerous daunting challenges that will require innovations, new technologies, and new ways of approaching agriculture if the food, feed, and fiber needs of the global population are to be met. [2]

The present NRC committee concurs with this finding as it relates to animal agriculture—and particularly agrees with characterizing the challenges as daunting. We recognize that many supporters of organic animal agriculture may disagree with the committee's finding that, although organic animal agriculture can contribute to local sustainability, such as by decreasing energy needs for transportation and by lessening the concentration of waste in a single location, in its present form it cannot be scaled up to meet current or future demands for animal protein. It is our hope that the committee's work will help the dedicated scientists, technicians, and animal agriculturists achieve their mutual goal of nutritiously feeding the world's population.

In this report, Chapter 1 first describes the challenge that the committee addressed. Chapter 2 discusses the challenge of global food security, Chapter 3 explores key issues for animal sciences research considerations for the United States, Chapter 4 examines global considerations for animal agricultural research, Chapter 5 addresses capacity building and infrastructure for research in food security and animal sciences, and Chapter 6 provides the committee's findings and recommendations for the field described in Chapters 3, 4, and 5.

This report has been reviewed in draft form by individuals chosen for their diverse perspectives and technical expertise, in accordance with procedures approved by the National Academies' Report Review Committee. The purpose of this independent review is to provide candid and critical comments that will assist the institution in making its published report as sound as possible and to ensure that the report meets institutional standards for objectivity, evidence, and responsiveness to the study charge. The review comments and draft manuscript remain confidential to protect the integrity of the process.

We wish to thank the following individuals for their review of this report: Donald Beitz, Iowa State University; Jason Clay, World Wildlife Fund; Russell Cross, Texas A&M University; Terry Etherton, The

---

[2] NRC (National Research Council). 2010. Toward Sustainable Agricultural Systems in the 21st Century. Washington, DC: The National Academies Press.

Pennsylvania State University; Terry McElwain, Washington State University; John Patience, Iowa State University; Jimmy Smith, International Livestock Research Institute; Ellen Silbergeld, Johns Hopkins University; Henning Steinfeld, Food and Agriculture Organization; and Peter Vitousek, Stanford University.

Although the reviewers listed above have provided many constructive comments and suggestions, they were not asked to endorse the conclusions or recommendations, nor did they see the final draft of the report before its release. The review of this report was overseen by R. James Cook, Washington State University, and Johanna Dwyer, Tufts Medical Center. Appointed by the National Academies, they were responsible for making certain that an independent examination of this report was carried out in accordance with institutional procedures and that all review comments were carefully considered. Responsibility for the final content of this report rests entirely with the authoring committee and the institution.

The report would not have been possible without the sponsors of this study, including the Association of American Veterinary Medical Colleges, the Bill & Melinda Gates Foundation, the Innovation Center for U.S. Dairy, the National Cattlemen's Beef Association, the National Pork Board, Tyson Foods, Inc., the U.S. Department of Agriculture (USDA), and the U.S. Poultry & Egg Association.

The committee gratefully acknowledges the following individuals for making presentations to the committee: Catherine Woteki, USDA; Kathy Simmons, National Cattlemen's Beef Association; Ying Wang, Innovation Center for U.S. Dairy; Chris Hostetler, National Pork Board; Kevin Igli, Tyson Foods, Inc.; Ted Mashima, Association of American Veterinary Medical Colleges; John Glisson, U.S. Poultry & Egg Association; Donald Nkrumah, Bill & Melinda Gates Foundation; Mary Beck, Mississippi State University; Molly Brown, National Aeronautics and Space Administration; Dennis Treacy, Smithfield Foods, Inc.; Trey Patterson, Padlock Ranch; Jason Clay, World Wildlife Fund; Suzanne Bertrand, International Livestock Research Institute; Laurie Hueneke, National Pork Producers Council; Mario Herrero, Commonwealth Scientific and Industrial Research Organisation; Henning Steinfeld, FAO; Russell Cross, Texas A&M University; Montague W. Demment and Anne-Claire Hervy, Association of Public and Land-grant Universities; Jude Capper, Montana State University; Raymond Anthony, University of Alaska, Anchorage; Randy Brummett, The World Bank; and Clare Narrod, University of Maryland. The committee would also like to recognize Keith Fuglie of USDA's Economic Research Service, who provided valuable data that greatly informed committee deliberations.

The deliberations on its broad task occurred on a fast-track schedule that could not have been met without the exceptional support of the committee staff. Staff members who contributed to this effort are Jennifer Saunders, senior program officer of the Science and Technology for Sustainability Program responsible for our study; Robin Schoen, director of the Board on Agriculture and Natural Resources; Kara Laney, program officer; Dominic Brose, program officer; Emi Kameyama, program associate; Brent Heard, sustainability fellow; Adriana Courembis, financial associate; Karen Autrey, report review associate; Marina Moses, director (through May 2014); Dylan Richmond, research assistant (through August 2014); Christine Johnson, sustainability fellow (through July 2014); Hope Hare, administrative assistant; Mirsada Karalic-Loncarevic, manager of the Technical Information Center; Marilyn Baker, director for reports and communication; and Richard Bissell, executive director, Policy and Global Affairs Division.

We thank especially the members of the committee for their tireless efforts throughout the development of this report.

Bernard D. Goldstein, Chair
Committee on the Considerations for
the Future of Animal Science Research

Contents

SUMMARY 1
    The Committee's Task 3
    Committee's Approach to the Task 4
    Recommendations 7

1 INTRODUCTION: OVERVIEW OF THE CHALLENGES FACING
   THE ANIMAL AGRICULTURE ENTERPRISE 19
    Challenges Toward the Year 2050 19
    Committee's Approach to the Task 24
    Defining Agricultural Sustainability 28
    The Role of and Need for R&D in Animal Sciences 40
    Structure of the Report 43
    References 44

2 GLOBAL FOOD SECURITY CHALLENGE: SUSTAINABILITY
   CONSIDERATIONS 51
    Future Demand for Animal Protein 51
    20th Century Expansion in Productivity 52
    The Way Forward 54
    Challenges to Sustainable Productivity 55
    The Predicament of Sustainable Animal Agriculture 79
    References 81

3 ANIMAL AGRICULTURE RESEARCH NEEDS: U.S.
   PERSPECTIVE 95
    Introduction 95
    Research Considerations for Sustainability 104
    References 189

4 GLOBAL CONSIDERATIONS FOR ANIMAL AGRICULTURE
   RESEARCH 215
    Introduction 215
    References 289

5 CAPACITY BUILDING AND INFRASTRUCTURE FOR
   RESEARCH IN FOOD SECURITY AND ANIMAL SCIENCES 311
    Introduction 311
    References 346

6 RECOMMENDATIONS 351

INDEXES

A COMMITTEE ON CONSIDERATIONS FOR THE FUTURE OF ANIMAL SCIENCE RESEARCH BIOGRAPHICAL INFORMATION 363
B STATEMENT OF TASK.. 369
C GLOSSARY 371
    References 380
D KEY STRATEGIES INVOLVING ANIMAL AGRICULTURE BEING FOCUSED ON BY USDA RESEARCH, EDUCATION, AND ECONOMICS (REE)... 383
E USDA ARS PROPOSED FY 2015 PRIORITIES 385
F ANIMAL HEALTH PRIORITIES FROM A 2011 USDA NIFA WORKSHOP... 387
    By Industry 387
    By Research Discipline 390
G RESULTS OF A USDA ARS- AND NIFA-SPONSORED WORKSHOP ON ANIMAL HEALTH 395
H SUMMARY OF NOAA/USDA FINDINGS ON ALTERNATIVE FEEDS FOR AQUACULTURE 398
I GOALS FOR PRIORITIES IDENTIFIED BY THE EU ANIMAL TASK FORCE.. 401
J USDA NIFA INVESTMENT IN ANIMAL SCIENCE BY SPECIES (SUBJECT OF INTEREST)... 405
K ANIMAL SCIENCE INVESTMENT BY KNOWLEDGE AREA.. 409
L USDA ARS ANIMAL/AGRICULTURE RESEARCH FY 2010–FY 2014 413

Abbreviations and Acronyms

ACDC	Agricultural Communications Documentation Center
ADSA	American Dairy Science Association
AFO	animal feeding operation
AFRI	Agriculture and Food Research Initiative
AI	artificial insemination
APLU	Association of Public and Land-grant Universities
ARS	Agricultural Research Service (U.S. Department of Agriculture)
ASAS	American Society of Animal Science
ASC	Aquaculture Stewardship Council
ATIP	Agricultural Technology Innovation Partnership
AVMA	American Veterinary Medical Association
BANR	National Research Council's Board on Agriculture and Natural Resources
BFT	biofloc technology
BSE	bovine spongiform encephalopathy
CAADP	Comprehensive Africa Agriculture Development Programme
CAC	Codex Alimentarius Commission
CAFO	concentrated animal feeding operation
CAP	Coordinated Agricultural Project (USDA)
CAST	Council for Agricultural Science and Technology
CDC	Centers for Disease Control and Prevention
CE	Cooperative Extension
CFD	computational fluid dynamics
CGIAR	Consultative Group on International Agricultural Research
CH_4	methane
CO_2	carbon dioxide
COOL	Country of Origin Labeling

CRADA	Cooperative Research and Development Agreement
CRIS	Current Research Information System (USDA)
CSES	Coalition for a Sustainable Egg Supply
CSREES	Cooperative State Research, Education, and Extension Service (USDA)
DDGS	distillers dried grains with solubles
DNA	deoxyribonucleic acid
DOE	U.S. Department of Energy
EADD	East Africa Dairy Development Project
ECF	East Coast fever
EFSA	European Food Safety Authority
EPA	U.S. Environmental Protection Agency
ERS	Economic Research Service (USDA)
ESMIS	Economics, Statistics, and Market Information System (USDA)
ET	embryonic transfer
EU	European Union
FAEIS	Food and Agricultural Education Information System (USDA)
FAIR	Farm Animal Integrated Research
FAO	Food and Agriculture Organization of the United Nations
FASEB	Federation of American Societies for Experimental Biology
FASS	Federation of Animal Science Societies
FCR	feed conversion ratio
FDA	Food and Drug Administration
FFAR	Foundation for Food and Agriculture Research
FPCM	fat- and protein-corrected milk
FSIS	Food Safety and Inspection Service (USDA)
FVE	Federation of Veterinarians of Europe
GAFSP	Global Agriculture and Food Security Program
GATT	General Agreement on Tariffs and Trade
GCM	general circulation model
GDP	gross domestic product

GEBV	genomic estimated breeding values
GHG	greenhouse gas
GI	gastrointestinal
GLOBIOM	Global Biosphere Management Model
GSI	Global Salmon Initiative
HACCP	hazard analysis and critical control points
HIV/AIDS	human immunodeficiency virus/acquired immune deficiency syndrome
HPAI	highly pathogenic avian influenza
ICAR	International Committee for Animal Recording
ILRI	International Livestock Research Institute
ILSI	International Life Sciences Institute
IPCC	Intergovernmental Panel on Climate Change
IPPC	International Plant Protection Convention
ISEAL	International Social and Environmental Accreditation and Labelling
IVF	in vitro fertilization
IWG-A	Interagency Working Group on Aquaculture
IWG-OA	Interagency Working Group on Ocean Acidification
LCA	life cycle assessment
LEAP	Livestock Environmental Assessment and Performance
MAP	modified atmosphere packaging
MERS	Middle East respiratory syndrome
MOET	multiple ovulation embryo transfer
MRP	multistate research program
N2O	nitrous oxide
NAGRP	National Animal Genome Research Program
NAHMS	National Animal Health Monitoring System
NASA	National Aeronautics and Space Administration
NGO	nongovernmental organization
NIFA	National Institute of Food and Agriculture (USDA)
NIH	National Institutes of Health
NMSP	Nutrient Management Spear Program

NOAA	National Oceanic and Atmospheric Administration
NRC	National Research Council
NRI	National Research Initiative (USDA)
NSF	National Science Foundation
OECD	Organisation for Economic Co-operation and Development
OIE	World Organization for Animal Health
OSHA	Occupational Safety and Health Administration (U.S. Department of Labor)
PAS	partitioned aquaculture systems
PCAST	President's Council of Advisors on Science and Technology
POP	persistent organic pollutants
PPP	public-private partnerships
PSA	Poultry Science Association
rBGH	recombinant bovine growth hormone
rbST	recombinant bovine somatotropin
rpST	recombinant porcine somatotropin
RAS	recirculating aquaculture system
REDNEX	reduce nitrogen excretion by ruminants
REE	Research, Education, and Economics (USDA)
R&D	research and development
SAES	state agricultural experiment stations
SARS	severe acute respiratory syndrome
SI	sustainable intensification
SNP	single nucleotide polymorphisms
SPF	specific-pathogen-free
SPS	sanitary and phytosanitary measures
TAD	transboundary animal diseases
TFP	total factor productivity
THI	thermal heat index
TSE	transmissible spongiform encephalopathy
UHT	ultra-high temperature
UN	United Nations
UNDP	United Nations Development Programme

USAID	U.S. Agency for International Development
USDA	U.S. Department of Agriculture
USDHHS	U.S. Department of Health and Human Services
USGCRP	U.S. Global Changes Research Program
USMEF	U.S. Meat Export Federation
vCJD	variant Creutzfeldt-Jakob disease
VFD	Veterinary Feed Directive
WAAP	World Association of Animal Production
WCED	United Nations World Commission on Environment and Development
WHO	World Health Organization
WTO	World Trade Organization
WWF	World Wildlife Fund

Summary

Reinvigorating animal agricultural research is essential to sustainably address the global challenge of food security[1]. The global demand for food from animal agriculture is anticipated to nearly double by 2050. Increased demand is due, in part, to a predicted increase in world population from 7.2 billion to between 9 billion and 10 billion people in 2050. The United Nations Food and Agriculture Organization (FAO) estimates that there will be a 73 percent increase in meat and egg consumption and a 58 percent increase in dairy consumption over 2011 levels worldwide by the year 2050. The increase in population puts additional pressure on the availability of land, water, and energy needed for animal and crop production. During this period, it is also anticipated that there will be significant growth in per capita animal-source food consumption related to increasing income and urbanization in developing

---

[1] When using the term *animal agriculture,* the committee is referring to livestock, poultry and aquaculture in total. *Livestock* includes cattle, sheep, horses, goats, and other domestic animals ordinarily raised or used on the farm. Domesticated fowl are considered poultry and not livestock (29 CFR § 780.328). Aquaculture, also known as fish or shellfish farming, refers to the breeding, rearing, and harvesting of plants and animals in all types of water environments including ponds, rivers, lakes, and the ocean (National Oceanic and Atmospheric Administration. 2014. What is Aquaculture? Online. Available at http://www.nmfs.noaa.gov/aquaculture/what_is_aquaculture.html. Accessed September 15, 2014).The committee uses the term *animal sciences* to refer to all disciplines currently contributing to animal food production systems. These disciplines are generally housed in departments focused on conventional animal sciences, animal husbandry, food sciences, dairy husbandry, poultry husbandry, veterinary science, veterinary medicine, and agricultural economics.
As defined by the 1996 World Food Summit, *food security* exists when all people, at all times, have physical and economic access to sufficient, safe, and nutritious food to meet their dietary needs and food preferences for an active and healthy life (Food and Agriculture Organization of the United Nations. 1996. World Food Summit Plan of Action, Paragraph 1 in Rome Declaration on World Food Security. World Food Summit 13-17 November 1996, Rome, Italy. Online. Available at http://www.fao.org/docrep/003/w3613e/w3613e00.HTM. Accessed August 14, 2014).

countries. Global environmental challenges, including global climate change, and the growing threat of disease transmission to and from agricultural animals add further challenges to sustainably meeting the demand for animal agriculture in the year 2050. Even in a stable world, the animal agricultural research enterprise would be significantly challenged to help rectify the current unequal distribution of animal food and to address the need to integrate social science research to better understand and respond to changing consumer preferences. Further, there is a high likelihood of a major threat to animal agriculture that no one currently predicts and for which a vibrant animal research enterprise will be central to an effective response to ensure global food security. Without additional investment, the current global animal agricultural research will not meet the expected demand for animal food products in 2050.

Agricultural sustainability, a focus of this report, has been defined by the National Research Council as having four generally agreed-upon goals: satisfy human food, feed, and fiber needs and contribute to biofuel needs; enhance environmental quality and the resource base; sustain the economic viability of agriculture; and enhance the quality of life for farmers, farm workers, and society as a whole. Sustainability is best evaluated not as a particular end state, but rather as a process that moves agricultural systems along a trajectory toward greater sustainability on each of the four goals.

The field of animal agriculture faces numerous challenges in meeting global food security in the context of the three pillars of sustainability (environment, economy, and society). Environmental considerations for animal agriculture include global environmental change, land- and water-use constraints, and impacts on biodiversity, among others. Several economic considerations include meeting increased demand, trade issues, and production growth. Social considerations include animal welfare, equity (e.g., fair labor practices, including agricultural worker health and protection of vulnerable human populations and rural communities), corporate social responsibility, business ethics, the naturalness of food products, and the use of biotechnology in food production. Whatever the definition of sustainability and however it is applied to animal agriculture, a key to ensuring a sustainable food system is a holistic systems approach. These approaches have become more important as the planet's resources necessary to sustain an increasing population are increasingly connected and are being further challenged by global environmental changes.

These challenges point to the need for more animal science research that improves integration in areas where these needs overlap, including research on food science, socioeconomics, and environmental sciences. A new roadmap for animal science research is needed that focuses on animal production in a way that informs and is informed by advances in biological sciences and by the broader socioeconomic and environmental conditions of the new century. Animal agriculture and animal protein production will substantially increase to meet demand from global population growth, but must do so in the context of sustainability.

THE COMMITTEE'S TASK

Recognizing the gap between the animal agricultural research enterprise and the challenges related to global food security, an ad hoc committee of experts was convened to prepare a report to identify critical areas of research and development (R&D), technologies and resource needs for research in the field of animal agriculture, both nationally and internationally (see Appendix B). The committee was asked to assess global demand for products of animal origin in 2050 within the framework of ensuring global food security; evaluate how environmental changes and limited natural resources may impact the ability to sustainably meet future global demand for animal products in a wide variety of production systems in the United States and internationally; and identify factors that may impact the ability of the United States to sustainably meet demand for animal products, including the need for trained human capital, product safety and quality, effective communication, and adoption of new knowledge, information, and technologies. The committee was also tasked with identifying the needs for human capital development, technology transfer and information systems for emerging and evolving animal production systems in developing countries; identifying the resources needed to develop and disseminate this knowledge and these technologies; and describing the evolution of sustainable animal production systems relevant to production and production efficiency metrics in the United States and in developing countries.

COMMITTEE'S APPROACH TO THE TASK

The task given to the committee was based on three underlying assumptions, which the committee did not reexamine in depth. First, global animal protein consumption will continue to increase based on population growth and on increased per capita animal protein consumption. There is a wealth of literature to support this assumption. Second, restricted resources (e.g., water, land, energy, and capital) and global environmental change will drive complex agricultural decisions that affect research needs. If the natural system is defined as one that is unaffected by humans, then agriculture is inherently disruptive. Third, current and foreseeable rapid advances in basic biological sciences provide an unparalleled opportunity to maximize the yield of investments in animal science R&D. The task to the committee was to identify what research is needed to achieve the goal of providing adequate, safe, and affordable nutritious food to the global population, taking into account several critical issues that factor into this process, such as public understanding and values, food safety concerns, poverty, trade barriers, socioeconomic dynamics, and health and nutrition (Figure S-1). The committee operated under a fast-track approach that began in March 2014 and concluded with the submission of the revised report in December 2014. This approach constrained deliberations to those areas clearly within the boundaries of the task. The report points to many directions that can be expanded upon and advanced in future deliberations on the subject of animal agricultural research needs. The committee recognizes that it is not unusual for an NRC committee charged with evaluating research to find that the area has been relatively underfunded. Accordingly, the committee went out of its way to develop analyses that evaluated this contention, which are detailed in Chapter 5.

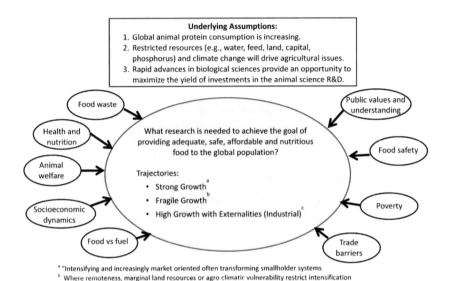

FIGURE S-1 Overview of inputs affecting animal agricultural needs.
SOURCE: Committee generated.

The committee engaged in a fast-track process that included meetings in March, May, July and September 2014. Data-gathering sessions that were open to the public were held during both the March and May meetings. The July meeting was a week-long intensive session, which included extensive reviews of relevant literature, deliberation, and drafting of report text; the September meeting was a closed writing session. During the March meeting, the committee heard from each of the study sponsors, including the Association of American Veterinary Medical Colleges, the Bill & Melinda Gates Foundation, the Innovation Center for U.S. Dairy, the National Cattlemen's Beef Association, the National Pork Board, Tyson Foods, Inc., the U.S. Department of Agriculture (USDA), and the U.S. Poultry & Egg Association. It also heard from speakers from various nongovernmental organizations (NGOs), industry, and federal agencies providing their perspectives on the need for animal science research. Presenters at the May meeting represented academia, NGOs, and federal agencies, and addressed topics such as ethical considerations in animal science research, sustainable aquaculture, funding equity for the field, and the impact of climate change on animal agriculture.

The committee reviewed and built upon a large body of written material on animal science issues, including literature that informed the committee on research needs for the field. The available data included other NRC reports, FAO reports, published research articles, and both U.S. and international governmental reports. The committee also reviewed many other documents related to USDA's budget and functioning. The committee recognized that there are public organizations and literature advocating reduction in the amount of animal protein consumed with the rationale that this would improve our health and well-being and reduce the impact of these agricultural systems on the environment (e.g., a meatless Monday campaign has gained support from many governmental and nonprofit organizations). The committee does not, however, address these issues because it was specifically tasked with identifying research needs for animal agriculture in light of the projected global increase in human consumption of animal protein. The committee does note that as long as animal protein continues to be consumed, there is value to R&D that improves the efficiency of its production.

Finally, in considering the charge to discuss global issues in animal agriculture, the committee noted that there is marked variability among and within different countries in their animal agricultural practices and needs. These reflect a variety of factors including climate, soil characteristics, and cultural practices. It has been estimated that for billions of people, particularly in sub-Saharan Africa, South and Southeast Asia, and other developing countries, raising food animals fulfills a social need beyond the provision of food[2]. In the United States, there exists a wide range of agricultural systems that reflect both geographical and cultural factors related to the different groups of immigrants who settled in different parts of the country.

Many agrarian societies are being rapidly affected by globalization of food supplies and urbanization, which are trends that are anticipated to continue. Some countries, such as Brazil, have rapidly emerging

---

[2] Herrero, M., D. Grace, J. Njuki, N. Johnson, D. Enahoro, S. Silvestri, and M. C. Rufino. 2013a. The roles of livestock in developing countries. Animal 7(S1):3-18. Herrero, M., P. Havlik, H. Valin, A. Notenbaert, M. C. Rufino, P. K. Thornton, M. Blummel, F. Weiss, D. Grace, and M. Obersteiner. 2013b. Biomass use, production, feed efficiencies and greenhouse gas emissions from global livestock systems. Proceedings of the National Academy of Sciences 110(52):20888-20893.

economies characterized by great increases in agricultural productivity that match or exceed that of the United States or Europe. Accordingly, the traditional distinction between developed and developing countries is overly simplistic. In this report, the committee provides an overview of the extensive variability in animal agriculture among developed and developing countries and among individual countries. This provides a basis for discussion of research and general needs for human capital development, technology transfer, and information systems. The committee does not go into detail for each country or region beyond the United States, but notes that there is much overlap between research needs in the United States and those in developing countries.

RECOMMENDATIONS

Animal production and the science that informs it are confronted by an emerging and globally complex set of conditions in the 21st century that generate new challenges for sustainable animal production, which in turn requires rethinking about the overall nature of animal science. These challenges include, but are not limited to, growing demand for animal products by an increasingly affluent, global population approaching 10 billion people; the globalization of food systems that cross continents with consequences for individual countries and regional concerns about food security; the intensification of production systems in the context of societal and environmental impacts; the development and maintenance of sustainable animal production systems in the face of global environmental change; and the multiple decadal stagnation in research funding for animal production. As described throughout this report, a new roadmap for animal science research is required. The findings and recommendations[3] described below will help to inform this new roadmap.

The breadth of the committee's task led to many recommendations being developed. The committee twice deliberated on prioritization of these recommendations. Early in the process the committee chose a limited number of broad and high-level overarching recommendations,

---

[3] Finding as used in the report is an actionable discovery, circumstance, revelation, issue, or fact drawn from evidence or analysis, and a recommendation is a proposal for action to be taken by leaders in government, NGOs, academia, and/or industry, as appropriate, to address one or more findings.

which were then refined in subsequent meetings and are described immediately below. At its last meeting, the committee chose its highest priorities from among all of the possible recommendations. These recommendations appear after the overarching recommendations and are specific to what the committee identified as key areas in animal agriculture in both the United States and globally. In addition to its recommendations, the committee identified complementary priorities for research, research support, and infrastructure, which can be found in Chapters 3-5.

Ideally, NRC committee recommendations should include an action statement specifying the agency or organization that should follow up. This works well if there is an individual sponsor with a single short-term task; however, the breadth of the tasks and the multiplicity of overlapping national and international public and private organizations involved in sponsoring or performing animal research limited the committee's ability to specify action pathways. Sorting out responsibilities for moving ahead is part of the reason that the committee has recommended the development of a U.S. Animal Science Strategic Plan under the leadership of the USDA.

Overarching Recommendations

Two central issues have guided National Research Council and other reports regarding the setting of research agendas for animal agriculture in recent years: productivity and sustainability. The committee built on these reports and emphasized the importance of research to sustainably and efficiently increase animal agricultural productivity. The committee's deliberations resulted in the following overarching recommendations:

- To achieve food security, research efforts should be improved through funding efforts that instill integration rather than independence of the individual components of the entire food chain. Success can only be achieved through strong, overarching, and inter- and transdisciplinary research collaborations involving both the public and private sectors. Animal science research should move toward a systems approach that emphasizes efficiency and quality of production to meet food security needs. The recently created Foundation for Food and Agricultural Research (FFAR) needs to incorporate holistic approaches to animal productivity and sustainability (Chapter 5).

- Continuing the research emphasis on improving animal productivity is necessary; however, concomitant research on the economic, environmental, and social sustainability nexuses of animal production systems should also be enhanced. Both public and private funding agencies should incorporate inter- and transdisciplinary approaches for research on animal productivity and sustainability (Chapters 3 and 5).
- There is a need to revitalize research infrastructure (human and physical resources), for example, through a series of strategic planning approaches, developing effective partnerships, and enhancing efficiency. In the United States, the committee recommends that USDA and the newly created FFAR spearhead the formation of a coalition to develop a U.S. Animal Science Strategic Plan or Roadmap for capacity building and infrastructure from 2014 to 2050. The coalition should be broad based and include representation from relevant federal agencies; colleges and universities that are involved in research, teaching, and outreach activities with food animals; NGOs; the private sector; and other relevant stakeholders. Areas of focus should include assessment of resource needs (human and physical infrastructure) to support the current and emerging animal science research enterprise; strategies to increase support for research, outreach, and instructional needs via formula funding, competitive funding, and public-private partnerships; curriculum development and delivery; evaluation of factors affecting hiring, retention, and diversity in the animal sciences; and mechanisms for research, priority setting to meet emerging, local, regional, national, and global needs (Chapter 5).
- Socioeconomic/cultural research is essential to guide and inform animal scientists and decision makers on appropriately useful and applicable animal science research as well as communication and engagement strategies to deal with these extensive challenges. Engagement of social scientists and researchers from other relevant disciplines should be a prerequisite as appropriate for integrated animal science research projects, such as National Institute of Food and Agriculture (NIFA) Coordinated Agricultural Project grants, to secure funding and approval of such projects (Chapters 3 and 5).
- For research in sustainable intensification of animal agriculture to meet the challenge of future animal protein needs, it is necessary to effectively close the existing broad communication gap between the public, researchers, and the food industries. This will require

research to better understand the knowledge, opinions, and values of the public and food system stakeholders, as well as the development of effective and mutually respectful communication strategies that foster ongoing stakeholder engagement. A coalition representing universities, federal agencies, industry, and the public should be formed to focus on communications research with the goals of enhancing engagement, knowledge dissemination, stakeholder participation, and informed decision making. Communications programs within agriculture schools, or in collaboration with other university components, such as schools of public health, could conduct this type of research (Chapters 3 and 5).

- The United States should expand its involvement in research that assists in the development of internationally harmonized standards, guidelines, and regulations related to both the trade in animal products and protection of the consumers of those products (Chapter 4).

Many of the recommendations and priorities discussed in each of the chapters are based on a central theme of the need for strategic planning to meet the challenges of the increased animal agricultural demand that is projected through 2050. These recommendations and priorities include planning for research in the United States and in developing countries and reconsideration of education and training in animal agriculture in the United States, particularly at the university level. These strategic planning activities should be guided by the need for systems approaches that integrate the many scientific disciplines and governmental and nongovernmental stakeholders involved in achieving the goal of food security based on sustainable animal agriculture.

Recommendations for U.S. Animal Agriculture

The committee developed several recommendations that are of high priority for reinvigorating the field of animal agriculture in the United States.

Public Funding

In view of the anticipated continuing increased demand for animal protein, growth in U.S. research related to animal agricultural productivity is imperative. Animal protein products contribute over $43 billion annually to the U.S. agricultural trade balance. Animal agriculture accounts for 60 to 70 percent of the total agricultural economy. In the past two decades, public funding, including formula funding and USDA Agricultural Research Service/National Institute of Food and Agriculture funding, of animal science research has been stagnant in terms of real dollars and has declined in relation to the research inflation rate. A 50 percent decline in the rate of increase in U.S. agricultural productivity is predicted if overall agricultural funding increases in normative dollars continue at the current rate, which is less than the expected rate of inflation of research costs. If funding does meet the rate of research cost inflation, however, a 73 percent increase in overall agricultural productivity between now and 2050 is projected and a 1 percent increase in inflation-adjusted spending is projected to lead to an 83 percent increase.

Despite documenting the clear economic and scientific value of animal science research in the United States, funding to support the infrastructure and capacity is evidently insufficient to meet the needs for animal food; U.S.-based research will be needed to address sustainability issues and to help developing countries sustainably increase their own animal protein production and/or needs. Additionally, animal science research and practices in the United States are often adopted, to the extent possible, within developing countries. Thus, increases in U.S. funding will favorably impact animal production enterprises in developing countries.

With the lack of increase in public funding of animal science research, private/industry support has increased. The focus of industry funding is more toward applied areas that can be commercialized in the short term. Many of these applications are built on concepts developed from publicly funded basic research. With the increased animal protein demands, especially poultry, more publicly funded basic research is needed.

RECOMMENDATION 3-1: To meet current and future animal protein demand, and to sustain corresponding infrastructure and capacity, public support for animal science research (especially basic research) should be restored to at least past levels of real dollars and maintained at a rate that meets or exceeds the annual rate of research inflation. This is especially critical for those species (i.e., poultry) for which the consumer demand is projected to significantly increase by 2050 and for those species with the greatest opportunity for reducing the

environmental impact of animal agriculture (Section 3-1 in Chapter 3).

Productivity and Production Efficiency

Regarding productivity and production efficiency, the committee finds that increasing production efficiency while reducing the environmental footprint and cost per unit of animal protein product is essential to achieving a sustainable, affordable, and secure animal protein supply. Technological improvements have led to system/structural changes in animal production industries whereby more efficient food production and less regional, national, and global environmental impact have been realized.

> **RECOMMENDATION 3-2: Support of technology development and adoption should continue by both public and private sectors. Three criteria of sustainability—(1) reducing the environmental footprint, (2) reducing the financial cost per unit of animal protein produced, and (3) enhancing societal determinants of sustainable global animal agriculture acceptability—should be used to guide funding decisions about animal agricultural research and technological development to increase production efficiency (Section 3-2 in Chapter 3).**

Breeding and Genetic Technologies

Further development and adoption of breeding technologies and genetics, which have been the major contributors to past increases in animal productivity, efficiency, product quality, environmental, and economic advancements, are needed to meet future demand.

> **RECOMMENDATION 3-3: Research should be conducted to understand societal concerns regarding the adoption of these technologies and the most effective methods to respectfully engage and communicate with the public (Section 3-3 in Chapter 3).**

Nutritional Requirements

The committee notes that understanding the nutritional requirements of the genetically or ontogenetically changing animal is crucial for optimal productivity, efficiency, and health. Research devoted to an understanding of amino acid, energy, fiber, mineral, and vitamin nutrition has led to technological innovations such as production of individual amino acids to help provide a diet that more closely resembles the animal's requirements, resulting in improved efficiency, animal health, and environmental gains, as well as lower costs; however, much more can be realized with additional knowledge gained from research.

> **RECOMMENDATION 3-4: Research should continue to develop a better understanding of nutrient metabolism and utilization in the animal and the effects of those nutrients on gene expression. A systems-based holistic approach needs to be utilized that involves ingredient preparation, understanding of ingredient digestion, nutrient metabolism and utilization through the body, hormonal controls, and regulators of nutrient utilization. Of particular importance is basic and applied research in keeping the knowledge of nutrient requirements of animals current (Section 3-4 in Chapter 3).**

Feed Technology

Potential waste products from the production of human food, biofuel, or industrial production streams can and are being converted to economical, high-value animal protein products. Alternative feed ingredients are important in completely or partially replacing high-value and unsustainable ingredients, particularly fish meal and fish oil, or ingredients that may otherwise compete directly with human consumption.

> **RECOMMENDATION 3-6.1: Research should continue to identify alternative feed ingredients that are inedible to humans and will notably reduce the cost of animal protein production while improving the environmental footprint. These investigations should include assessment of the possible impact of changes in the protein product on the health of the animal and the eventual human consumer, as well as the environment (Section 3-6.1 in Chapter 3).**

Animal Health

The subtherapeutic use of medically important antibiotics in animal production is being phased out and may be eliminated in the United States. This potential elimination of subtherapeutic use of medically important antibiotics presents a major challenge.

> **RECOMMENDATION 3-7: There is a need to explore alternatives to the use of medically important subtherapeutic antibiotics while providing the same or greater benefits in improved feed efficiency, disease prevention, and overall animal health (Section 3-7 in Chapter 3).**

Animal Welfare

Rising concern about animal welfare is a force shaping the future direction of animal agricultural production. Animal welfare research, underemphasized in the United States compared to Europe, has become a high-priority topic. Research capacity in the United States is not commensurate with respect to the level of stakeholder interest in this topic.

> **RECOMMENDATION 3-8: There is a need to build capacity and direct funding toward the high-priority animal welfare research areas identified by the committee. This research should be focused on current and emerging housing systems, management, and production practices for food animals in the United States. FFAR, USDA-AFRI, and USDA-ARS should carry out an animal welfare research prioritization process that incorporates relevant stakeholders and focuses on identifying key commodity-specific, system-specific, and basic research needs, as well as mechanisms for building capacity for this area of research (Section 3-8 in Chapter 3).**

Climate Change

Although there is uncertainty regarding the degree and geographical variability, climate change will nonetheless impact animal agriculture in diverse ways, from affecting feed quality and quantity to causing environmental stress in agricultural animals. Animal agriculture affects and is affected by these changes, in some cases significantly, and must adapt to them in order to provide the quantity and affordability of animal protein expected by society. This adaptation, in turn, has important implications for sustainable production. The committee finds that adaptive strategies will be a critical component of promoting the resilience of U.S. animal agriculture in confronting climate change and variability.

> **RECOMMENDATION 3-11.2: Research needs to be devoted to the development of geographically appropriate climate change adaptive strategies and their effect on greenhouse gas (GHG) emissions and pollutants involving biogeochemical cycling, such as that of carbon and nitrogen, from animal agriculture because adaptation and mitigation are often interrelated and should not be independently considered. Additional empirical research quantifying GHG emissions sources from animal agriculture should be conducted to fill current knowledge gaps, improve the accuracy of emissions inventories, and be useful for improving and developing mathematical models predicting GHG emissions from animal agriculture (Section 3-11.2 in Chapter 3).**

Socioeconomic Considerations

Although socioeconomic research is critical to the successful adoption of new technologies in animal agriculture, insufficient attention has been directed to such research. Few animal science departments in the United States have social sciences or bioethics faculty in their departments who can carry out this kind of research.

> **RECOMMENDATION 3-12: Socioeconomic and animal science research should be integrated so that researchers, administrators, and decision makers can be guided and informed in conducting and funding effective, efficient, and productive research and technology transfer (Section 3-12 in Chapter 3).**

Communications: The committee recognizes a broad communication gap related to animal agricultural research and objectives between the animal science community and the consumer. This gap must be bridged if animal protein needs of 2050 are to be fulfilled.

> **RECOMMENDATION 3-13: There is a need to establish a strong focus on communications research as related to animal science research and animal agriculture, with the goals of enhancing knowledge dissemination, respectful stakeholder participation and engagement, and informed decision making (Section 3-13 in Chapter 3).**

Recommendations for Global Animal Agriculture

Overall, the committee strongly supports an increase in funding of global animal research both by governments and the private sector. The committee also identified several recommendations directed toward global animal agriculture.

Infrastructural Issues

The committee notes that per capita consumption of animal protein will be increasing more quickly in developing countries than in developed countries through 2050. Animal science research priorities have been proposed by stakeholders in high-income countries, with primarily U.S. Agency for International Development, World Bank, Food and Agriculture Organization, Consultative Group on International Agricultural Research, and nongovernmental organizations individually providing direction for developing countries. A program such as the Comprehensive Africa Agriculture Development Programme (CAADP) demonstrates progress toward building better planning in agricultural development in developing countries, through the composite inclusion of social, environmental, and economic pillars of sustainability.

In addition, for at least the last two decades, governments worldwide have been reducing their funding for infrastructure development and training for animal sciences research. Countries and international funding agencies should be encouraged to adapt an integrated agriculture research system to be part of a comprehensive and holistic approach to agriculture production. A system such as CAADP can be adapted for this purpose.

> **RECOMMENDATION 4-1: To sustainably meet increasing demands for animal protein in developing countries, stakeholders at the national level should be**

involved in establishing animal science research priorities (Section 4-1 in Chapter 4).

Technology Adoption

The committee finds that proven technologies and innovations that are improving food security, economics, and environmental sustainability in high-income countries are not being utilized by all developed or developing countries because in some cases they may not be logistically transferrable or in other ways unable to cross political boundaries. A key barrier to technological adoption is the lack of extension to smallholder farmers about how to utilize the novel technologies for sustainable and improved production as well as to articulate smallholder concerns and needs to the research community. Research objectives to meet the challenge of global food security and sustainability should focus on the transfer of existing knowledge and technology (adoption and, importantly, adaptation where needed) to nations and populations in need, a process that may benefit from improved technologies that meet the needs of multiple, local producers. Emphasis should be placed on extension of knowledge to women in developing nations.

RECOMMENDATION 4-5.2: Research devoted to understanding and overcoming the barriers to technology adoption in developed and developing countries needs to be conducted. Focus should be on the educational and communication role of local extension and advisory personnel toward successful adoption of the technology, with particular emphasis on the training of women (Section 4-5.2 in Chapter 4).

Animal Health

Zoonotic diseases account for 70 percent of emerging infectious diseases. The cost of the six major outbreaks that have occurred between 1997 and 2009 was $80 billion. During the last two decades, the greatest challenge facing animal health has been the lack of resources available to combat several emerging and reemerging infectious diseases. The current level of animal production in many developing countries cannot increase and be sustained without research into the incidence and epidemiology of disease and effective training to manage disease outbreaks, including technically reliable disease investigation and case findings. Infrastructure is lacking in developing countries to combat animal and zoonotic diseases, specifically a lack of disease specialists and diagnostic laboratory facilities that would include focus on the etiology of diseases. There is a lack of critical knowledge about zoonoses presence, prevalence, drivers, and impact. Recent advances in technology offer opportunities for improving the understanding of zoonoses epidemiology and control.

RECOMMENDATION 4-7.1: Research, education (e.g., training in biosecurity), and appropriate infrastructures should be enhanced in developing countries to alleviate the problems of animal diseases and zoonoses that result in enormous losses to animal health, animal producer livelihoods, national and regional economies, and human health (Section 4-7.1 in Chapter 4).

As discussed above, in addition to the recommendations presented in this summary, the committee identified complementary priorities for research, research support, and infrastructure which specifically address the contents of and can be found in chapters 3-5.

1

Introduction: Overview of the Challenges Facing the Animal Agriculture Enterprise

CHALLENGES TOWARD THE YEAR 2050

Reinvigorating animal agricultural research is essential to sustainably address the global challenge of food security[1]. The demand for food from animal agriculture is anticipated to nearly double by 2050. Increased demand is due, in part, to a predicted increase in world population from 7.2 billion to between 9 and 10 billion people in 2050 (United Nations, 2013). The increase in population puts additional pressure on the availability of land, water, and energy needed for animal and crop agriculture. During this period, it is also anticipated that there will be

---

[1] When using the term *animal agriculture* the committee is referring to livestock, poultry, and aquaculture in total. *Livestock* includes cattle, sheep, horses, goats, and other domestic animals ordinarily raised or used on the farm. Domesticated fowl are considered poultry and not livestock (29 CFR § 780.328). Aquaculture, also known as fish or shellfish farming, refers to the breeding, rearing, and harvesting of plants and animals in all types of water environments including ponds, rivers, lakes, and the ocean (NOAA, 2014). The committee uses the term *animal sciences* to refer to all disciplines currently contributing to animal food production systems. These disciplines are generally housed in departments focused on conventional animal sciences, animal husbandry, food sciences, dairy husbandry, poultry husbandry, veterinary science, veterinary medicine, and agricultural economics.
As defined by the 1996 World Food Summit, *food security* exists when all people, at all times, have physical and economic access to sufficient, safe, and nutritious food to meet their dietary needs and food preferences for an active and healthy life (FAO, 1996).

significant growth in per capita animal meat consumption related to increasing urbanization and income in developing countries. Global environmental challenges, including global climate change, and the growing threat of disease transmission to and from agricultural animals add further challenges to sustainably meeting the demand for animal agriculture in 2050. Even in a stable world, the animal agricultural research enterprise would be significantly challenged to help rectify the current unequal distribution of animal calories and the need to integrate social science research so as to better understand and respond to changing consumer preferences. Furthermore, a vibrant animal research enterprise will be central to an effective response to potential threats to animal agriculture to ensure global food security (Box 1-1 provides a brief summary of challenges facing sustainable animal agriculture described in this chapter and throughout the report). Without additional investment, the current U.S. animal agricultural research enterprise will have difficulty meeting the expected demand for animal food products in 2050.

BOX 1-1
Selected Challenges to Meeting Sustainable Animal Agriculture by 2050

- Growth in demand for animal protein due to:
  - Population growth
  - Increasing global affluence
  - Increase in per capita animal protein intake
- Impact of global environmental change on:
  - Climate
  - Habitats
  - Animal feedstocks
- Water and land scarcity
- Changes in consumer preferences
- Changes in national and international regulatory requirements reflecting public concerns about animal agriculture practices
- Role of trade barriers and other governmental actions on animal agriculture in different regions of the world
- Health considerations, such as emerging infectious diseases and foodborne pathogens
- Lack of research funding in the future

INTRODUCTION

Animal protein currently provides 13 percent of the calories produced globally from agriculture and represents 26 percent of the world's dietary protein (Fraser, 2014). In the United States, 133.9 lbs of animal protein were consumed in 2012, and animal products accounted for over half of the value of agricultural production (USDA ERS, 2014)[2]. Animal protein continues to be a significant part of the American diet, with more than 37 million tons of meat consumed annually (U.S. Census Bureau, 2011). In the United States alone, animal products account for over half of the value of agricultural products (USDA, 2014b). The United States has the largest fed-cattle industry in the world and is the world's largest producer of high-quality[3], grain-fed beef. The farm value of milk production in the United States is second only to beef among food animal industries and equal to corn. The U.S. swine industry has seen a rapid shift to fewer and larger operations associated with technological change and an evolving industry structure that has led to pork production accounting for about 10 percent of the world's supply and the United States becoming the largest pork exporter in 2005 (America's Pork Checkoff, 2009). The U.S. poultry industry is the world's largest producer and second largest exporter of poultry meat, and is a major egg producer. Poultry products will continue to increase the amount and share of the animal protein market desired by the American consumer as well as the export market (USDA, 2014b). In 2014, the value of U.S. food animal production was projected to be about $185 billion and the value of crop production projected to be $195 billion, representing 42 and 44 percent, respectively, of the total agricultural sector value (USDA ERS, 2014). The United States, however, is allocating less than 0.20 percent, including both National Institute of Food and Agriculture and USDA Agricultural Research Service appropriations, of the U.S. food animal production value back into publicly funded animal science research.

Worldwide, the Food and Agriculture Organization (FAO) estimates that there will be a 73 percent increase in meat and egg consumption and a 58 percent increase in dairy consumption over 2011 levels by the year 2050 (McLeod, 2011). This increase will not be evenly distributed,

---

[2] Animal protein in this context is defined as red and white meat including poultry, dairy and its products, eggs and their products, and all fish and shellfish.
[3] The committee defines high-quality meat in terms of providing protein and caloric value. The committee recognizes that the organoleptic properties of meat may lead to overeating, obesity, and attendant diseases, but discussing these factors is beyond the charge to the committee.

however; models indicate that between 2000 and 2050, North America and Europe will see little growth in animal protein consumption, whereas consumption in Asia and Africa will more than double. Food animal consumption in Latin America and the Caribbean will also increase significantly (Rosegrant et al., 2009). Governments are grappling with how to address this increased demand, particularly given finite natural resources.

Aquaculture also critically contributes to the world's food supply, and demand continues to increase as incomes rise. The FAO reports that over the past five decades, the world fish food supply has outpaced global population growth and has come to constitute an important source of nutritious food and animal protein for much of the world's population (FAO, 2014). Of particular note is the growth in global trade in fishery products. According to the FAO, developing economies, whose exports represented just 34 percent of world trade in 1982, saw their share rise to 54 percent of total fishery export value by 2012 (FAO, 2014). For many developing nations, fish trade represents a significant source of foreign currency earnings in addition to its important role in income generation, employment, food security, and nutrition. Aquaculture's important role as the fastest growing food production sector in the world to complement other animal agricultural production sectors. Globally, the specific growth rate of aquaculture since its emergence about 60 years ago is approximately 7.4 percent, whereas all other livestock has averaged about 2.6 percent during that same period. The virtually untapped oceanic shelf resources offer possibilities that are not available for the terrestrial-based animal agriculture. Only about 0.05 percent of the available shelf area is currently used for farming, and adverse environmental impacts associated with mariculture are less than the terrestrial-based counterparts. Also, because of its infancy, it has had to address unique issues that are specific to its rapid growth. The feed efficiency of fish, crustaceans, and mollusks is a unique attribute that is a very important contributor to sustainable intensification. The FAO also reports that developed countries continue to dominate world imports of fish and fishery products, although their share has decreased in recent years.

Although investment in agricultural research and development (R&D) continues to be one of the most productive investments, with rates of return between 30 and 75 percent, it has been neglected, particularly in low-income countries (FAO, 2009a). The remarkable advances in animal agriculture in recent years have been a result of R&D

and the adoption of new technologies, particularly in areas such as food safety, genetics and breeding, reproductive efficiencies, nutrition, and disease control (Roberts et al., 2009). These have led to major productivity gains in various species.

Animal science research has improved animal productivity and thus decreased the costs of animal products to consumers, increased food safety and food security, decreased environmental impacts of livestock and poultry production, and addressed public concerns about animal welfare.

Despite the demonstrated importance of animal agricultural research to current global food security, the field faces several significant impediments that limit the ability to sustainably increase productivity to meet future global demand. Recognizing this gap between the animal agricultural research enterprise and the challenges related to global food security, the National Research Council (NRC) convened an ad hoc committee of experts to prepare a report that identifies critical areas of R&D, technologies, and resource needs for research in the field of animal agriculture, both nationally and internationally. Specifically, the committee was tasked to prepare a report to identify critical areas of R&D, technologies, and resource needs for research in the field of animal agriculture nationally and internationally. Specifically, the committee was asked to assess global demand for products of animal origin in 2050 within the framework of ensuring global food security; evaluate how climate change and limited natural resources may impact the ability to meet future global demand for animal products in sustainable production systems, including typical conventional, alternative, and evolving animal production systems in the United States and internationally; and identify factors that may affect the ability of the United States to meet demand for animal products, including the need for trained human capital, product safety and quality, effective communication, and adoption of new knowledge, information and technologies. The committee was also tasked with identifying the needs for human capital development, technology transfer, and information systems for emerging and evolving animal production systems in developing countries; identifying the resources needed to develop and disseminate this knowledge and these technologies; and describing the evolution of sustainable animal production systems relevant to production and production efficiency metrics in the United States and in developing countries.

COMMITTEE'S APPROACH TO THE TASK

The committee's task was based upon three underlying assumptions which the committee did not reexamine in depth. First, global animal protein consumption will continue to increase based on population growth and increased per capita animal protein consumption. There is a wealth of literature to support this assumption, some of which is cited above. Second, restricted resources (e.g., water, land, energy, and capital) and global environmental change will drive complex agricultural decisions that affect research needs. Third, current and foreseeable rapid advances in basic biological sciences provide an unparalleled opportunity to maximize the yield of investments in animal science R&D (Figure 1-1; Research, Education and Economics Task Force, 2004). The committee was tasked with identifying the research needed to achieve the goal of providing adequate, safe, and affordable nutritious food to the global population, taking into account critical issues such as public understanding and values, food safety, poverty, trade barriers, socioeconomic dynamics, and health and nutrition. This report is a result of the committee's deliberations based on these assumptions and contributing factors. The committee's goal was to identify where research should focus so that "sustainable intensification" can be achieved and the protein needs of the projected global population in 2050 can be met. The success of intensive sustainable aquaculture is based on increasing production while simultaneously minimizing/eliminating negative impacts on the environment. The committee, therefore, has recommended critical areas of research based on the quantitative reduction of impacts such as carbon footprint and waste. The recommendations provided should guide agencies in the establishment of research priorities (targets) that have quantitative-based goals in mind.

The committee operated under a fast-track approach that began in March 2014 and concluded with the submission of the revised report in November 2014. This approach constrained deliberations to those areas clearly within the boundaries of the task; however, the report presents many areas that can be expanded upon and advanced in future deliberations on the subject of animal agricultural research needs. The committee recognizes that it is not unusual for an NRC committee charged with evaluating research to find that the area has been relatively underfunded. Accordingly, the committee went out of its way to develop analyses that evaluated this contention, which are detailed in Chapter 5.

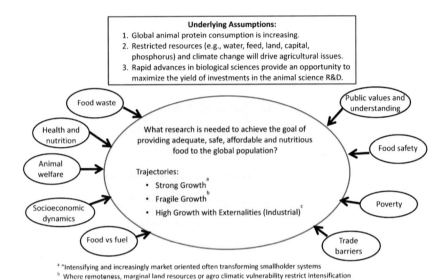

FIGURE 1-1 Overview of inputs affecting animal agricultural needs.
SOURCE: Committee generated.

The committee held meetings in March, May, July, and September 2014. Data-gathering sessions that were open to the public were held during both the March and May meetings. The July meeting was a week-long intensive session, which included extensive reviews of relevant literature, deliberation, and drafting of report text; the September meeting was a closed writing session. During the March meeting, the committee heard from each of the study sponsors: the Association of American Veterinary Medical Colleges, the Bill & Melinda Gates Foundation, the Innovation Center for U.S. Dairy, the National Cattlemen's Beef Association, the National Pork Board, Tyson Foods, Inc., the U.S. Department of Agriculture, and U.S. Poultry & Egg Association. The committee also heard from various nongovernmental organizations (NGOs), industry, and federal agencies providing their perspectives on the need for animal science research. Presenters at the May meeting represented academia, NGOs, and federal agencies, and addressed such topics as ethical considerations in animal science research, sustainable aquaculture, funding equity for the field, and the impact of climate change on animal agriculture. The committee reviewed and built upon a large body of written material on animal science issues, including literature that informed the committee on research needs for the field. The available data included other NRC reports, FAO reports, published research articles, and both U.S. and international governmental reports. The committee also reviewed many other documents related to USDA's budget and activities.

The committee recognizes that there are public organizations and literature advocating for reducing and/or eliminating the amount of animal protein in our diets with the rationale that this would improve our health and well-being and reduce the impact of these agricultural systems on the environment (e.g., a meatless Monday campaign has gained support from many governmental and nonprofit organizations [Righter, 2012]). The committee does not, however, address these issues because it was specifically tasked with identifying research needs for animal agriculture in light of the projected global increase in human consumption of animal protein. The committee notes that under any scenario, there is value to R&D that improves the efficiency of animal protein production.

The committee discussed at length the charge given to it within the context of the expertise of its membership and time constraints regarding the delivery of the report. The critical needs for research in animal agriculture are expansive and go well beyond the focus of this report. The committee adhered to its main charge, which was to consider meeting the animal product demands for 2050 in a sustainable way across different production systems with strong consideration of trained human capital, product safety, and effective technologies. We interpreted the charge to focus on production rather than, for example, the ethical goal of food equity. Nevertheless, the report points to nonproduction issues throughout.

The committee's task was not to set out research directions for studies on the social and policy effects of animal food production per se, but to make recommendations specifically for animal science research. Also, the expertise of the committee members did not allow for specific recommendations to be made about the highest-priority social sciences research topics.

In considering the charge to discuss global issues in animal agriculture, the committee noted that there is marked variability among and within different countries in their animal agricultural practices and needs, and much overlap with research needs in the United States. These reflect a variety of factors including climate, soil characteristics, and cultural practices. Prominent issues in the United States are overconsumption and greenhouse gas (GHG) emissions, whereas in developing countries, public health issues, food security, undernutrition, and adaptation to climate change take priority. For perhaps a billion people, particularly in sub-Saharan Africa, South and Southeast Asia, and other developing areas, the raising of food animals fulfills a social need beyond the provision of food (Herrero and Thornton, 2013; Herrero et al., 2013a). In the United States, there exists a wide range of agricultural systems that reflect geographical as well as cultural factors related to the different groups of immigrants who settled in different parts of the country.

Many agrarian societies are rapidly being affected by globalization of food supplies and urbanization, which are trends that are anticipated to continue. Some countries, such as Brazil, have rapidly emerging economies characterized by great increases in agricultural productivity that match or exceed that of the United States or Europe. Accordingly, the traditional distinction between developed and developing countries is overly simplistic. In this report, the committee provides a brief overview of the extensive variability in animal agriculture among developed and developing countries and individual countries. This provides a basis for discussion of research and general needs for human capital development, technology transfer, and information systems. The committee does not go into details for each country or region beyond the United States.

The committee believes that the developed and non-developed dichotomy is not as useful as in the past. Gradations of economies exist that cloud the distinction and possible generalities for either type. In terms of animal agriculture in particular, sharp distinctions are difficult. For example, Brazil is now the global leader in soybean production by which it has expanded its cattle industry. Argentina has similarly expanded its cattle industry, and India is now expected to be the world's leader in dairy production.

Finally, in considering rural land use, the committee recognizes that to some extent the United States and developing countries are going in opposite directions. The United States is characterized by loss of agricultural land to urban sprawl. The U.S. population is moving from inner cities to suburbs in which people live in individual homes on quarter-acre plots and in which farms are converted into shopping malls and schools with large asphalt parking areas. In contrast, the developing world continues to be characterized by urbanization caused by the move of rural subsistence agricultural workers to cities. This difference is accentuated by the shift in the United States from an economy based on factories, which requires large numbers of workers to be concentrated in a specific area, to a knowledge-based economy in which productive labor can use modern Internet-based technology with less need to congregate in central locations. Europe differs from the United States by having stronger rules to protect its farmland from encroachment by cities, which is an approach buttressed by trade barriers to protect otherwise marginal agricultural producers.

In this introductory chapter, the committee explores the role of research in understanding and meeting global food demand, and broadly discusses the role of sustainability and of systems approaches in considering animal science research needs. Major uncertainties, potential opportunities, and likely hindrances related to research needs for animal agriculture are briefly reviewed. Further chapters expand on these subjects.

DEFINING AGRICULTURAL SUSTAINABILITY

Sustainability has progressed from a goal first enunciated in the 1980s in the landmark report Our Common Future (WCED, 1987) to specific actionable frameworks that can be used to guide decision making (NRC, 2011, 2013; see also Alliance for Sustainable Agriculture[4] and Sustainability Consortium[5]). It is interpreted variously by different research communities. The 1987 "Bruntland Report" focuses on sustainable development and defines it as meeting the needs of the present without compromising the ability of future generations to meet their own needs (WCED, 1987). This definition mirrors that of the 1969 U.S. National Environmental Policy Act, which described the need to "create and maintain conditions, under which humans and nature can exist in productive harmony, that permit fulfilling the social, economic, and other requirements of present and future generations" (42 U.S.C. § 4331(a)). That policy expresses what is now described as sustainability, and has been cited previously in NRC reports (NRC, 2011, 2013). With the emergence of earth system, ecosystem, and natural capital research, the role of the environment in the sustainability concept was enlarged to include the provisioning of humankind without threatening the functioning of the earth system, in which "provisioning humankind"

---

[4] http://allianceforsustainability.com.
[5] http://www.sustainabilityconsortium.org.

includes socioeconomic development, akin to the Bruntland report (NRC, 1999; Kates et al., 2001). The key shift, therefore, is the maintenance of functioning ecosystems, landscapes, and the earth system to provide the environmental services that nature provides and humankind wants.

Agricultural sustainability, a focus of this report, is defined by the NRC (2010) as having four generally agreed-upon goals consistent with the visions of sustainability as noted above:
- Satisfy human food, feed and fiber needs, and contribute to biofuel needs.
- Enhance environmental quality and the resource base.
- Sustain the economic viability of agriculture.
- Enhance the quality of life for farmers, farm workers, and society as a whole.

Sustainability is best evaluated not as a particular end state, but as a process that moves farming systems along a trajectory toward greater sustainability on each of the four goals.

Davidson (2002) similarly notes that "agriculture must be internally sustainable, externally sustainable, and also serve as a resource that is available to support other sectors of the economy and society. A system that pays heed to each of these three areas is likely to be able to meet the needs of the present without compromising the ability of future generations to meet their own needs." Internal sustainability refers to preserving the agricultural resource base, including avoiding degradation of soil and water. It includes responding to the threats of animal diseases, the usual vagaries of climate and market forces, and loss of land and water to nonagricultural uses. Also included is the maintenance of the human capital necessary to sustain agricultural communities through succeeding generations. External sustainability refers to avoiding the externalities of agricultural production imposed on other natural resources and the environment and on the nonagricultural society. Responsive sustainability requires agriculture to be sufficiently vibrant and resilient in the face of crises and opportunities in other sectors of the economy, including global climate change. This third heading of responsive sustainability requires that agriculture, including animal production systems, must be sufficiently dynamic and flexible to respond to crises in other societal sectors.

By definition, sustainability is forward looking and addresses intergenerational and longer timescales in research and planning that operate at the ecosystem level, whereby imbalances are avoided or minimized. Under sustainable practices, the impacts of animal agriculture positively contribute to the provision of ecosystem services. These timescales and the range of factors considered—economic, social, and environmental—in sustainability approaches raise a series of issues about appropriate metrics for gauging and comparing animal agricultural productivity, which are considered in depth in Chapter 3. Sustainability also has implications across geographical areas, which can differ depending upon social and environmental issues. Approaches that can increase local sustainability may have adverse impacts on the region, and effective approaches to regional sustainability may have adverse global impacts.

An important issue in considering challenges to sustainability is the dynamic interrelation between various seemingly disparate actions. In Box 1-2, the committee describes the documented impact of overfishing off of the west coast of Africa, leading to a price increase for fish in the local market and ultimately to an increase in bushmeat hunting. This had an impact on biodiversity, endangered species, and a wide variety of processes pertinent to human health and sustainability.

BOX 1-2
Broad Implications of Animal Protein Availability:
Impact on Endangered Species and Bushmeat Hunting

Overfishing off the west coast of Africa leading to price increases for fish in local markets has been causally associated with an increase in bushmeat hunting (Brashares et al., 2004). Concerns include loss of tropical biodiversity (Waite, 2007), an increased likelihood of the spread of existing or emerging infectious diseases to humans, particularly simian viruses (Wolfe et al., 2004), and a possible climate threat due to the loss of carbon storage from large trees whose seeds are usually spread by large-bodied vertebrates (Brodie and Gibbs, 2009). This is further evidence that the demand for animal protein is leading to the loss of wild animals, a process known as defaunation (Dirzo et al., 2014). The role that bushmeat may have played in the Ebola virus disease outbreak in West Africa can be found in Box 4-6.

Whatever the definition, and however applied to animal agriculture, a key to ensuring a sustainable food system is a holistic systems approach (Chapter 3). Holistic systems approaches have become more important as the planet's resources necessary to sustain an increasing population are increasingly connected and are further being challenged by global environmental changes. Below is a summary of the sustainability challenges facing the field of animal agriculture in the context of meeting global food security challenges and the three pillars of sustainability (environmental, economic, and social).

Environmental Considerations for Animal Agriculture

If the natural ecosystem is defined as one that is unaffected by humans, then agriculture is inherently disruptive. Livestock is the largest land-use sector on Earth, occupying 30 percent of the Earth's ice-free surface (Steinfeld et al., 2006; Reid et al., 2008). According to Steinfeld (2014), 50 percent of the arable land in industrialized countries is used to grow animal feed. Livestock utilizes one-third of global cropland for animal feed production, is responsible for 72 percent of deforestation, and as a sector consumes 32 percent of global freshwater (Herrero, 2014). The water used by the livestock sector is over 8 percent of global human water use (Steinfeld et al., 2006). Given this context, the increased demand for animal protein has important implications for natural resources.

Animal agriculture also has implications for global environmental change, which refers to the totality of changes, both natural and anthropogenic in origin, under way in the earth system, from ecosystems to climate change. Animal agriculture affects these changes, in some cases significantly, and must adapt to them in order to provide the quantity and affordability level of animal products expected by society. For example, animal agriculture is responsible for about 14.5 percent of global GHG emissions, according to the FAO (Gerber et al., 2013). Major contributors are nitrous oxide (N_2O) from manure storage and application (25 percent), carbon dioxide (CO_2) from deforestation (34 percent), and methane (CH_4) from enteric fermentation of ruminants (26 percent). The 2.7 billion tons of CO_2 from animal agriculture is equivalent to "9 percent of all global anthropogenic CO_2 emissions; the 2.2 billion tons of CH_4 emissions from animal agriculture represent 37 percent of global anthropogenic CH_4 emissions; and the 2.2 billion tons of N_2O emissions comprise 65 percent of total global anthropogenic N_2O emissions" (Steinfeld, 2014). There are also significant variations in the emissions from different species; these differences are further pronounced when comparing developed and developing nations (de Vries and de Boer, 2010). FAO reported in 2011 that 44 percent of agriculture-related GHG outputs occurred in Asia, 25 percent in the

Americas, 15 percent in Africa, 12 percent in Europe, and 4 percent in Oceania (Tubiello et al., 2014). The environmental and resource impact of animal agriculture has long-run economic implications all along the animal supply chain from producers to animal product consumers. Economic modeling of this impact can inform decision-makers about the needed technical and policy responses. For example, reducing the GHG emissions of beef cattle through investments in technology could improve not only the bottom line of the rancher, but also the whole economic sustainability of the industry.

Sustainable Intensification

The lower global climate change footprint of U.S. animal agriculture compared to historic conditions is related to the intensification of animal agriculture (Chapter 3)[6]. Briefly, as detailed below, the ability to concentrate animal agriculture and to enhance productivity has been made feasible by public-funded and industry-funded research, ranging from enhancing the birth number and health of young animals to improvements in the selection and delivery of feed. Research has also been of importance in mitigating the threats to land, water, and health caused by wastes and to other concerns, such as animal welfare associated with concentrated animal feeding operations; however, these issues persist and are the cause of increasing consumer concerns. Because of the multiple roles of food animals in the developing world, incentives for sustainable intensification may also include minimizing the costs of animal protein production as a way to maintain an effective food animal population. Although sustainable intensification has been adopted as a policy goal for national and international institutions, it has also been subject to criticism. Box 1-3 includes areas that interface with sustainable intensification and ways that shared agendas might best be pursued (Garnett et al., 2013).

---

[6]According to the U.S. Global Climate Change Program, global change refers to "changes in the global environment that may alter the capacity of the Earth to sustain life. Global change encompasses climate change, but it also includes other critical drivers of environmental change that may interact with climate change, such as land use change, the alteration of the water cycle, changes in biogeochemical cycles, and biodiversity loss" (U.S. Global Change Research Program, 2014).

> **BOX 1-3**
> **Policy Goals Interfacing with Sustainable Intensification (SI)**
>
> Identifying the need for increased food security amid increasing global populations and wealth, Garnett et al. (2013) provided a nuanced discussion of SI as a policy objective. These authors say that SI has been adopted as a policy goal by many prominent institutions, but also attracts serious criticism. They adopt a broad perspective for analysis which attempts to see how SI interfaces with other policy objectives and to what extent it represents a step toward attaining food security. They discuss the four key premises underlying SI, the first of which is the need for an increase in agricultural production, which increases yields in many low-income countries and also considers environmental sustainability. Second, because the increasing area of land used in agriculture has a negative environmental impact, this higher production must be attained through higher yields. A third premise is that food security requires altering business-as-usual practices, because environmental sustainability is as important to this goal as raising agricultural productivity. Fourth, although SI is a goal to be attained, it does not specify the means through which it can be reached. Different approaches, ranging from conventional to organic to technological, can be tested with respect to attaining SI (Garnett et al., 2013).
>
> The authors also consider five other policy goals that must interface with SI in its implementation. The first of these is biodiversity and land use, because through land and water contamination "agriculture is a greater threat to biodiversity than any other human activity." This speaks to the importance of exploring "land-sharing" and "land-sparing" processes that can help SI systems accommodate for the trade-offs between increased yields and environmental concerns. Accommodating animal welfare goals in limiting the negative effects (such as disease) associated with intensification policies as well as objectives for human nutrition through the food available for consumption are also identified as important considerations. Finally, accommodating economic support for rural economies and sustainable development goals, such as targeting investment in agriculture as a mechanism for economic growth and facilitating the social and production capital necessary to meet sustainability objectives, is identified as an important policy consideration when implementing SI. Garnett et al. (2013) identify SI as a "new, evolving concept," whose "meaning and objectives [are] subject to debate and contest." Because SI is being recommended and pursued by many groups, policy considerations such as those identified by the autors must be considered and explored as we implemtn this concept into development plans.

Despite these criticisms, given animal agriculture's immense environmental impact amid increases in demand for animal protein, sustainable intensification has become a potential means for reaching production goals while preserving environmental quality. The Intergovernmental Panel on Climate Change argues that the sustainable intensification of agriculture can decrease GHG emissions per unit of agricultural product (IPCC, 2014), and a report by FAO has acknowledged that "intensification—in terms of increased productivity both in food animal production and in feed crop agriculture—can reduce greenhouse gas emissions from deforestation and pasture degradation" (Steinfeld et al., 2006). Evidence in the study of agricultural production supports these claims, with intensification resulting in net avoided emissions of 161 gigatons of carbon between 1961 and 2005, despite increased fertilizer production and use (Burney et al., 2010). Evidence as to the environmental success of sustainable intensification has also been observed in the U.S. beef industry, with the carbon footprint per billion kilograms of beef being reduced by 16.3 percent when comparing 2007 and 1977 values (Capper, 2011). Similarly, the U.S. dairy industry has reduced feed use by 77 percent, land use by 90 percent, and water use by 65 percent, and has achieved a 63 percent decrease in GHG emissions per kilogram of milk from 1944 to 2007 (Capper et al., 2009).

Economic Considerations for Animal Agriculture

The economic importance of animal agriculture cannot be overstated. This sector contributes 40 percent of the global value of agricultural output and supports the livelihoods and food security of almost a billion people (FAO, 2009b). It represents 1.5 percent of world gross domestic product (GDP), and in industrialized countries livestock production comprises more than half of the agriculture-related GDP (Herrero et al., 2013b). This sector has approximately $1.4 trillion in assets and employs 1.3 billion people, including from 400 million to over 700 million of the world's poor who rely on animals for meat, milk, and fertilizer (Herrero et al., 2013a). In the United States, the annual economic value of livestock and poultry sales exceeds $183 billion (USDA ERS, 2014). In the European Union, the livestock sector contributes about 130 billion euros annually to the European economy (Animal Task Force, 2013).

Animal protein production is expected to grow from 2014 to 2023, albeit at a slower rate of 1.6 percent per year, compared to the previous decade (OECD-FAO, 2014). Demand for meat of all types is expected to increase by 53 million tons over this time period, with 58 percent of this increase coming from the Asia and Pacific region, and 18 percent from Latin America and the Caribbean contributing. The developed countries of North America and Europe are expected to contribute 15 percent to this growth, and Africa is projected to contribute 7 percent. Meat trade is projected to grow slower than in the past decade, and in global terms just over 11 percent of meat output will be traded. The most significant import demand growth originates from Asia, which represents the greatest share of additional imports for all meat types (OECD-FAO, 2014). By the end of this decade, poultry meat production is projected to overtake pork production, making poultry the number one animal protein source globally.

America's freshwater and marine aquaculture industry meets only 5 to 7 percent of U.S. demand for seafood (NOAA, 2012). Products of marine aquaculture in the United States represent 10 percent of the total domestic production (206,767 tonnes). In 2012, the U.S. seafood trade deficit surpassed $10 billion for the first time (NOAA, 2012) with the United States being the second highest importer of seafood in the world with a mean annual increase of 5.1 percent from 2002 to 2012. In 2013, the total value of aquaculture sales from U.S. production was $1.37 billion, up from $1.09 billion in 2005, (USDA, 2014a), a 25.7 percent increase. This most recent value of U.S.-produced food fish (fish, crustaceans, and mollusks) is approximately 1 percent of the total value globally ($137.7 billion). Key findings from an OECD-FAO (2014) report projecting global meat consumption and production growth can be found in Box 1-4 (excluding aquaculture). Global aggregate production and demand for meat products (excluding aquaculture) can be found in Table 1-1.

BOX 1-4
Key Findings from OECD-FAO Agricultural Outlook (2014):
Animal Agriculture from 2014 to 2023

- "Nominal meat prices are expected to remain high throughout the outlook period. Animal feed costs remain above historic norms and rising costs related to other inputs such as energy, labor, water and land will also support higher prices. In real terms, however, meat prices have already, or will soon, peak, and will decline moderately by 2023."

- "Global meat production is projected to rise by 1.6 percent annually over the Outlook period, down from 2.3 percent in the last ten years. Driven largely by demand preferences, poultry meat will become the largest meat sector by 2020. Over the projection period poultry meat production will capture almost half of the increase in global meat production by 2023, compared to the base period."

- "Global meat consumption per capita is expected to reach 36.3 kg in retail weight by 2023, an increase of 2.4 kg compared to the base period. This additional consumption will mostly (72 percent) consist of poultry, followed by pig, sheep and bovine meat. Consumption growth in developed countries will be slower than that of the developing countries, but in absolute terms, at 69 kg per capita, will remain more than double that in developing countries by the end of the projection period."

- "Meat trade is projected to grow slower than in the past decade and in global terms just over 10.6 percent of meat output will be traded. The most significant import demand growth originates from Asia, which represents the greatest share of additional imports for all meat types."

TABLE 1-1 Past and Future Aggregate Production and Consumption for Meat Products.

	Production					Consumption				
	1,000 tonnes	Annual Growth Rates, %				1,000 tonnes	Annual Growth Rates, %			
	2005/2007	1961-2007	1991-2007	2005/2007-2030	2005/2007-2050	2005/2007	1961-2007	1991-2007	2005/2007-2030	2005/2007-2050
World										
Bovine	63,583	1.5	0.9	1.3	1.2	62,321	1.5	0.9	1.3	1.2
Ovine	12,876	1.7	1.6	1.6	1.5	12,670	1.7	1.8	1.7	1.5
Pig meat	99,917	3.1	2.3	1.2	0.8	99,644	3.1	2.3	1.2	0.8
Poultry meat	81,994	5.2	4.4	2.1	1.8	81,545	5.1	4.4	2.1	1.8
Total meat	258,370	2.9	2.4	1.5	1.3	256,179	2.9	2.5	1.6	1.3
Developing countries										
Bovine	34,122	2.9	3.0	2.0	1.8	31,975	3.1	2.7	2.2	1.9
Ovine	9,462	3.1	3.4	1.9	1.7	9,695	3.3	3.3	2.0	1.8
Pig meat	60,483	5.7	3.7	1.5	1.1	60,584	5.6	3.7	1.6	1.1
excl. China	15,504	3.9	3.5	2.0	1.7	16,053	3.8	3.6	2.1	1.8
Poultry meat	44,880	7.4	6.4	2.8	2.4	44,543	7.3	6.2	2.9	2.4
Total meat	148,946	4.8	4.2	2.1	1.7	146,797	4.9	4.1	2.2	1.8
excl. China	80,660	3.7	3.7	2.3	2.1	78,958	3.8	3.4	2.5	2.2
excl. China and Brazil	59,957	3.3	3.0	2.3	2.4	64,357	3.6	3.3	2.7	2.4

	Production					Consumption				
	1,000 tonnes	Annual Growth Rates, %				1,000 tonnes	Annual Growth Rates, %			
	2005/2007	1961-2007	1991-2007	2005/2007-2030	2005/2007-2050	2005/2007	1961-2007	1991-2007	2005/2007-2030	2005/2007-2050
Total meat by region										
Sub-Saharan Africa	6,802	2.5	3.0	2.9	2.9	7,334	2.8	3.4	3.2	3.0
Near East/North Africa	8,918	4.5	3.9	2.4	2.2	10,292	4.3	3.7	2.7	2.3
Latin America and Caribbean	40,585	3.7	4.5	1.7	1.3	34,557	3.8	3.6	1.7	1.3
excl. Brazil	19,882	2.6	3.1	1.9	1.6	19,955	3.1	3.4	2.0	1.6
South Asia	7,180	2.8	1.6	4.4	4.0	6,685	2.6	1.2	4.5	4.2
East Asia	85,121	6.5	4.5	1.9	1.4	86,806	6.5	4.7	1.9	1.4
excl. China	16,834	4.7	3.2	2.3	2.1	18,967	4.8	3.7	2.4	2.0
Developed countries	109,424	1.6	0.6	0.7	0.5	109,382	1.5	0.7	0.6	0.4

SOURCE: Alexandratos and Bruinsma (2012). Reprinted with permission of FAO.

Social Considerations for Animal Agriculture

It is becoming increasingly apparent that for research advances in animal productivity to be useful, consideration must be given to the social norms of the communities and countries in which they are to be applied. Studies have indicated that there are a host of issues beyond food safety and quality that influence the acceptability, and hence the sustainability, of existing and new animal agricultural practices. These factors include environmental, economic, and social concerns (NRC, 2010), with the latter playing an increasingly important role not only with respect to the regulatory decision-making process, but in terms of shaping consumer and supply-chain purchasing decisions in many countries (Mench, 2008; Matthews and Hemsworth, 2012). Social concerns include those related to animal welfare, equity (e.g., fair labor practices, including agricultural worker health and safety and protection of vulnerable human populations and rural communities), corporate social responsibility and business ethics, food security, agricultural and food traditions, naturalness of food products, and the use of biotechnology in food production (NRC, 2010; Anthony, 2012; Niles, 2013; Lister et al., 2014).

Many animal agricultural issues also interface with human health, starting with sufficient availability to avoid famine and undernutrition to concerns about overconsumption and the relationship of eating meat to diseases prevalent in both the developed and developing world (Keats and Wiggins, 2014). Disease transmission from animals to humans is another main issue of interest to human health. Outcomes of research relating micronutrient intake to positive and negative effects on human health may guide breeding practices, including genetic modification of animals to achieve dietary goals (e.g., lower-cholesterol eggs). An issue particularly worthy of additional research is improving response to the development of antibiotic-resistant microorganisms that threaten human and animal health.

Animal agricultural productivity also depends upon the health of the agriculture workforce. Agriculture has historically been a dangerous trade and remains so today (OSHA, 2014). U.S. agricultural workers have seven times the annual death rate of the average U.S. workforce (24.9 vs. 3.5 deaths per 100,000 people). This does not include commercial fishery workers who are at the highest fatality risk of almost any industry (126 deaths per 100,000 people). Animal production agricultural workers' injury rate (6.7 per 100 people) was higher than that of crop production workers (5.5 per 100 people) (OSHA, 2014). Globally, agricultural worker health risks are often compounded by the lack of technology to replace unsafe practices, the lack of personal protective equipment, and the frequency of child labor. Research on the issue of worker health is usually the province of health agencies and is divorced from standard agricultural research funding considerations. In the United States, the National Institute of Occupational Safety and

Health funds programs related to agricultural worker health. The National Institutes of Health Fogarty Center has funded global agricultural health programs pairing U.S. academic programs with developing countries, in part to train future researchers.

The Millennium Development Goals are particularly valuable in providing metrics that permit measurement of outcomes whereby assessment of impacts on human nutrition, poverty, health, and overall socioeconomic welfare at the community and household levels can be confidently assessed. Related to animal health and agriculture, 13 zoonoses are responsible for a staggering 2.2 billion human illnesses and 2.3 million deaths per year, mostly in low-income and middle-income countries. Of the post-2015 development goals, several proposed goals include elimination of extreme poverty and sustainable social, economic, and environmental development (Kelly et al., 2014).

THE ROLE OF AND NEED FOR R&D IN ANIMAL SCIENCES

Advances in animal agriculture have been a result of R&D and new technologies, particularly in areas such as food safety, genetics and breeding, reproductive efficiencies, nutrition, disease control, biotechnology, and the environment (Roberts et al., 2009). Major productivity gains in various species are also attributed to R&D in this field. Developments in reproductive technologies will continue to allow acceleration of genetic selection (Hume et al., 2011). The neglect of investment in animal agriculture documented by the committee in Chapter 5 runs contrary to the significant economic value and high rate of return of this sector to the United States and globally. The committee recognizes that there are various sources of funding for R&D in the field, including private funding, but also notes the essential role that public funding plays in addressing longer-term research needs, and particularly in supporting research that addresses public goods. The committee has therefore focused its assessment on federal sources in the United States. As such, the United States is allocating less than 0.20 percent of the U.S. agricultural value in public funding for animal science research.

Additionally, although budgets were relatively flat in terms of nominal dollars for public investment in animal science research during 2004-2012, the annual value in U.S. exports for beef increased from $0.8 to $5.5 billion (USMEF, 2012a), pork increased from $2.2 to $6.3 billion (USMEF, 2012c), lamb increased from $12.3 to $26.2 million (USMEF,

INTRODUCTION

2012b), broiler exports increased from $1.8 billion (2006) to over $4 billion (2012) (USDA ERS, 2013), and dairy exports increased from $1.29 billion (2006) to $5.28 billion (2013) (U.S. Dairy Export Council, 2014). The increased value in U.S. exports demonstrates the importance of animal agriculture to the U.S. economy and the need to continue to invest in the research that has driven gains in this important industry.

The FAO stated that "the livestock sector requires renewed attention and investments from the agricultural research and development community and robust institutional and governance mechanisms that reflect the diversity within the sector...The challenges posed by the livestock sector cannot be solved by a single string of actions or by individual actors alone. They require integrated efforts by a wide range of stakeholders" (FAO, 2009b).

Impediments and Opportunities: Reinvigorating Animal Sciences Research

The committee highlights in the report the limitations to funding facing animal science research. It also identifies several other impediments to developing research that optimizes sustainable animal agriculture domestically and globally, and for which the committee based its subsequent recommendations. These include inadequate research infrastructure, including personnel, facilities and other organizational issues; political and social impediments; insufficient collaboration among government, industry, and academia, different disciplines, and basic and applied sciences and technologies; and problems in technology transfer. Although good metrics are available for quantifying research outcomes, there is a need to evolve a better process for identifying and funding future research needs. Finally, there is a lack of strategic planning for the field that cuts across the different dimensions of sustainability, considers various timescales and intergenerational issues, and carefully considers the implications of actions across local, regional, and global dimensions.

Despite these impediments, the committee notes many opportunities to improve agricultural productivity in both the United States and the developing world. These include advances in general biology pertinent to better understanding animal growth, lactation, and welfare; breeding and growth techniques; genetics, including improved growth characteristics and protection against diseases; technological advances such as minimizing animal production wastes, including recycling, improving animal welfare, and minimizing spoilage of food (e.g., through better packaging); advances in developing research approaches that involve cooperation between scientists and multiple stakeholders, especially smallholder food animal producers; and finally advances in social sciences, such as improved communication among the public, the food animal industry, and scientists (Box 1-5).

BOX 1-5
Opportunities to Improve Research That Optimizes Sustainable Animal Agriculture

- Advances in general biology and other basic research pertinent to understanding animal growth and welfare
- Advances in breeding and growth techniques
- Advances in nutrition and management
- Advances in genetics
    - Improve growth characteristics
    - Protect against diseases (e.g., reduce antibiotic use)
    - Identify and select for traits in animals that increase their adaptability and resilience to climate change and variability
    - Identify and select for traits in animals and in gut microbiomes that increase animal nutrient and energy utilization, and decrease nutrient excretion
- Advances in technology
    - Minimize animal production wastes and improve nutrient recycling in animal and plant agriculture
    - Minimize environmental and resource use footprints
    - Improve animal welfare
    - Protect against disease
    - Minimize spoilage of food (e.g., through better packaging)
- Advances in social sciences
    - Improve communication among the public, the food animal industry and scientists
    - Improve understanding of the economic and social drivers that govern (impact) food animal development
    - Improve understanding and development of policy tools that optimize animal food production

To keep to the statement of task, the committee had to focus on higher-level issues that were crosscutting to many developing regions of the world and presents these challenges throughout Chapter 4. The committee had to prioritize what it could address within the report and could not address all differences in trends across all different regions, trends, and species around the world. It chose to present larger regions and systems, such as increasing production in poultry across Africa and aquaculture across Asia, discussed in Chapter 4, that would have impacts across larger regions. The committee acknowledges that there are regional differences in per capita animal product consumption patterns that are important to consider; however, it was not able to discuss this in great detail.

The committee is cognizant of the inherent uncertainties in any research planning effort that attempts to look beyond a few years, let alone one that is given a target date 36 years from the present. Any such planning effort is subject to uncertainties in demand, global climate change and other environmental impacts, social and regulatory issues affecting acceptability of animal agricultural practice and consumer preferences, trade issues, and pertinent basic scientific advances. The committee's highest confidence is in the prediction that animal agricultural productivity will be significantly impacted by at least one major factor that is not foreseen today, which itself is a reason to have available a vibrant animal agricultural research enterprise capable of responding to this unforeseen threat (Box 1-6). The remainder of the report, including the committee's findings, recommendations, and research priorities, is the committee's attempt to embrace the opportunities available to enhance and reinvigorate the field, while recognizing these complex uncertainties.

STRUCTURE OF THE REPORT

The remainder of the report is organized into five chapters. Chapter 2 provides a broad overview of the sustainability challenges associated with feeding a population of 10 billion people in 2050, particularly given the increased demand for animal protein. Chapter 3 discusses research needs for the field of animal agriculture from the U.S. perspective, including priorities identified by previous entities (other examples of these research priorities are included in Appendixes E-I). Chapter 4 describes research needs at the global level. In its deliberations, the committee came to the recognition that scarcity of human resources capable of meeting the challenge was a major factor underlying all animal agricultural research planning for the future. On the basis of this finding, the committee decided to write a separate chapter that highlights this issue rather than separately consider this issue within each of the chapters. Accordingly, Chapter 5 focuses on the capacity-building and infrastructure needs for research in food security and animal sciences in the United States. Chapter 6 provides a summary of the committee's recommendations.

BOX 1-6
Major Uncertainties in the Assessment of Research Needs and Opportunities in Sustainable Animal Agriculture for 2050

- The extent of the impact of global environmental change on animal foodstocks and habitats
- The rate of development of new scientific understanding or new technologies pertinent to animal food production research
- The impact of overlapping social movements related to animal welfare, organic foods, and vegetarianism on the consumption of animal protein
- The impact of science-based health information on consumer preferences
- The rate of population growth and of growth of per capita protein consumption, including the extent of sustainable economic development and its role in protein consumption
- The state of the economic system, including the extent of institution or removal of trade barriers and the free exchange of animal protein among countries and trading groups
- The impact of major new diseases or the spread of existing diseases on animal health

---

REFERENCES

Alexandratos, N., and J. Bruinsma. 2012. World Agriculture Towards 2030/2050: The 2012 Revision. EAS Working Paper 12-03. Rome: FAO. Online. Available at http://www.fao.org/docrep/016/ap106e/ap106e.pdf. Accessed June 16, 2014.

America's Pork Checkoff. 2009. Quick Facts: The Pork Industry at a Glance. National Pork Board. Online. Available at http://www.porkgateway.org/filelibrary/piglibrary/references/npb%20quick%20%20facts%20book.pdf. Accessed December 16, 2014.

Animal Task Force. 2013. Research & Innovation for a Sustainable Livestock Sector in Europe. White Paper. Online. Available at http://www.animaltaskforce.eu/Portals/0/ATF/documents%20for%20scare/ATF%20white%20paper%20Research%20priorities%20for%20a%20sustainable%20livestock%20sector%20in%20Europe.pdf. Accessed June 16, 2014.

Anthony, R. 2012. Building a sustainable future for animal agriculture: An environmental virtue ethic of care approach within the philosophy of technology. Journal of Agricultural and Environmental Ethics 25(2):123-144.

Brashares, J. S., P. Arcese, M. K. Sam, P. B. Coppolillo, A. R. E. Sinclair, and A. Balmford. 2004. Bushmeat hunting, wildlife declines, and fish supply in West Africa. Science 306(5699):1180-1183.

Brodie, J. F., and H. K. Gibbs. 2009. Bushmeat hunting as climate threat. Science 326(5951):364-365.

Burney, J. A., S. J. Davis, and D. B. Lobell. 2010. Greenhouse gas mitigation by agricultural intensification. Proceedings of the National Academy of Sciences of the United States of America 107(26):12052-12057.

Capper, J. L. 2011. The environmental impact of beef production in the United States: 1977 compared with 2007. Journal of Animal Science 89(12):4249-4261.

Capper, J. L., R. A. Cady, and D. E. Bauman. 2009. The environmental impact of dairy production: 1944 compared with 2007. Journal of Animal Science 87(6):2160-2167.

Davidson, H. H. 2002. Agriculture. Pp. 347-368 in Stumbling Toward Sustainability, J. C. Dernbach, ed. Washington, DC: Environmental Law Institute.

de Vries, M., and I. J. M. de Boer. 2010. Comparing environmental impacts for livestock products: A review of life cycle assessments. Livestock Science 128(1-3):1-11.

Dirzo, R, H. S. Young, M. Galetti, G. Ceballos. N. J. B. Issac, and B. Collen. 2014. Defaunation in the Anthropocene. Science 34(6195):401-406.

FAO (Food and Agriculture Organization of the United Nations). 1996. World Food Summit Plan of Action, Rome Declaration on World Food Security. Online. Available at http://www.fao.org/docrep/003/w3613e/w3613e00.HTM. Accessed August 14, 2014.

FAO. 2009a. How to Feed the World in 2050. Rome: FAO. Online. Available at http://www.fao.org/fileadmin/templates/wsfs/docs/expert_paper/How_to_Feed_the_World_in_2050.pdf. Accessed August 28, 2014.

FAO. 2009b. The State of Food and Agriculture: Livestock in Balance. Rome: FAO. Online. Available at http://www.fao.org/docrep/012/i0680e/i0680e.pdf. Accessed August 28, 2014.

FAO. 2014. The State of World Fisheries and Aquaculture: Opportunities and Challenges. Rome: FAO.

Fraser, N. 2014. Sustainable livestock. For people, for the planet. Paper presented at the International Meat Secretariat (IMS) World Meat Congress, June 16, 2014, Beijing. Online. Available at http://www.meat-ims.org/wp-content/uploads/2014/06/Plenary-IMS-Beijing-2014-Fraser.editsJD-1-2.pdf. Accessed August 28, 2014.

Garnett, T., M. C. Appleby, A. Balmford, I. J. Bateman, T. G. Benton, P. Bloomer, B. Burlingame, M. Dawkins, L. Dolan, D. Fraser, M. Herrero, I. Hoffmann, P. Smith, P. K. Thornton, C. Toulmin, S. J. Vermeulen, and H. C. J. Godfray. 2013. Sustainable intensification in agriculture: Premises and policies. Science 341(6141):33-34.

Gerber, P. J., H. Steinfeld, B. Henderson, A. Mottet, C. Opio, J. Dijkman, A. Falcucci, and G. Tempio. 2013. Tackling Climate Change through Livestock: A Global Assessment of Emissions and Mitigation Opportunities. Rome: FAO. Online. Available at http://www.fao.org/docrep/018/i3437e/i3437e.pdf. Accessed August 28, 2014.

Herrero, M. 2014. Livestock in the developing world: Searching for sustainable solutions. Presentation at the First Meeting on Considerations for the Future of Animal Science Research, March 10, 2014, Washington, DC.

Herrero, M., and P. K. Thornton. 2013. Livestock and global change: Emerging issues for sustainable food systems. Proceedings of the National Academy of Sciences of the United States of America 110(52):20878-20881.

Herrero, M., D. Grace, J. Njuki, N. Johnson, D. Enahoro, S. Silvestri, and M. C. Rufino. 2013a. The roles of livestock in developing countries. Animal 7(S1):3-18.

Herrero, M., P. Havlík, H. Valin, A. Notenbaert, M. C. Rufino, P. K. Thornton, M. Blummel, F. Weiss, D. Grace, and M. Obersteiner. 2013b. Biomass use, production, feed efficiencies and greenhouse gas emissions from global livestock systems. Proceedings of the National Academy of Sciences of the United States of America 110(52):20888-20893.

Hume, D. A., C. B. A. Whitelaw, and A. L. Archibald. 2011. The future of animal production: improving productivity and sustainability. Journal of Agricultural Science 149(S1):9-16.

IPCC (Intergovernmental Panel on Climate Change). 2014. Climate Change 2014: Impacts, Adaptation, and Vulnerability. Cambridge, UK: Cambridge University Press.

Kates, R. W., W. C. Clark, R. Corell, J. M. Hall, C. C. Jaeger, I. Lowe, J. J. McCarthy, H. J. Schellnhuber, B. Bolin, N. M. Dickson, S. Faucheux, G. C. Gallopin, A. Grübler, B. Huntley, J. Jäger, N. S. Jodha, R. E. Kasperson, A. Mabogunje, P. Matson, H. Mooney, B. Moore, T. O'Riordan, and U. Svedin. 2001. Sustainability science. Science 292(5517):641-642.

Keats, S., and S. Wiggins. 2014. Future Diets: Implications for Agriculture and Food Prices. Overseas Development Institute Report. Available: http://www.odi.org/sites/odi.org.uk/files/odi-assets/publications-opinion-files/8776.pdf. Accessed October 13, 2014.

Kelly, A., B. Osburn, and M. Salman. 2014. Veterinary medicine's increasing role in global health. 2(7):e379-e380. Online. Available at http://www.thelancet.com/pdfs/journals/langlo/PIIS2214-109X(14)70255-4.pdf. Accessed February 4, 2015.

Lister, G., G. Tonsor, M. Brix, T. Schroeder, and C. Yang. 2014. Food Values Applied to Livestock Products. Online. Available at http://www.agmanager.info/livestock/marketing/WorkingPapers/WP1_Food Values-LivestockProducts.pdf. Accessed August 15, 2014.

Matthews, L. R., and P. H. Hemsworth. 2012. Drivers of change: Law, international markets, and policy. Animal Frontiers 2(3):40-45.

McLeod, A., ed. 2011. World Livestock 2011: Livestock in Food Security. Rome: FAO. Online. Available at http://www.fao.org/docrep/014/i2373e/i2373e.pdf. Accessed September 25, 2014.

Mench, J. A. 2008. Farm animal welfare in the U.S.A.: Farming practices, research, education, regulation, and assurance programs. Applied Animal Behaviour Science 113(4):298-312.

Niles, M. T. 2013. Achieving social sustainability in animal agriculture: Challenges and opportunities to reconcile multiple sustainability goals. Pp. 193-210 in Sustainable Animal Agriculture, E. Kebreab, ed. Oxon, UK: CAB International.

NOAA (National Oceanic and Atmospheric Administration). 2012. Fisheries of the United States, 2012: A Statistical Snapshot of 2012 Fish Landings. Online. Available at http://www.st.nmfs.noaa.gov/Assets/commercial/fus/fus12/FUS_2012_factsheet.pdf. Accessed September 3, 2014.

NRC (National Research Council). 1999. Our Common Journey: A Transition Toward Sustainability. Washington, DC: National Academies Press.

NRC. 2010. Toward Sustainable Agricultural Systems in the 21st Century. Washington, DC: The National Academies Press.

NRC. 2011. Sustainability and the U.S. EPA. Washington, DC: The National Academies Press.

NRC. 2013. Sustainability for the Nation: Resource Connection and Governance Linkages. Washington, DC: The National Academies Press.

OECD-FAO (Organisation for Economic Co-operation and Development and Food and Agriculture Organization). 2014. OECD-FAO Agricultural Outlook: 2014-2023. Online. Available at http://www.oecd.org/site/oecd-faoagriculturaloutlook. Accessed September 25, 2014.

OSHA (U.S. Department of Labor Occupational Safety and Health Administration). 2014. Agricultural Operations. Online. Available at https://www.osha.gov/dsg/topics/agriculturaloperations/index.html. Accessed September 3, 2014.

Reid, R. S., K. A. Galvin, and R. S. Kruska. 2008. Global significance of extensive grazing lands and pastoral societies: An introduction. Pp. 1-24 in Fragmentation in Semi-arid and Arid Landscapes. New York: Springer.

Research, Education and Economics Task Force. 2004. National Institute for Food and Agriculture: A Proposal. Washington, DC: U.S. Department of Agriculture. Online. Available at http://www.google.com/url?sa=t&rct=j&q=&esrc=s&source=web&cd=1&ved=0CB4QFjAA&url=http%3A%2F%2Fwww.ars.usda.gov%2Fsp2userfiles%2Fplace%2F00000000%2Fnational.doc&ei=qb7HVOHaDYfEgwTg44OYDg&usg=AFQjCNGqOs-G_HpC3Y97nH9sz9Fsa9RAnQ&sig2=sdWMQJnqFAXTEkxogBOtMA&bvm=bv.84349003,d.eXY. Accessed February 4, 2015.

Righter, A. 2012. Survey Results from Sodexo's Meatless Monday Initiative. Center for a Livable Future. Online. Available at http://www.livablefutureblog.com/2012/04/survey-results-sodexo-meatless-monday. Accessed September 24, 2014.

Roberts, R. M., G. W. Smith, F. W. Bazer, J. Cibelli, G. E. Seidel, Jr., D. E. Bauman, L. P. Reynolds, and J. J. Ireland. 2009. Farm animal research in crisis. Science 324(5926):468-469.

Rosegrant, M. W., M. Fernandez, A. Sinha, J. Alder, H. Ahammad, C. de Fraiture, B. Eickhout, J. Fonseca, J. Huang, O. Koyama, A. M. Omezzine, P. Pingali, R. Ramirez, C. Ringler, S. Robinson, P. Thornton, D. van Vuuren, and H. Yana-Shapiro. 2009. Looking into the future for agriculture and AKST (Agricultural Knowledge Science and Technology). Pp. 307-376 in Agriculture at a Crossroads, B. D. McIntyre, H. R. Herren, J. Wakhungu, and R. T. Watson, eds. Washington, DC: Island Press.

Steinfeld, H. 2014. Animal agriculture and climate change. Presentation at the Second Meeting on Considerations for the Future of Animal Science Research, May 13, Washington, DC.

Steinfeld, H., P. Gerber, T. Wassenaar, V. Castel, M. Rosales, and C. de Haan. 2006. Livestock's Long Shadow: Environmental Issues and Options. Rome: FAO. Online. Available at http://www.fao.org/docrep/010/a0701e/a0701e00.htm. Accessed August 15, 2014.

Tubiello, F. N., M. Salvatore, R. D. Cóndor Golec, A. Ferrara, S. Rossi, R. Biancalani, S. Federici, H. Jacobs, and A. Flammini. 2014. Agriculture, Forestry and Other Land Use Emissions by Sources and Removals by Sinks: 1990-2011 Analysis. ESS Working Paper 2. Rome: FAO. Online. Available at http://www.fao.org/docrep/019/i3671e/i3671e.pdf. Accessed September 25, 2014.

United Nations. 2013. World Population Prospects: The 2012 Revision, Key Findings and Advance Tables. New York: United Nations. Online. Available at http://esa.un.org/wpp/documentation/pdf/WPP2012_%20KEY%20FINDINGS.pdf. Accessed August 15, 2014.

U.S. Census Bureau. 2011. The 2011 Statistical Abstract: International Statistics.

U.S. Dairy Export Council. 2014. Export Trade Data. Online. Available at https://www.usdec.org/Why/content.cfm?ItemNumber=82452. Accessed August 15, 2014.

USDA. 2014a. Census of Aquaculture (2013). Volume 3, Special Studies. Part 2.

USDA. 2014b. USDA Agricultural Projections to 2023. Long-Term Projections Report OCE-2014-1. Prepared by the Interagency Agricultural Projections Committee. Available at http://www.ers.usda.gov/publications/oce-usda-agricultural-projections/oce141.aspx.

USDA ERS (U.S. Department of Agriculture Economic Research Service). 2013. Statistics & Information. Online. Available at http://www.ers.usda.gov/topics/animal-products/poultry-eggs/statistics-information.aspx. Accessed October 31, 2014.

USDA ERS. 2014. U.S. and State-Level Farm Income and Wealth Statistics (Includes the U.S. Farm Income). Online. Available at http://www.ers.usda.gov/data-products/farm-income-and-wealth-statistics/us-and-state-level-farm-income-and-wealth-statistics-(includes-the-us-farm-income-forecast-for-2014).aspx. Accessed August 13, 2014.

U.S. Global Change Research Program. 2014. Glossary. Online. Available at http://www.globalchange.gov/climate-change/glossary. Accessed September 2, 2014.

USMEF (U.S. Meat Export Federation). 2012a. Total U.S. Beef Exports 2003-2012 (Including Variety Meat). Online. Available at http://www.usmef.org/downloads/Beef-2003-to-2012.pdf. Accessed August 15, 2014.

USMEF. 2012b. Total U.S. Lamb Exports 2003-2012 (Including Variety Meat). Online. Available at http://www.usmef.org/downloads/Lamb-2003-to-2012.pdf. Accessed August 15, 2014.

USMEF. 2012c. Total U.S. Pork Exports 2003-2012 (Including Variety Meat). Online. Available at http://www.usmef.org/downloads/Pork-2003-to-2012.pdf. Accessed August 15, 2014.

Waite, T. A. 2007. Revisiting evidence for sustainability of bushmeat hunting in West Africa. Environmental Management 40(3):476-480.

WCED (United Nations World Commission on Environment and Development). 1987. Our Common Future. Oxford, UK: Oxford University Press.

Wolfe, N. D., A. T. Prosser, J. K. Carr, U. Tamoufe, E. Mpoudi-Ngole, J. N. Torimiro, M. LeBreton, F. E. McCutchan, D. L. Birx, and D. S. Burke. 2004. Exposure to nonhuman primates in rural Cameroon. Emerging Infectious Diseases 10(12):2094-2099.

2

Global Food Security Challenge: Sustainability Considerations

FUTURE DEMAND FOR ANIMAL PROTEIN

A growing consensus is forming around the prediction that the global human population will reach nearly 10 billion people by 2050 (FAO, 2009a). Providing adequate and nutritious food for such a large population highlights the importance of the world's agriculture system. Indeed, the Food and Agricultural Organization of the United Nations (FAO) projects that food production will have to increase by 70 percent over the same time frame (FAO, 2009a). Note that as with any prediction, there are numerous underlying assumptions and uncertainties associated with the reported number; however, it is almost certain that there will be a need for global food production to increase substantially in the foreseeable future. Projected increases in income globally will increase demands for not only more food but for better quality food, leading to an increased intake of animal protein (FAO, 2009a; Masuda and Goldsmith, 2010). The demand for more high-quality foods will have to be met by increases derived from plant and animal production systems.

As a result, animal agriculture in the 21st century faces increasing and persistent challenges to produce more animal protein products in the context of an emerging, globally complex set of conditions for sustainable animal production. This, in turn, requires the rethinking of the very nature of animal science. In addition to the increasing demand for animal products in the context of globalization of food systems, these challenges include, but are not limited to, consequences for individual country and regional concerns about food security, such as the impact of geopolitical strife on food production and distribution, the intensification of production systems in the context of societal and environmental impacts, the development and maintenance of sustainable animal production systems in the face of global environmental change, and the multidecadal decrease in public funding in real dollars for animal science in the United States and variable funding worldwide.

These challenges point to the need for animal science research that improves the integration among its disciplinary components, including food science and the socioeconomic and environmental sciences with which these various challenges overlap. A new roadmap for animal science research is required that focuses on animal production but intimately informs and is informed by the broader socioeconomic and environmental conditions of the new century. Thus, animal agriculture and animal protein production must substantially increase in production and efficiency, but in the context of sustainability. Interestingly, agriculture, and particularly crop production, has before faced similar questions related to global production and efficiency. The outcome for that challenge in the 1960s was predicated on the genetic improvement in major crop staples, often referred to as the Green Revolution (IFPRI, 2002). As a result, cereal production in Asia more than doubled between 1970 and 1995 (IFPRI, 2002). The environmental and socioeconomic consequences of that effort are still being debated (IFPRI, 2002), given the high input demands for hybrid crops and the difficulty that marginal, smallholder farmers have in participating in their production. Today's agricultural researchers and practitioners can learn from the successes and mistakes of the Green Revolution to design and conduct even more successful research that optimizes animal protein production while minimizing environmental, social, and economic impacts. This is the principal challenge for today's researchers and policy makers.

20th CENTURY EXPANSION IN PRODUCTIVITY

Agricultural research has made significant strides in the last century regarding productivity in the United States. Farm efficiencies, including animal production, have improved. For example, a fourfold increase in milk yield per cow was obtained between 1944 and 2007. The dairy milk industry has achieved a 59 percent increase in total milk production (53 billion kg in 1944 vs. 84 billion kg in 2007) while also decreasing the national dairy herd from 25.6 to 9.2 million cattle (Capper et al., 2009).

Similarly, beef cattle productivity increased significantly between 1977 and 2007. The average slaughter weight (607 kg in 2007 vs. 468 kg in 1977) and growth rate (1.18 kg/day in 2007 vs. 0.72 kg/day in 1977) resulted in the total average days from birth to slaughter being reduced from 609 days (1977) to 485 days (2007) (Capper, 2013). The pork industry increased the number of hogs marketed from 87.6 million in 1959 to 112.6 million in 2009 from a breeding herd that decreased in size by 39 percent over the same time period (Cady et al., 2013). In the last 50 years, the poultry industry has made tremendous progress. According to Ferket (2010), in 1957, a 42-day-old broiler weighed 540 g with a feed conversion rate (FCR) of 2.35. In 2010, a broiler of the same age weighed 2.8 kg with an FCR under 1.70. Not only has the growth performance of broilers improved significantly during the past 50 years, but also its conformational structure has changed. Similar changes have occurred in turkeys. In 1966, an 18-week-old turkey tom weighed about 8 kg with an FCR of 3.0. in 2010, a tom of the same age could weigh over 19 kg with an FCR under 2.55. Layer performance has also changed significantly from 1958 until the present day. Over the past 50 years, egg production per hen has increased over 64 percent, egg mass per hen by 83 percent, and the amount of feed consumed per gram of egg produced decreased by over 20 percent (Ferket, 2010). Crop yields, some of which serve animal agriculture, have undergone similar increases. Average corn yields increased from approximately 1.6 tons/ha in the first third of the 20th century to approximately 9.5 tons/ha in 2009 (Edgerton, 2009). Aquaculture production increased 12 percent in volume and 19 percent in value from 1998 to 2008, principally from the shellfish sector, which includes mollusks and crustaceans (Olin, 2012).

Outside of a few intensive production systems, the developing world has not witnessed similar growth in animal agricultural productivity across its many sectors. Productivity in developing countries, particularly sub-Saharan Africa and parts of Southeast Asia, is far below world averages and too low to support expanding local demand (Sanchez, 2010). Agriculture in the developing world faces inadequate inputs and infrastructure, insufficient agricultural research focused on local environmental concerns, and competition from specialized commercial production destined for distant markets (Sanchez et al., 2007). Although technologies that may be of value to animal agriculturalists in developing countries exist or are under development, low levels of use have resulted in a wide yield gap (Table 2-1; NRC, 2009). The exception is aquaculture, where gains have been made and now constitutes 49 percent of the entire amount of seafood (excluding plants) consumed in the world and is expected to reach 62 percent by 2030 (FAO, 2014b). Commercial aquaculture production increased from 4.8 to 66.6 million tons between 1980 and 2012, a 13.9-fold increase (FAO, 2014b). As a result of these and other gains, the food supply in the United States and many other parts of the world has been affordable and abundant. The volume of global aquaculture production has risen dramatically since the 1970s, largely due to productivity growth facilitated by improved research and development (R&D), input factors, and technical efficiency in aquaculture farming (Asche, 2013).

TABLE 2-1 Average Animal Product Output in 2007 (kg of meat or milk per animal)

Country	Beef & Buffalo	Poultry Meat	Goat	Cattle Milk
United States	357	1.77	29	9,219
Tanzania	107	0.91	12	173
China	135	1.41	14	3,109
Kenya	150	1.16	12	565
Nigeria	130	1.00	13	240

SOURCE: Gapminder (http://www.gapminder.org).

THE WAY FORWARD

Responding to a perceived crisis in U.S. agriculture and recognizing that continued innovation is key to meeting this crisis, the U.S. 2004 Danforth Task Force recommended that the U.S. Department of Agriculture (USDA) create two new programs: the National Institute of Food and Agriculture (NIFA), subsequently created by the Food, Conservation and Energy Act of 2008, and the Agriculture and Food Research Initiative (AFRI). NIFA, mirroring earlier recommendations of the USDA, proposed a de-emphasis on agricultural research on productivity, efficiency, and innovation, and the creation of activities on renewable energy, obesity, human disease prevention of zoonotic disease (e.g., avian influenza), and environmental impact of agriculture.

Internationally funded research in the developing world focuses on improved productivity or public health considerations (e.g., zoonotic issues). Examples of agencies that fund international research are the World Bank, United Nations Development Programme, U.S. Agency for

International Development, Rockefeller Foundation, and Bill & Melinda Gates Foundation. The majority of this funding is through financial assistance, donations, and loans, and focuses on production and social well-being, which includes translational research (research to application) and education and training activities.

The direction of international research indicates correctly that sustainable animal agriculture is intertwined with food security (NRC, 2012). As defined by the 1996 World Food Summit, food security refers to conditions in which "all people, at all times, have physical and economic access to sufficient, safe, and nutritious food to meet their dietary needs and food preferences for an active and healthy life" (FAO, 1996). Addressing food security requires the consideration of factors other than production, such as access to food (i.e., entitlements), as part of meeting sustainable animal production goals. After all, famine and malnutrition, including "food deserts" in the United States, have long occurred and continue to persist in the midst of plentiful food (Sen, 1983; Baro and Deubel, 2006). Concerns about the prevalence of food deserts prompted a 2009 Institute of Medicine workshop exploring the negative public health effects of food deserts in low-income areas in the United States (IOM and NRC, 2009). Improving entitlements, which falls beyond the research domain of animal agriculture, would help to ameliorate food security problems in 2050, but does not eliminate the need for significant advances in sustainable production.

Today's research must carefully consider environmental, health and disease, sociocultural considerations, community welfare, animal welfare, economic and policy constraints and other factors. The needs for current and future global food security cannot be met without greater emphasis and expansion of R&D devoted to productivity, efficiencies, and innovation in animal agriculture. The challenges of meeting these sustainability goals are massive and far beyond the domain of corporate interests. The committee agrees with the conclusion of the 2004 Danforth report that publicly sponsored research is essential to continued agricultural innovation.

CHALLENGES TO SUSTAINABLE PRODUCTIVITY

Global Environmental Change

Global environmental change refers to the totality of changes, both natural and anthropogenic in origin, under way in the earth system from ecosystems to climate change. Animal agriculture affects this change through the landscapes it consumes and the biogeochemical cycles it affects, and is also affected by these changes, in some cases significantly, and must adapt to them in order to provide the quantity and affordability of animal protein expected by society. This adaptation, in turn, has important implications for sustainable production.

Approximately 17 billion food animals globally occupy 30 percent of the ice-free land surface of Earth, resulting in about 72 percent of deforestation worldwide and consuming 32 percent of freshwater globally (Reid et al., 2008; FAO, 2009b; Nepstad et al., 2011). Currently, food animals contribute 14.5 percent of global greenhouse gas (GHG) emissions, according to the FAO (Gerber et al., 2013). Enteric fermentation from ruminants is the second largest global source of methane (Makkar and Vercoe, 2007). Land degradation from overgrazing rangelands has long been noted across many different socioeconomic and environmental conditions, leading to soil erosion and soil nutrient loss, reduced feed stocks, and habitat changes, among other impacts (Havstad et al., 2007; Jun Li et al., 2007). The impacts of some aquaculture practices have negatively affected wild fish stocks, prompting development of certification programs to help ameliorate such impacts (Naylor et al., 1998). These are only a few of the environmental consequences of animal agriculture (Wirsenius, 2003; Steinfeld et al., 2006a; Galloway et al., 2010), which also include environmental tradeoffs between land or water use for animal agricultural and nonagricultural purposes (e.g., urbanization and conservation), preservation, and maintenance of biotic diversity and functioning ecosystems and landscapes (Reid et al., 2008). Animal agriculture is increasingly confronted by these and other environmental issues.

By the end of this century, the global average temperature is likely to increase by between 2.6 and 4.8°C (NAS and RS, 2014). Significant temperature increases and rising drought are projected for much of the global land acres currently devoted to food animal production and crops used to support production worldwide (Parry et al., 2007; Thornton et al., 2009; Meehl et al., 2013). Overall, global climate change occurring in tandem with land-use changes raises a series of water and feed availability and quality problems for animal agriculture. Among these are the tradeoffs on environmental (ecosystem) services at large due to water and land demands of various systems of meat, egg, and milk production (Herrero et al., 2009; Thornton, 2010). For aquaculture, freshwater systems will be subject to many of the changes anticipated for terrestrial systems such as flood, drought, and water availability. Marine-based systems will be subject to changes in sea surface temperatures, ocean acidification, and other related marine environmental variables, such as wind velocity, currents, and wave action (Handisyde et al., 2006). These changes will impact all types of animal agricultural production systems as well as the livelihoods of the communities they support.

The exact impact on food animal production and the future of animal agriculture as a whole are still the subjects of debate (Thornton, 2010). Some contend that by 2050, food animal production will reach or exceed certain environmental thresholds (Pelletier and Tyedmers, 2010). However, it is also contended that food animal production can be increased while attenuating environmental impact (Steinfeld and Gerber, 2010), possibly as a result of agricultural land use reaching a peak due to more land-efficient production processes (Ausubel et al., 2012). It is also believed that policies can be effective in decreasing meat consumption (Myers, 2014) among groups whose current consumption levels are high. Animal science research funding is needed because the true outcomes of these factors will reflect the capacity of animal agriculture to incorporate variation in geographic location and production types to adapt to global environmental change. The FAO, for example, found significant regional variation in GHG emission intensities from dairy, ranging from 1.3 CO_2-eq/kg FPCM (fat- and protein-corrected milk) in North America (average milk yield of 8,900 kg/year) to 7.5 kg CO_2-eq/kg FPCM in sub-Saharan Africa (average milk yield of 300 kg/year) (FAO, 2010). Hall et al. (2011) found that within a single aquaculture species group globally, impacts could vary by 50 percent or more. Adaptive capacities for much of sub-Saharan Africa and Southeast Asia are especially problematic for land-based systems of production, such as crop production (Parry et al., 2007; Lobell et al., 2008; Thornton et al., 2009), although environmental problems encountered in these areas attributable to climate change will be experienced throughout tropical regions globally (Parry et al., 2004; Morton, 2007; Thornton et al., 2009). These problems include negative consequences of climate warming for the quantity and quality of feeds owing to lower herbage growth, pasture compositional changes, increasing drought, and nitrogen leaching from the increased intensity of rainfall; increasing animal heat stress; reduction in water availability; and changing pest and disease vector challenges. In contrast, increased length of the growing season may favor food animal production at higher latitudes (Baker et al., 1993; FAO, 2007). Freshwater and marine-based aquaculture at temperate latitudes would probably decrease substantially due to detrimental physiological demands caused by increases in water temperatures from climate warming.

The goal of reducing animal agriculture's ecological footprint in the face of increasing production has progressively gained more attention (Clay, 2009; Chicago Council on Global Affairs, 2013). How this goal might be achieved will vary by the type of production system; systems vary from high to low input (intensive to extensive) in terms of the amount of capital and technological input needed, output per unit area, and time required to manage animals (Bertrand, 2014). Thorough analysis is necessary before viewing any production system as more or less sustainable, given the many environmental (ecosystem) services, species types, and social and livelihood implications. Generally, however, intensive systems for feed and animal production tend to produce fewer emissions and use less land per unit of production (Burney et al., 2010; O'Mara, 2011; Herrero et al., 2013). For example, the U.S. dairy industry has reduced feed use by 77 percent, land use by 90 percent, and water use by 65 percent and has achieved a 63 percent decrease in GHG emissions per kilogram of milk (Capper et al., 2009). The U.S. beef industry has reduced feed (19 percent), land use (33 percent), water use (12 percent) and GHG emissions (16 percent) per kilogram of beef over the referenced 30-year time period (Capper, 2011). The U.S. pork industry has achieved similar improvements, reducing water use by 41 percent, land use by 78 percent, and GHG emissions by 35 percent per pound of meat (Boyd and Roger, 2012). Similarly, the environmental footprint of the U.S. egg industry per kilogram of eggs produced is 65 percent lower in acidifying emissions, 71 percent lower in eutrophying emissions, 71 percent lower in GHG emissions, and 31 percent lower in cumulative energy demand than in 1960 (Pelletier et al., 2014).

In extensive animal agriculture, on the other hand, high constraints to capital and technology, as well as production goals that often differ from those in the commercial sector (Netting, 1993), results in lower production and, in some cases, increases footprint measures, such as land area per food animal head (Capper 2013; Herrero et al., 2013). Various assessments for such systems in Africa envision shifts in food animal portfolios (e.g., cattle to sheep and goats or the reverse) to offset projected climate change impacts on livelihoods (Seo and Mendelsohn, 2008), although quantification of the environmental consequences of these shifts has received minimal research attention. In 2008, 39 percent of all aquaculture was extensive, 42 percent was semi-intensive, and 20 percent was intensive (based on the weight of harvested fish). It has been projected that the available food needs for a global population of 9.2 billion people in 2050 will require a slightly greater than 100 percent increase in aquaculture production from 67 to 140 tonnes. This need can only be accomplished through the transition from extensive to semi-intensive systems and some semi-intensive systems to intensive systems (Waite et al., 2014).

Overall, the data demonstrate that intensive food animal systems reduce resource use, waste output, and GHG emissions per unit of food compared to extensive systems. This observation, however, has a major caveat. Although intensive systems allow for more efficiencies and often enhance the local economy and increase employment, they may have greater effects on local air, water, and land quality, especially where such systems cluster locationally (Atkinson and Watson, 1996; Eghball et al., 1997; Monaghan et al., 2005). The environmental impacts resulting from mismanagement of wastes include, among others, excess nutrients, such as nitrogen and phosphorus in water, which can contribute to low levels of dissolved oxygen. Other examples include detrimental effects on water quality from the runoff from confinement consolidation and intensification production. These concerns highlight the importance of proper management, location, and permitting and monitoring, which can reduce, minimize, or eliminate these impacts (Hribar, 2010; EPA, 2014). Although these footprint metrics for high-input systems have decreased substantially (Naylor et al., 2009), the overall magnitude of demand points to the need for sustained attention to the issues indicated by these metrics. The use of alternative feedstuffs, both plant and animal derived, continue to be assessed as viable production alternatives to help mitigate these environmental impacts. Various calculations indicate, however, that appropriate intensive animal production could yield significant environmental mitigation consequences for all agriculture globally (Herrero et al., 2009; Thornton and Herrero, 2010; Havlík et al., 2014). The movement from extensive to semi-intensive and intensive aquaculture production globally includes systems that employ recirculation of water yields, and could increase water-use efficiency in the aquaculture enterprise. This reduction also decreases the conflict of water use with other agricultural production systems. Addressing these impacts and lowering the ecological footprint of animal agriculture will increasingly become important because high-input animal agriculture in certain parts of the world can produce significant, detrimental consequences elsewhere in the world for community livelihoods, food security, and access to resources, such as water (McLeod, 2011).

Health and Disease

Animal Health and Disease

Considerable challenges to the health and well-being of animals and humans are presented by animal disease pathogens. Understanding and developing effective measures to control animal infectious diseases are sometimes problematic mainly due to the lack of direct and continuous monitoring of the animal status as would otherwise occur with humans. The majority of these pathogens also have the ability to have a long period of carrier status in host animals (i.e., animals show no clinical signs but are able to transmit pathogens). For example, the bacterium Salmonella enterica may develop carrier status in an animal host, and such carriers typically excrete high levels of bacteria during recovery from enteric or systemic disease without showing clinical signs (Stevens et al., 2009). The carrier state may exist for the lifetime of the animal host with bacterial species such as S. enterica serovar Dublin. Therefore, preventive measures are critically important components of the first line of defense. There is a need for exploration and deployment of state-of-the art approaches for early diagnosis and surveillance to provide a network of global intelligence on their spread and an assessment of risk. The delivery of effective vaccine strategies for the control of major animal pathogens will be especially important, and finding new and better vaccines able to deliver long-lasting and durable protective immunity will be needed to be effective against multiple strains or variants.

One main goal of animal health is in food provision and food safety. During the last four decades, there have been several emerging and new health events that have received public attention. Most of these health events were linked to animal diseases or originated in animal products, including avian influenza, bovine spongiform encephalopathy, West Nile fever, sudden acute respiratory syndrome, HIV/AIDS, and Ebola virus. Because of the extensive involvement of animals and their products in these events, animal scientists and animal production sectors have been involved in measures to minimize the spread and impact of these diseases. The public health sector, particularly within central government or international agencies, has maintained the lead in the effort to control or eradicate these diseases. Nevertheless, the prevention effort requires major involvement of animal health officers and others in related industries because the roots of most of these diseases extend to animal populations, particularly to food-producing animals (Salman, 2009).

The Agreement on the Application of Sanitary and Phytosanitary Measures (SPS Agreement) was finalized as part of the Uruguay Round of Multilateral Trade Negotiations General Agreement on Tariff and Trade (GATT, signed in Marrakesh in April 1994). Subsequent to its approval, the World Trade Organization (WTO) was established. The SPS Agreement's main intent was to provide guidelines and provisions to member countries to facilitate trade while taking measures to protect human, animal, or plant life or health. The agreement advocates the use of international standards from the World Organization for Animal Health (OIE), the Codex Alimentarius, and the International Plant Protection Convention (IPPC) as the basis for recommended standards (Zepeda et al., 2001). It has become clear that the health status of animals and their products plays a major role in import and export regulations. This type of requirement for trade has placed pressure on the animal health structure both nationally and internationally. Animal scientists, particularly veterinary professionals, throughout the world are faced with having to fulfill a crucial role in protecting their country's animal health status, providing sound surveillance information on the occurrence of diseases within their territories, and conducting scientifically valid risk analyses to establish justified import requirements (Salman, 2009). Population-based approaches for disease management require scientifically sound research prior to application of available options.

Animal diseases are severely affecting the production of food animals and disrupting regional and international trade in animals and animal products. Such diseases as hand, foot, and mouth disease, African swine fever, blue tongue, and classical swine fever are eminent transboundary animal diseases. These diseases are notorious for their ability to severely affect, and indeed disrupt, regional and international trade in animals and animal products. They are also notorious for the enormous financial damage and ethical violations caused when introduced into countries that are free from these diseases. The burden of these diseases involving the loss of animals and biological diversity and the lowering of production efficiency is generally much less well known or is underestimated. Declines in biodiversity have been identified as potentially increasing infectious disease transmission among ecological populations in certain cases (Keesing et al., 2010). Furthermore, these diseases threaten food security and the livelihoods of smallholders and prevent animal husbandry sectors from developing their economic potential (Kelly et al., 2013).

Emerging and reemerging animal diseases that were considered under control are threatening trade and the disruption of animal protein distribution. The appearance of the porcine epidemic diarrhea (PED) virus, for example, in North America presents significant challenges for producers and animal health officers. The disease is a viral infection affecting swine populations with significant impact on production with potential transmission through the fecal-oral route. The introduction of this virus to North America is evidence of the need for biosecurity that is more comprehensive to include all possible ways for pathogens and harmful materials to enter the food chain. The rapid spread of PED among intensive swine premises in North America should be considered a challenge to researchers to explore better options for monitoring and surveillance systems (Box 2-1).

BOX 2-1
Porcine Epidemic Diarrhea (PED) Virus

The PED virus has affected European swine for over 40 years, but only appeared in the United States in 2013 (Kerr, 2013). PED is spread between animals through fecal to oral transmission and can result in pigs suffering from vomiting, diarrhea, dehydration, and poor appetite, among other symptoms, which can result in high death rates, especially for younger animals (Kerr, 2013). PED cannot be transmitted to humans, but has the potential to create significant economic losses due to animal sickness and death in the pork industry (USDA, 2014). In a Chinese outbreak of PED in 2010, death rates of approximately 100 percent were recorded for piglets (Sun et al., 2012). Because of the "hardy and virulent nature" of the PED virus (Canadian Food Inspection Agency, 2014), the pork industry has cause for concern regarding the effects of this illness. Although a PED vaccine exists and has been used in some countries, the vaccine may not be fully effective against all disease strains of the virus (Kerr, 2013). Research toward mitigating the effects of PED and preventing infections are of critical importance to the pork industry.

Human Health and Disease

The World Health Organization (WHO) has recognized that the worldwide upswing in resistance to antibiotics is based on a combination of factors that includes "overuse in many parts of the world by both human and animals, particularly for minor infections, and misuse due to lack of access to appropriate treatment" (WHO, 2001). According to the U.S. Centers for Disease Control and Prevention (CDC), antibiotic use in people is a primary factor (CDC, 2013a), and the most acute problem is in hospitals. And the most resistant organisms in hospitals are emerging because of "poor antimicrobial stewardship among humans" (CDC, 2013b). Antibiotics are among the most commonly prescribed drugs used in human medicine; however, CDC estimates that up to 50 percent of all antibiotics prescribed are not needed or are not optimally effective as prescribed.

The U.S. Food and Drug Administration (FDA) recently took the major step of issuing guidance documents to promote the judicious use of antibiotics in agriculture. In particular, Guidance No. 213 established procedures for voluntarily phasing out growth promotion indications for medically important antibiotics in alignment with FDA's Guidance No. 209 and published proposed changes to the Veterinary Feed Directive (VFD) regulation. The VFD regulation mandates the rules and responsibilities of licensed veterinarians in prescribing and administering medically important antibiotics in feed. Guidance No. 209 adopted principles that the use of medically important antimicrobial drugs in food-producing animals should be limited to uses that are considered necessary for ensuring animal health, and should include veterinary oversight or consultation. There have been reports that many leaders in industry have moved forward to implement Guidance Nos. 209 and 213 (FDA, 2014).

In addition, The President's Council of Advisors on Science and Technology recently proposed additional findings and recommendations concerning how the U.S. government can best combat the rise of antibiotic resistance (PCAST, 2014b). This report, released simultaneously with a National Strategy on Combating Antibiotic Resistant Bacteria (PCAST, 2014a) and with an Executive Order (White House, 2014), made recommendations in three areas, including improving surveillance of the rise of antibiotic resistant bacteria and acting on surveillance data to implement appropriate infection control; increasing the longevity of current antibiotics and scaling up proven interventions to decrease the rate at which microbes develop resistance to current antibiotics; and increasing the rate at which new antibiotics, as well as other interventions, are discovered and developed.

Related to animal agriculture, the report reiterates support for FDA's Guidance Nos. 209 and 213 described above and directed the agency to proceed with their implementation. It adds that USDA, through its Cooperative Extension Service, should establish and lead a national education program to help food animal producers comply with these guidances. The report adds that FDA should also assess progress by monitoring changes in total sales of antibiotics in animal agriculture and in use of antibiotics, where possible (PCAST, 2014b).

Related to research, the report recommends

> expanding fundamental research relevant to developing new antibiotics and alternatives for treating bacterial infections, including requesting dedicated funds for the National Institutes of Health (NIH) and FDA to support fundamental research aimed at understanding and overcoming antibiotic resistance, and for Defense Advanced Research Projects Agency (DARPA) and Defense Threat Reduction Agency (DTRA) to support non-traditional approaches to overcoming antibiotic resistance. (PCAST, 2014b).

It also recommends "developing alternatives to antibiotics in agriculture, noting that USDA should develop, in collaboration with NIH and the agriculture industry, a comprehensive R&D strategy to promote the fundamental understanding of antibiotic resistance and the creation of alternatives to or improved uses of antibiotics in food animals" (PCAST, 2014b). The report states that one mechanism that should be employed is a USDA multidisciplinary innovation institute (PCAST, 2014b).

Regarding coordination, the report recommends establishing an interagency Task Force on Combating Antibiotic Resistant Bacteria that should include members from all relevant agencies, as well as establishing a President's Advisory Council on Combating Antibiotic Resistant Bacteria comprising nonfederal experts. Related to international coordination, the report notes that the federal government should support development of the WHO Global Action Plan and continue to elevate the issue of antibiotic resistance to the level of a global priority by encouraging or requiring, as appropriate, coordination among countries for surveillance, reporting, research, antibiotic stewardship, and development of new and next generation drug and diagnostics development (PCAST, 2014b).

Animal Health

During the last few decades, various regions of the world were infected with serious zoonotic diseases such as immunodeficiency viruses, SARS, MERS, and reemerging diseases such as tuberculosis, undulant fever and Rift Valley fever. Furthermore, there were 2.4 billion human illnesses and 2.2 million deaths per year from foodborne illness, and more than 1.7 million deaths from HIV/AIDS in 2011. Global outbreaks of influenza, for example, have occurred periodically in the human population. The viruses of the outbreaks in the 20th century were avian in origin and arose through mutational events. In particular, recent evidence of direct bird-to-human transmission has increased global concerns over the pandemic potential of these viruses (Dinh et al., 2006). Escherichia coli O157:H7 is another major public health concern in North America and other parts of the world. The feces of animals, particularly cattle, are considered the primary source of these bacteria, and major routes of human infection include consumption of food and water contaminated with feces and, to a lesser extent, contact with live animals. Human infections are often asymptomatic or result in uncomplicated diarrhea, but may progress to bloody diarrhea, kidney disease, and death (Griffin and Tauxe, 1991).

Food Safety

Food safety concerns pertinent to foods of animal origin often relate to Salmonella, parasite infections, antibiotic residues, Listeria, Campylobacter, Staphylococcus, and Clostridium. According to a National Animal Health Monitoring System fact sheet that detailed results of a Salmonella prevalence study, only 38 percent of sampled farms with finishing hogs had samples that were positive when tested for Salmonella (USAPHIS, 1997). Of the 38 percent of Salmonella-positive farms, the level of bacterial shedding in finishing hogs was low at an estimated 6 percent. In another study, 9.6 percent of uncooked or unprocessed samples obtained from retail stores were contaminated with Salmonella (Duffy et al., 2000). Although proper handling and cooking will prevent the consumer from becoming infected, it is of primary importance that animal science researchers work to identify means to reduce foodborne pathogens.

Antimicrobials

Antimicrobials and improvement in vaccine efficiency have saved millions of human lives. Over the course of the 20th century, deaths from infectious diseases declined markedly and contributed to a 29-year increase in life expectancy (CDC, 1999). They also play an important role in modern agriculture and in enhancing food security by preventing disease and improving food safety for humans. Antimicrobials are also used in animal agriculture to alter the animal's gut microflora and decrease the level of pathological bacteria present for production purposes. This aids feed conversion and hastens growth. Prudent and judicious use of antimicrobials is an important piece of the sustainability challenge. Animal agriculture, however, is at risk now of losing this progress through bacterial resistance. The CDC and WHO have identified antibiotic resistance as one of the greatest threats to human health worldwide. Resistance is now spreading to the point where there is a rise in so-called superbugs (CDC 2013a).

The use of antimicrobials in animal agriculture has come under intense scrutiny in recent years with some charging that use in animals is impacting human health. Although use of antimicrobials in animals (food animals and companion animals) is a factor that contributes to the wider pool of resistance, the problem is not limited to modern agricultural practices and stems from a number of factors, including environmental, biological, and management practices in both agriculture and human health care (Box 2-2; Salman et al., 2008). A recent OECD paper (Rushton et al., 2014) noted that the use of low-level antibiotics enhances the overall productivity of livestock systems, but could in the long term lead to antimicrobial resistance and to animal diseases that could lessen this productivity effect. The authors note the concern of health authorities about antimicrobial-resistant organisms and call for greater cooperation among national and international agricultural and health organizations in improving preventive approaches that do not depend on antibiotics used for human health. They also call for greater cooperation in gathering data and developing effective policies that balance the benefits and risks of low-level antibiotics in livestock system productivity and animal and human health.

The committee examined practices related to antibiotic use in other countries and data relating to those practices, for example, Denmark, which instituted a compulsory ban in 2000 and regularly reports on antibiotic use and resistance patterns.

Among the problems is that the number of new antimicrobials developed and approved has decreased steadily over the past three decades, from over 15 in the mid-1980s to just 2 since 2008. The drug development pipeline is failing for a variety of reasons—including the

> **BOX 2-2**
> **United Kingdom Report on the Use of Antimicrobials in Animals**
>
> "Increasing scientific evidence suggests that the clinical issues with antimicrobial resistance that we face in human medicine are primarily the result of antibiotic use in people, rather than the use of antimicrobials in animals. Nevertheless, use of antimicrobials in animals (which includes fish, birds, bees, and reptiles) is an important factor contributing to the wider pool of resistance which may have long term consequences" (UK Department of Health, 2013).

scientific difficulty of discovering new drugs, challenging pathways for regulatory approval, and low return on investment. As a result, pharmaceutical companies have decreased or eliminated their investments in antibiotic drug development (PCAST, 2012). In the United States, unlike traditional animal agriculture, the farming of aquatic species has come under strict regulations concerning the use of antimicrobials. Use is highly controlled and is based on identification of the disease and whether there is an antimicrobial that will be effective against the disease. Other regulatory restrictions include treating the population for an appropriate period of time and then adhering to an appropriate withdrawal time.

Most aquaculture occurs in Asia, and use of antimicrobials is indiscriminate, most often having no foundation for choice other than convenience. In addition, any attempt at enforcement of violations is difficult. The European Union places a particular emphasis on regulating the importation of aquaculture products from Asian countries, particularly those who frequently use antimicrobial and other additive materials (EC, 2014). Application of control measures will initially require trained professionals to provide the proper diagnosis and surveillance. Standards of administration need to be developed and will require international organizations to implement a cooperative effort. Because most of the seafood consumed in the United States is imported, the question of inspection protocol for imports continues to persist. The Farm Animal Integrated Research (FAIR) effort in 2012, which was undertaken to identify key priorities and strategies for the future, noted that the "the use of antimicrobials in animal agriculture has been the source of much controversy in recent years as critics express concerns about antimicrobial resistance. Investments in research will be essential to more clearly understand these issues and provide decision makers with science-based information to develop better-informed policies" (FASS, 2012).

Sociocultural Considerations

As discussed above, animal science research has contributed to remarkable increases in the production efficiency of agricultural animals. It is becoming increasingly apparent that in order for research advances in animal productivity to be useful, consideration must be given to the social norms of communities and countries in which they are to be applied. Although surveys in developed countries show consistently that food quality and safety are the most important considerations for consumers (Lusk and Briggeman, 2009), there are a host of other issues that influence the acceptability, and hence the sustainability, of existing and new animal agriculture practices. These factors include environmental, economic, and social concerns (NRC, 2010), with the latter playing an increasingly important role not only with respect to the regulatory decision-making process, but in terms of shaping consumer and supply-chain purchasing decisions in many countries (Mench, 2008; Matthews and Hemsworth, 2012). Social concerns include those related to animal welfare, equity (e.g., fair labor practices, protection of vulnerable human populations and rural communities, agricultural worker health and safety), corporate social responsibility and business ethics, food security, agricultural and food traditions, and naturalness of food products (NRC, 2010; Anthony, 2012; Niles, 2013; Lister et al., 2014).

Social concerns cannot be adequately addressed using only traditional animal sciences research approaches (Box 2-3). Dealing with the complex issues associated with the values surrounding agricultural sustainability requires transdisciplinary approaches that bring together experts in the natural sciences, social sciences, and humanities to address wicked problems (Peterson, 2013). A critical element of addressing social challenges is to determine what the public knows ("objective knowledge") and believes they know ("subjective knowledge") about animal agriculture technologies and practices (Costa-Font et al., 2008). Increasing public understanding of animal agriculture and aquaculture technologies is important. Public perceptions of risks and benefits of particular technologies are dynamic and factual information can shape those perceptions as long as it is conveyed by sources that are trusted

(Costa-Font et al., 2008) and in language that resonates with the various stakeholders. Perceptions are also shaped by cultural values and can be influenced by information that creates uncertainty about the safety or importance of the innovation in question, even if that information comes from individuals or organizations with vested interests (Oreskes and Conway, 2010). Research on climate change provides some insights on which animal agricultural technology might build. It indicates that information about technology is most likely to be accepted if it is perceived as being relevant to the recipient, based on adequate scientific information, produced in a way that is unbiased and respectful of stakeholders' divergent values and beliefs, and fair in its treatment of opposing views and interests (Cash et al., 2003).

BOX 2-3
Genetically Engineered Salmon

The potential fate of genetically engineered salmon in the marketplace represents one example of the effects social concerns can have on the adoption of new technologies in animal agriculture (Van Eenennaam and Muir, 2011). In 1995, a U.S. company called AquaBounty applied for FDA approval to market Atlantic salmon transgenic for expression of a Chinook salmon growth hormone gene. The insertion of this gene leads to a doubling of fish growth rate and the potential for increasing the affordability and availability of salmon. During the two decades in which the application was under consideration, public debate grew increasingly heated, with more than 1.5 million people writing letters in opposition to FDA approval of what became dubbed the "Frankenfish." Concerns were raised about the welfare of the transgenic salmon themselves, their potential to escape and cause the collapse of wild salmon populations, food safety issues, negative impacts on commercial salmon fishermen, and, quoting Alaska Senator Lisa Murkowski, "messing with Mother Nature" (Welch, 2014). In 2014, the Senate Appropriations Committee passed an amendment requiring labeling of transgenic salmon if they are approved by FDA for human consumption, although it appears that approval may be moot—a coalition of 30 consumer, food safety, environmental, sustainable agriculture, public health, and animal health and welfare organizations have successfully lobbied 65 major supermarkets to pledge that they will not sell it. The approval of the application is still pending.

Community Welfare

The number of hungry and malnourished people, due to the recent sharp increase in food prices both nationally and internationally, has increased the awareness of policy makers and of the general public to the fragility of the global food system. Poverty is associated with endemic and chronic diseases, poor nutrition, environmental challenges, and a lack of leadership in understanding public health and food safety. Poverty affects one out of seven people in the world today. Approximately 2.6 billion people worldwide receive less than $2 a day and 1 billion people live on $1.25 a day and suffer from chronic hunger. Globally, 25 percent of children are born malnourished with compromised growth, have weakened cognitive and immune systems, and risk early childhood death (FAO, 2009a; Lam et al. 2013).

The greatest concentration of poverty is in sub-Saharan Africa and in South Asia, where subsistence farming is associated with mixed farming of both food animals and crops. In addition, many people are nomadic, following their animals as they seek water and forage on open ranges in desert regions. It is estimated that there will be a major wave of farmers leaving rural areas and migrating to urban areas over the next three decades, spurred by global and regional climate change impacts. Such a transformation is challenging for the world as there are significant concerns about whether there will be adequate and safe food, nutrition, and water resources by midcentury (FAO, 2009a). Food animals are crucial to the lives of poor farmers and their children. In developing countries, animal agriculture contributes to employment and GDP, but poverty and malnutrition persist. The production of food animals as an animal protein source increasingly expands beyond household and community food animal assets into commercial production. Animal health issues confront the nearly 1 billion poor producers with consequences for their livelihoods and human health (Kelly et al., 2014).

Animal Welfare

Concerns about animal welfare increasingly shape the acceptability and adoption of food animal production technologies. Although concerns about animal welfare are not new, it was not until the 1960s that significant public concern about animal welfare associated with the intensification of animal agriculture began to be apparent. These concerns centered mainly on confinement rearing and led to an influential British government committee report that stated that both the physical and mental states of animals were important for welfare and recommended that all farm animals should be provided at least sufficient space to "stand up, lie down, turn around, groom themselves and stretch their limbs," referred to as the "Five Freedoms" (Brambell, 1965). Over time, this list of freedoms was expanded to include freedom from fear, distress, discomfort, pain, injury, disease, hunger, and thirst as well as freedom to express normal behaviors via provision of appropriate facilities and social companions. These Five Freedoms have been accepted by legislative and standards-setting bodies in many countries as ethical principles that underlie the care and treatment of farm animals. They also form the basis for a joint statement by the American Veterinary Medical Association (AVMA) and the Federation of Veterinarians of Europe (FVE) describing the roles of veterinarians with respect to animal welfare (AVMA, 2014).

As public interest in and concern about animal welfare increases globally, emphasis on regulation, voluntary standards setting, and the development of animal welfare labeling programs continues to increase. Many countries now have regulations, codes of practice, or standards covering animal transport and humane slaughter, and an increasing number also have standards covering the breeding, housing and management of farm animals (FAO, 2014a). The most far-reaching regulations are in the European Union, which has banned several common animal housing systems because they involve a high degree of confinement; many countries within the European Union have also banned or are planning to ban painful procedures such as castration or beak trimming, or at least require that pain relief be provided when these procedures are performed. Although the main focus has been on food animals and poultry, there has recently been growing interest in the welfare of farmed fish, with particular emphasis on reducing pain and stress during handling and slaughter (Branson, 2008).

National and multinational retailers have also played a significant role with respect to animal welfare in developed countries via their purchasing preferences and specifications (Veissier et al., 2008). A major thrust in the last few years has been the development of animal welfare standards that can be audited to provide assurance that they are being followed. In the United States there is now a mix of these standards that have been developed by different sectors including industry, retailers, and independent animal welfare certification programs (Mench, 2008). Animal welfare research has provided an important underpinning for these kinds of standards. However, much scientific information is still lacking (Rushen et al., 2011), and this information needs to be generated in order to ensure that standards are scientifically based while still addressing evolving social concerns about the treatment of animals. It is clear that animal welfare concerns can rapidly drive major changes in agricultural animal production methods, as is currently happening for egg production in the European Union and United States (Box 2-4). The challenge going forward will be to ensure that new or modified housing systems and management practices developed to improve animal welfare also contribute to maintaining or enhancing other aspects of the sustainability of animal agriculture (Tucker et al., 2014).

BOX 2-4
Egg Production in Transition

Egg production provides a case study of the way in which social issues can affect animal production systems and, consequently, the research needed to ensure that new systems are sustainable. Eggs are a primary and inexpensive source of animal protein in both developing and developed countries, with global production of nearly 65 tonnes annually from approximately 5 billion hens. Most egg-laying hens are housed in conventional cages. This housing system began to be adopted on a large scale in the 1950s because it reduced the potential for transmission of soilborne parasites, facilitated egg collection, improved egg cleanliness, and was economically efficient; however, public pressure is now propelling the egg industry in developed countries to move away from conventional cages because of concerns about crowding and behavioral restriction of the hens (Mench et al., 2011). This pressure and the resulting mechanisms for change have come from different sources in different regions and countries, illustrating the complex social matrix surrounding animal agriculture. One approach has been legislative, with the European Union and New Zealand banning conventional cages in 2012. Australia has not regulated egg production methods, but a large proportion of consumers there now preferentially purchase free-range eggs, which is moving the egg industry toward alternative production systems to retain market share (Mench et al., 2011). In contrast, in the United States the main egg producer group, the United Egg Producers, has announced their members' intentions to discontinue the use of conventional cages despite the fact that more than 90 percent of consumers purchase conventional cage eggs. This has been driven by anticipation of

> consumer dissatisfaction with the current production system, resulting in state-by-state regulation, as has already occurred in six states. Retailers have also played a role in other countries. In Germany, for example, the major retailers decided to purchase only cage-free eggs, despite the fact that German producers were already transitioning to so-called enriched cages containing perches, nests and a scratch area, which were allowed under both European Union and German law. Animal welfare concerns are thus driving a global shift in egg production systems that will need to be accompanied by a new research effort focused on identifying optimal housing, genetics, and management for these systems, as well as improving other aspects of their sustainability such as those related to environmental impacts and food affordability (Mench et al., 2011).

Animal welfare is now becoming an increasingly prominent international issue, and as such, will potentially affect not only the United States, but animal production practices globally. The OIE began the process of developing global standards and also proposed a definition of animal welfare, which has been accepted by the 178 OIE member countries and territories:

Animal welfare is how an animal copes with the conditions in which it lives. An animal is in a good state of welfare if (as indicated by scientific evidence) it is healthy, comfortable, well nourished, safe, able to express innate behaviour, and if it is not suffering from unpleasant states such as pain, fear, and distress. Good animal welfare requires disease prevention and appropriate veterinary treatment, shelter, management and nutrition, humane handling and humane euthanasia or humane slaughter.... (OIE, 2014).

This definition echoes many of the points in the Five Freedoms. This definition, and the associated 10 General Principles for animal welfare adopted by the OIE (Box 2-5; Fraser et al., 2013) will be very important for global trade in animal products because the OIE is the WTO reference organization for standards setting for animal health. The OIE already has global animal welfare standards for transport and slaughter as well as the production of beef and chicken, which are the two most internationally traded commodities.

> **BOX 2-5**
> **World Animal Health Organization General Principles**
>
> 1. Genetic selection should always take into account the health and welfare of animals.
> 2. The physical environment, including the substrate (walking surface, resting surface, etc.) should be suited to the species and breed so as to minimize risk of injury and transmission of diseases or parasites to animals.
> 3. The physical environment should allow comfortable resting, safe and comfortable movement, including normal postural changes, and the opportunity to perform types of natural behaviour that animals are motivated to perform.
> 4. Social groupings of animal should be managed to allow positive social behaviour and minimise injury, distress and chronic fear.
> 5. Air quality, temperature and humidity in confined spaces should support good animal health and not be aversive to animals. Where extreme conditions occur, animals should not be prevented from using their natural methods of thermoregulation.
> 6. Animals should have access to sufficient feed and water, suited to the animals' age and needs, to maintain normal health and productivity and to prevent prolonged hunger, thirst, malnutrition or dehydration.
> 7. Diseases and parasites should be prevented and controlled as much as possible through good management practices. Animals with serious health problems should be isolated and treated promptly or killed humanely if treatment is not feasible or recovery is unlikely.
> 8. Where painful procedures cannot be avoided, the resulting pain should be managed to the extent that available methods allow.
> 9. The handling of animals should foster a positive relationship between humans and animals and should not cause injury, panic, lasting fear or avoidable stress.
> 10. Owners and handlers should have sufficient skill and knowledge to ensure that animals are treated in accordance with these principles.

Policy Constraints

Political factors and policy development have considerable influence on animal agriculture decision making and therefore necessarily affect the future research agenda. Requirements for animal production and trade in animal protein products vary widely from country to country. Additionally, policies that affect animal agriculture in the United States and other countries are dynamic and change over time. Likewise, research needs are not static and must consider the shifting political realities and controlling policy decisions.

Examples of policies that affect animal agricultural research needs are easy to find. The United States has an aggressive ethanol production mandate that has resulted in dramatic price increases in corn, with cascading impacts on land uses, especially in the Midwest (Donner and Kucharik, 2008; Wright and Wimberly, 2013). The increase in ethanol production in the United States compels the need for additional research in alternative feeds for animal production, a topic that would not be a top priority except that it arises from ethanol policy. Another example is the use of beta-agonists to promote growth and increase lean muscle in cattle, which are approved for use by the USDA in the United States (Texas A&M AgriLife Extension, 2013) and have been widely found in Brazilian beef even after a temporary ban (Beef Central, 2013). Internationally, however, China, Russia, Taiwan, and other countries have banned the use of beta-agonists (Centner et al., 2014). Differing international requirements create trade and production challenges to animal agriculture industries.

Protein trade is also subject to disparate phytosanitary inspection practices at international borders which may not be related to food safety but instead are designed to address unrelated political concerns. The trade implications of animal welfare standards are becoming obvious in Thailand, Argentina, and other rapidly developing countries, which are increasing their export markets for animal products by producing according to EU- or European country–specific standards (Bowles et al., 2005). In addition, some multinational companies are beginning to harmonize and enforce their animal welfare standards or preferences globally. For example, Unilever recently announced that after having significantly increased its purchase of cage-free eggs in Europe and the United States, it will purchase only cage-free eggs globally by 2020 (WorldPoultry, 2013). These types of changes could have significant effects on global development of animal agriculture as well as the level of food availability by influencing which producers can supply to those companies, potentially favoring larger, better-capitalized producers able not only to produce sufficient products but to commit resources toward ensuring compliance.

Although partly attributable to social habits and practices, food waste and food loss are other areas that are driven by policy choices. Food safety requirements and government-imposed shelf-life requirements related to mandatory expiration dates cause considerable loss of food regardless of actual health threats. The loss of this food volume has been estimated to be approximately one-third of food produced for human consumption (Gustavsson et al., 2011). Government-imposed political and policy choices certainly present an ongoing challenge concerning the future considerations and corresponding direction of animal research efforts going forward.

Economic Considerations

Even if a food animal system in some region of the world is considered to be sustainable in an environmental or social sense, it cannot thrive or survive unless that system is also economically sustainable. Animal industries face economic constraints from numerous sources, including domestic and international regulations and trade barriers, market structure, knowledge, access to resources, financing, access to veterinary care, cost of inputs, and technology (Box 2-6). These constraints are not uniform throughout the world and can vary even within regions. For example, producers in developed countries, utilizing more intensive systems may be constrained by government intervention in the markets, whereas smallholders in developing countries may be more constrained by risk and uncertainties in weather and market vagaries in the regions where they operate (Jarvis, 1986).

The sustainability of food animal production is particularly affected by existing economic policies (Schillhorn, 1999; Upton, 2004; Otte et al., 2012). Policies that enhance the efficient functioning of markets allow food animal producers to receive appropriate signals regarding resource allocation and costs (Upton, 2004). Food animal markets are sensitive to both economic and technical forces (Steinfeld et al., 2006b). Consequently, food animal development strategies and policies must be informed by a clear understanding of not only the various production constraints faced by producers in and across regions and countries but also the demand for different food animal products (Otte et al., 2012).

Although government intervention in markets is often distortive, some intervention is justified on various social grounds, such as protection of populations from animal diseases and communities from land grabbing, the restriction of imports to protect producers from international dumping, or the need to initiate long-term industry development (Upton, 2004). In general, food animal development in most countries can improve domestic welfare, help alleviate poverty, and reduce food security concerns if food animal markets operate free of distortive intervention and investments are made in complementary research, infrastructure, and animal health programs to support the sustainable growth of the industry (Jarvis, 1974, 1986; Schillhorn, 1999; Otte et al., 2012). Historically, research, education, and nondistorted market signals have proven fruitful in the development of sustainable food animal production.

BOX 2-6
Economic Forces in Animal Agriculture

The role of economic forces in constraining the ability of livestock industries to meet growing food demand and contribute to global food security can be illustrated by adapting the Mosher framework as elaborated by Timmer et al. (1983) to livestock. This framework considers the relationship between the actual performance of an agricultural system and its potential performance given technical and economic constraints to growth (i.e., a yield gap). Understanding the drivers of the yield gap can help guide technology and capacity development. The performance measure is yield (output per animal), which may vary widely across livestock on farms in a given region and across regions. Given some level of technology (which includes the breed), yield variability results from regional differences in environmental, social, climatic, political, and other conditions. Even within a region, however, yields will vary because some producers may have better access to pastures or water, commercial markets, or agricultural extension services, or may utilize more efficient management techniques. Whatever the reasons, the observed yield variations across livestock of a particular type are ranked from highest to lowest in the Mosher framework along the vertical axis of a graph (Figure 2-1). The horizontal axis is the percentage distribution of livestock of that type by yield level.

For beef cattle, curve *a* in Figure 2-1 represents the distribution of beef yields achieved by livestock raised in some region, referred to as the "achievement curve." The yields along the curve are ordered from highest to lowest so that, given the differences in yield performance of beef cattle raised, the achievement curve slopes downward from left to right. Curve *t* represents the technical ceiling to yield performance. The technical ceiling represents the maximum biologically possible yields that can be achieved given available technology. Curve *e* represents the economic ceiling to yield performance, which is the yield distribution possible when farmers maximize their profits and economic conditions are at their optimum given the social, political, cultural, environmental, infrastructure, and other exogenous conditions impacting the cattle production system. The position of curve *e* in the graph depends on how well economic systems function and the constraints imposed by the exogenous conditions. Obviously curve *e* cannot lie above curve *t*. How far below curve *t* it will lie is determined by the efficiency of the economic system in delivering inputs such as production financing, genetic material, and feedstuffs to producers and in determining prices, transporting, feeding, and processing live animals, and delivering meat products to end users. Lack of infrastructure, unfavorable policies relating to production, confined animal feeding, the cattle–environment interface, food safety regulations, and international trade, as well as restrictive land tenure arrangements and many other conditions increase the technical and economic yield gap. At the same time, droughts and other unfavorable weather conditions, lack of technical training and education, poor access to extension services, and other factors affect the ability of farmers to maximize profits and can exacerbate the technical and economic yield gap.

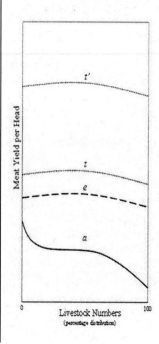

Investments in new technology,

such as improved genetics, improved feeding regimes to enhance yield gains, or improved livestock or pasture management techniques, can raise the technical ceiling from t to somewhere like t' creating the opportunity for increased profitability which may raise the economic ceiling (e curve), leading to the possibility of an increase in yield performance. However, unless investments are also made to relieve any constraints to an increase in the economic ceiling, actual yield performance will not increase. Improved technology may not raise productivity and profitability of cattle production in a region such as Africa if producers have poor access to commercial markets, production financing, education, and/or extension services and face inefficient delivery systems hampered by lack of refrigeration, poor quality roads, bridges, ports, and other critical infrastructure. Agricultural policies that discriminate against cattle production in favor of crops or environmental improvement can also hinder productivity. Thus, a consideration of the economic constraints to increased meat production must accompany investments in new technologies if the potential returns to those investments are to be realized.

SOURCE: Used with permission from the World Bank. Timmer et al., 1983. Food Policy Analysis.

THE PREDICAMENT OF SUSTAINABLE ANIMAL AGRICULTURE

Animal production systems are complex, and decisions about the direction of future research needed to improve their sustainability are influenced by an array of factors identified in this chapter. It seems unlikely that there is a "one-size-fits-all" solution to achieve sustainability that cuts across domains such as production systems, species, and regions with different climates, cultures, and socioeconomic conditions. In fact, these kinds of solutions are difficult to achieve even within particular domains, because sustainability is a "wicked problem." Wicked problems require a much different research approach than tame problems (e.g., determining the energy requirements of a growing steer) that research from the traditional animal science disciplines have served so well. Peterson (2013) characterizes the wicked problem of sustainability in animal agriculture as having four properties: (1) no definitive formulation of the problem exists; (2) its solution is not true or false, but rather better or worse; (3) stakeholders have radically different frames of reference concerning the problem; and (4) the underlying cause-and-effect relationships related to the problem are complex, systemic, and either unknown or highly uncertain. Systems approaches have been advocated as a means to address wicked problems (Williams and van't Hof, 2014).

Much of the sustainability predicament stems from differences in values and interests from the different stakeholders involved (Norton, 2012). As such, stakeholders often have divergent answers to the questions, "What is the current state of animal agriculture sustainability?" and "What should it be?" By failing to recognize different ways to formulate the problem and by advancing straight to solutions, intractable debates emerge (i.e., problem of legitimacy; Cash et al., 2003). Decisions about "what is sustainability" drive research questions, tools, and metrics. Additionally, stakeholders in animal agriculture come from different sections of the production chain (e.g., consumers vs. primary production vs. processors or retailers) and often use different vocabularies. Different vocabularies are also used in different animal sciences disciplines and related fields of study (e.g., agricultural economics), which can create miscommunications among those within and outside of academia.

The sustainability of animal agriculture is further complicated because it involves addressing other inherently challenging problems, such as climate change, animal welfare, and food security. There is a temptation to try to resolve these kinds of complex problems by developing quantitative models. However, unintended or unforeseen consequences and externalities associated with a change in animal production will always exist and accordingly fall outside the capabilities of any model or research framework employed. Furthermore, integrating sustainability metrics is also influenced by values, which inevitably leads to a decision about what metrics are more important and how much more important (e.g., is animal welfare more or less important than environmental impact in deciding which production system to use?). Many animal scientists are not trained to consider the role of their own and others' values in making decisions about sustainability issues related to the animal sciences. Ignoring differences in values, including differences in risk tolerance, creates significant challenges in making the problems associated with sustainability easier to solve. Incorporating quantitative approaches, such as participatory strategies (Swanson et al., 2011) into the research enterprise can help to address sustainability issues by involving a broad array of stakeholders in decision making (de Boer and Cornelissen, 2002; Mollenhorst and de Boer, 2004; NRC,

2011). Animal science research can contribute to assessments of sustainability by providing data that reduce the uncertainty about underlying causes and effects and define degrees of risk (Goldstein, 1999). Arriving at the "best" technical solution, however, involves increasing stakeholder participation and acknowledging the role of values in decision making and not just "educating the public" (Slovic, 1999). The involvement of social scientists in addressing these complex issues will lead to animal science research and technology transfer that is more relevant and consequential.

REFERENCES

Anthony, R. 2012. Taming the unruly side of ethics: Overcoming challenges of a bottom-up approach to ethics in the areas of climate change and food security. Journal of Agricultural and Environmental Ethics 25(6):813-841.

Asche, F. 2013. Demand and Productivity Growth as Drivers of a Successful Aquaculture Industry. Online. Available at http://www.acadcmia.edu/4112190/Demand_and_productivity_growth_as_drivers_of_a_successful_aquaculture_industry._Asche_July_23. Accessed August 29, 2014.

Atkinson, D., and C. A. Watson. 1996. The environmental impact of intensive systems of animal production in the lowlands. Animal Science 63(3):353-361.

Ausubel, J. H., I. K. Wernick, and P. E. Waggoner. 2012. Peak farmland and the prospect for land sparing. Population and Development Review 38(S1):217-238.

AVMA (American Veterinary Medical Association). 2014. The Veterinarian's Role in Animal Welfare. Online. Available at https://ebusiness.avma.org/files/productdownloads/VetsRoleinAW.pdf. Accessed August 17, 2014.

Baker, B., J. Hanson, R. Bourdon, and J. Eckert. 1993. The potential effects of climate change on ecosystem processes and cattle production on U.S. rangelands. Climatic Change 25(2):97-117.

Baro, M., and T. F. Deubel. 2006. Persistent hunger: Perspectives on vulnerability, famine, and food security in sub-Saharan Africa. Annual Reviews of Anthropology 35: 521-538.

Beef Central. 2013. Brazil's exports impacted by beta agonist use. Online. Available at http://www.beefcentral.com/trade/trade-brazils-exports-impacted-by-beta-agonist-use. Accessed October 16, 2014.

Bertrand, S. 2014. Animal research: Addressing the needs of the coming 50 years. Presentation at the First Meeting on Considerations for the Future of Animal Science Research, March 10. Washington, DC.

Bowles, D., R. Paskin, M. Gutierrez, and A. Kasterine. 2005. Animal welfare and developing countries: Opportunities for trade in high-welfare products from developing countries. Revue Scientifique et Technique 24(2):783-790.

Boyd, G., and C. Roger. 2012. A 50-Year Comparison of the Carbon Footprint of the U.S. Swine Herd: 1959-2009. Des Moines, IA: National Pork Board.

Brambell, F. W. R., ed. 1965. Report of the Technical Committee to Enquire into the Welfare of Animals Kept Under Intensive Livestock Husbandry Systems. London, UK: Her Majesty's Stationery Office.

Branson, E. J., ed. 2008. Fish Welfare. Hoboken, NJ: Wiley-Blackwell.

Burney, J. A., S. J. Davis, and D. B. Lobell. 2010. Greenhouse gas mitigation by agricultural intensification. Proceedings of the National Academy of Sciences of the United States of America 107(26):12052-12057.

Cady, R. A., G. Boyd, L. Wittig, G. Bryan, P. J. Holden, A. L. Sutton, and D. Anderson. 2013. A 50-year comparison of the environmental impact and resource use of the U.S. swine herd: 1959 vs. 2009. Proceedings of the ADSA-ASAS Joint Annual Meeting, July 8-12, Indianapolis, IN.

Canadian Food Inspection Agency. 2014. Porcine Epidemic Diarrhea (PED) Situation in Canada. Online. Available at http://www.inspection.gc.ca/animals/terrestrial-animals/diseases/other-diseases/ped/eng/1392762503272/1392762576176. Accessed September 10, 2014.

Capper, J. L. 2011. The environmental impact of beef production in the United States: 1977 compared with 2007. Journal of Animal Science 89(12):4249-4261.

Capper, J. L. 2013. Should we reject animal source foods to save the planet? A review of the sustainability of global livestock production. South African Journal of Animal Science 43(3):233-246.

Capper, J. L., R. Cady, and D. Bauman. 2009. The environmental impact of dairy production: 1944 compared with 2007. Journal of Animal Science 87(6):2160-2167.

Cash, D. W., W. C. Clark, F. Alcock, N. M. Dickson, N. Eckley, D. H. Guston, J. Jäger, and R. B. Mitchell. 2003. Knowledge systems for sustainable development. Proceedings of the National Academy of Sciences of the United States of America 100(14):8086-8091.

CDC (Centers for Disease Control and Prevention). 1999. Achievements in public health, 1900-1999: Control of infectious diseases. Morbidity and Mortality Weekly Report 48(29):621-629. Online. Available at http://www.cdc.gov/mmwr/PDF/wk/mm4829.pdf. Accessed August 17, 2014.

CDC. 2013a. Antibiotic Resistance Threats in the United States. Online. Available at http://www.cdc.gov/drugresistance/threat-report-2013/pdf/ar-threats-2013-508.pdf. Accessed August 16, 2014.

CDC. 2013b. CDC Telebriefing on Today's Drug-Resistant Health Threats. Press Briefing Transcript. Online. Available at http://www.cdc.gov/media/releases/2013/t0916_health-threats.html. Accessed August 16, 2014.

Centner, T. J., J. C. Alvey, and A. M. Stelzleni. 2014. Beta agonists in livestock feed: Status, health concerns, and international trade. Journal of Animal Science 92(9):4234-4240.

Chicago Council on Global Affairs. 2013. Advancing Global Food Security: The Power of Science, Trade, and Business. Chicago: Chicago Council on Global Affairs. Online. Available at http://www.thechicagocouncil.org/UserFiles/File/GlobalAgDevelopment/Report/2013_Advancing_Global_Food_Security.pdf. Accessed September 25, 2014.

Clay, J. 2009. Market Transformation and Reducing the Ecological Footprint: Making More from Less. Online. Available at http://awsassets.panda.org/downloads/jason_clay_guatemala.pdf. Accessed August 29, 2014.

Costa-Font, M., J. M. Gil, and W. B. Traill. 2008. Consumer acceptance, valuation of and attitudes towards genetically modified food: Review and implications for food policy. Food Policy 33(2):99-111.

de Boer, I. J. M., and A. M. G. Cornelissen. 2002. A method using sustainability indicators to compare conventional and animal-friendly egg production systems. Poultry Science 81(2):173-181.

Dinh, P. N., H. T. Long, N. T. K. Tien, N. T. Hien, L. T. Q. Mai, L. H. Phong, L. V. Tuan, H. V. Tan, N. B. Nguyen, P. V. Tu, N. T. M. Phuong, and the World Health Organization/Global Outbreak Alert and Response Network Avian Influenza Investigation Team in Vietnam. 2006. Risk factors for human infection with Avian Influenza A H5N1, Vietnam, 2004. Emerging Infectious Diseases 12(12):1841-1847.

Donner, S. D., and C. J. Kucharik. 2008. Corn-based ethanol production compromises goal of reducing nitrogen export by the Mississippi River. Proceedings of the National Academy of Sciences of the United States of America 105(11):4513-4518.

Duffy, E. A., K. E. Belk, J. N. Sofos, G. R. Bellinger, and G. C. Smith. 2000. United States retail pork microbiological baseline. Pp. 305-309 in Proceedings of the Pork Quality and Safety Summit. Des Moines, IA: National Pork Producers Council.

EC (European Commission). 2014. EU Import Conditions for Seafood and Other Fishery Products. Online. Available at http://ec.europa.eu/food/international/trade/im_cond_fish_en.pdf. Accessed August 29, 2014.

Edgerton, M. D. 2009. Increasing crop productivity to meet global needs for feed, food, and fuel. Plant Physiology 49(1):7-13.

Eghball, B., J. F. Power, J. E. Gilley, and J. W. Doran. 1997. Nutrient, carbon, and mass loss during composting of beef cattle feedlot manure. Journal of Environmental Quality 26(1):189-193.

EPA (U.S. Environmental Protection Agency). 2014. Region 7 Concentrated Animal Feeding Operations (CARFOs). Online. Available at http://www.epa.gov/region07/water/cafo/cafo_impact_environment.htm. Accessed September 10, 2014.

FAO (Food and Agriculture Organization of the United Nations). 1996. World Food Summit Plan of Action, Paragraph 1 in Rome Declaration on World Food Security. World Food Summit November 13-17. Rome: FAO. Online. Available at http://www.fao.org/docrep/003/w3613e/w3613e00.HTM. Accessed August 14, 2014.

FAO. 2007. Adaptation to Climate Change in Agriculture, Forestry and Fisheries: Perspective, Framework and Priorities. Rome: FAO. Online. Available at http://www.fao.org/nr/climpag/pub/adaptation_to_climate_change_2007.pdf. Accessed September 10, 2014.

FAO. 2009a. How to Feed the World in 2050. Rome: FAO. Online. Available at http://www.fao.org/fileadmin/templates/wsfs/docs/expert_paper/How_to_Feed_the_World_in_2050.pdf. Accessed August 28, 2014.

FAO. 2009b. The State of Food and Agriculture 2009: Livestock in the Balance. Rome: FAO. Online. Available at http://www.fao.org/docrep/012/i0680e/i0680e.pdf. Accessed August 28, 2014.

FAO. 2010. Greenhouse Gas Emissions from the Dairy Sector: A Life Cycle Assessment. Rome: FAO. Online. Available at http://www.fao.org/docrep/012/k7930e/k7930e00.pdf. Accessed September 25, 2014.

FAO. 2014a. Gateway to Farm Animal Welfare. Rome: FAO. Available at http://www.fao.org/ag/againfo/themes/animal-welfare/en. Accessed August 17, 2014.

FAO. 2014b. The State of World Fisheries and Aquaculture: Opportunities and Challenges. Rome: FAO. Available at http://www.fao.org/3/a-i3720e.pdf. Accessed September 25, 2014.

FASS (Federation of Animal Science Societies). 2012. FAIR (Farm Animal Integrated Research) 2012. Available at http://www.fass.org/docs/FAIR2012_Summary.pdf. Accessed August 5, 2014.

FDA (Food and Drug Administration). 2014. FDA Update on Animal Pharmaceutical Industry Response to Guidance #213. Available at http://www.fda.gov/AnimalVeterinary/SafetyHealth/AntimicrobialResistance/JudiciousUseofAntimicrobials/ucm390738.htm. Accessed December 19, 2014.

Ferket, P. R. 2010. Poultry nutrition moves towards higher standards. World Poultry. Online. Available at http://www.worldpoultry.net/Breeders/Nutrition/2010/5/Poultry-nutrition-moves-towards-higher-standards-WP007480W/. Accessed December 13, 2014.

Fraser, D., I. J. Duncan, S. A. Edwards, T. Grandin, N. G. Gregory, V. Guyonnet, P. H. Hemsworth, S. M. Huertas, J. M. Huzzey, D. J. Mellor, J. A. Mench, M. Spinka, and H. R. Whay. 2013. General principles for the welfare of animals in production systems: The underlying science and its application. Veterinary Journal 198(1):19-27.

Galloway, J., F. Dentener, M. Burke, E. Dumont, A. Bouwman, R. A. Kohn, H. A. Mooney, S. Seitzinger, and C. Kroeze. 2010. The impact of animal production systems on the nitrogen cycle. Pp. 83-95 in Livestock in a Changing Landscape: Drivers, Consequences, and Responses, H. Steinfeld, H. A. Mooney, F. Schneider, and L. E. Neville, eds. Washington, DC: Island Press.

Gapminder. 2014. Gapminder Agriculture. Online. Available at http://www.gapminder.org. Accessed August 29, 2014.

Gerber, P. J., H. Steinfeld, B. Henderson, A. Mottet, C. Opio, J. Dijkman, A. Falcucci, and G. Tempio. 2013. Tackling Climate Change Through Livestock: A Global Assessment of Emissions and Mitigation Opportunities. Rome: FAO. Online. Available at http://www.fao.org/docrep/018/i3437e/i3437e.pdf. Accessed August 28, 2014.

Goldstein, B. D. 1999. The precautionary principle and scientific research are not antithetical. Environmental Health Perspectives 107(12):594-595.

Griffin, P. M., and R. V. Tauxe. 1991. The epidemiology of infections caused by Escherichia coli O157:H7, other enterohemorrhagic E. coli, and the associated hemolytic uremic syndrome. Epidemiologic Reviews 13(1):60-98.

Gustavsson, J., C. Cederberg, U. Sonesson, R. van Otterdijk, A. Meybeck. 2011. Global Food Losses and Food Waste – Extent, Causes and Prevention. Rome: FAO. Online. Available at http://www.fao.org/docrep/014/mb060e/mb060e00.pdf. Accessed September 25, 2014.

Hall, S. J., A. Delaporte, M. J. Phillips, M. Beveridge, and M. O'Keefe. 2011. Blue Frontiers: Managing the Environmental Costs of Aquaculture. Penang, Malaysia: World Fish Center. Online. Available at http://www.worldfishcenter.org/sites/default/files/report.pdf. Accessed September 25, 2014.

Handisyde, N. T., L. G. Ross, M. C. Badjeck, and E. H. Allison. 2006. The Effects of Climate Change on World Aquaculture: A Global Perspective. Stirling, UK: Institute of Aquaculture. Online. Available at http://www.ecasa.org.uk/Documents/Handisydeetal.pdf. Accessed September 25, 2014.

Havlík, P., H. Valin, M. Herrero, M. Obersteiner, E. Schmid, M. C. Rufino, A. Mosnier, P. K. Thornton, H. Böttcher, R. T. Conant, S. Frank, S. Fritz, S. Fuss, F. Kraxner, and A. Notenbaert. 2014. Climate change mitigation through livestock system transitions. Proceedings of the National Academy of Sciences of the United States of America 111(10):3709-3714.

Havstad, K. M., D. P. C. Peters, R. Skaggs, J. Brown, B. Bestelmeyer, E. Frederickson, J. Herrick, and J. Wright. 2007. Ecological services to and from rangelands of the United States. Ecological Economics 64(2):261-268.

Herrero, M., P. K. Thornton, P. Gerber, and R. S. Reid. 2009. Livestock, livelihoods and the environment: Understanding the trade-offs. Current Opinion in Environmental Sustainability 1(2):111-120.

Herrero, M., P. Havlík, H. Valin, A. Notenbaert, M. C. Rufino, P. K. Thornton, M. Blummel, F. Weiss, D. Grace, and M. Obersteiner. 2013. Biomass use, production, feed efficiencies, and greenhouse gas emissions from global livestock systems. Proceedings of the National Academy of Sciences of the United States of America 110(52):20888-20893.

Hribar, C. 2010. Understanding Concentrated Environmental Health: Animal Feeding Operations and Their Impact on Communities. Bowling Green, OH: National Association of Local Boards of Health.

IFPRI (International Food Policy Research Institute). 2002. Green Revolution: Curse or Blessing? Online. Available at http://ifpri.org/pubs/ib/ib11.pdf. Accessed August 15, 2014.

IOM and NRC (Institute of Medicine and National Research Council). 2009. The Public Health Effects of Food Deserts: Workshop Summary. Washington, DC: The National Academies Press.

Jarvis, L.S. 1974. Cattle as capital goods and ranchers as portfolio managers: An application to the Argentine cattle sector. Journal of Political Economy 82(3):489-520.

Jarvis, L. S. 1986. Livestock Development in Latin America. Washington, DC: World Bank.

Jun Li, W., S. H. Ali, and Q. Zhang. 2007. Property rights and grassland degradation: A study of the Xilingol pasture, Inner Mongolia, China. Journal of Environmental Management 85(2):461-470.

Keesing, F., L. K. Belden, P. Daszak, A. Dobson, C. D. Harvell, R. D. Holt, P. Hudson, A. Jolles, K. E. Jones, C. E. Mitchell, S. S. Myers, T. Bogich, and R. S. Ostfeld. 2010. Impacts of biodiversity on the emergence and transmission of infectious diseases. Nature 468(7324):647-652.

Kelly, A. M., J. D. Ferguson, D. T. Galligan, M. Salman, and B. I. Osburn. 2013. One health, food security, and veterinary medicine. Journal of the American Veterinary Medical Association 242(6):739-743.

Kelly, A., B. Osburn, and M. Salman. 2014. Veterinary medicine's increasing role in global health. Lancet Global Health 2(7):e379-e380.

Kerr, S. 2013. Porcine Epidemic Diarrhea Virus Confirmed in U.S. Washington State University Extension Fact Sheet 1001-2013. Online. Available at http://eden.lsu.edu/Topics/AgDisasters/Documents/PED%20fact%20sheet.pdf. Accessed September 10, 2014.

Lam, H.-M., J. Remais, M.-C. Fung, L. Xu, and S. S.-M. Sun. 2013. Food supply and food safety issues in China. Lancet 381(9882):2044-2053.

Lister, G., G. Tonsor, M. Brix, T. Schroeder, and C. Yang. 2014. Food Values Applied to Livestock Products. Online. Available at http://www.agmanager.info/livestock/marketing/WorkingPapers/WP1_Food Values-LivestockProducts.pdf. Accessed August 15, 2014.

Lobell, D. B., M. B. Burke, C. Tebaldi, M. D. Mastrandrea, W. P. Falcon, R. L. Naylor. 2008. Prioritizing climate change adaptation needs for food security in 2030. Science 319(5863):607-610.

Lusk, J. L., and B. C. Briggeman. 2009. Food values. American Journal of Agricultural Economics 91(1):184-196.

Makkar, H. P., and P. E. Vercoe, eds. 2007. Measuring Methane Production from Ruminants. Dordrecht, The Netherlands: Springer.

Masuda, T., and P. D. Goldsmith. 2010. China's meat consumption: An income elasticity analysis and long-term projections. In Proceedings of Agricultural and Applied Economics Association 2010 AAEA, CAES, & WAEA Joint Annual Meeting, Agricultural & Applied Economics Association, July 25-27, Denver, CO. Online. Available at http://ageconsearch.umn.edu/bitstream/61601/2/Poster11972AAEA_Masuda Goldsmith20100503b.pdf. Accessed September 25, 2014.

Matthews, L. R., and P. H. Hemsworth. 2012. Drivers of change: Law, international markets, and policy. Animal Frontiers 2(3):40-45.

McLeod, A., ed. 2011. World Livestock 2011: Livestock in Food Security. Rome: FAO. Online. Available at http://www.fao.org/docrep/014/i2373e/i2373e.pdf. Accessed September 25, 2014.

Meehl, G. A., A. Hu, J. Arblaster, J. Fasullo, and K. E. Trenberth. 2013. Externally forced and internally generated decadal climate variability associated with the interdecadal Pacific oscillation. Journal of Climate 26:7298-7310.

Mench, J. A. 2008. Farm animal welfare in the U.S.A.: Farming practices, research, education, regulation, and assurance programs. Applied Animal Behaviour Science 113(4):298-312.

Mench, J. A., D. A. Sumner, and J. T. Rosen-Molina. 2011. Sustainability of egg production in the United States: The policy and market context. Poultry Science 90(1):229-240.

Mollenhorst, H., and I. J. M. de Boer. 2004. Identifying sustainability issues using participatory SWOT analysis—A case study of egg production in the Netherlands. Outlook on Agriculture 33(4):267-276.

Monaghan, R. M., R. J. Paton, L. C. Smith, J. J. Drewry, and R. P. Littlejohn. 2005. The impacts of nitrogen fertilisation and increased stocking rate on pasture yield, soil physical condition and nutrient losses in drainage from a cattle grazed pasture. New Zealand Journal of Agricultural Research 48(2):227-240.

Morton, J. F. 2007. The impact of climate change on smallholder and subsistence agriculture. Proceedings of the National Academy of Sciences of the United States of America 104(50):19680-19685.

Myers, B. 2014. Livestock's hoof print. Environmental Forum 31(2):34-39.

NAS and RS (National Academy of Sciences and The Royal Society). 2014. Climate Change: Evidence & Causes. Online. Available at http://dels.nas.edu/resources/static-assets/exec-office-other/climate-change-full.pdf. Accessed August 29, 2014.

Naylor, R. L., R. J. Goldburg, H. Mooney, M. Beveridge, J. Clay, C. Folke, N. Kautsky, J. Lubchenco, J. Primavera, and M. Williams. 1998. Nature's subsidies to shrimp and salmon farming. Science 282(5390):883-884.

Naylor, R. L., R. W. Hardy, D. P. Bureau, A. Chiu, M. Elliott, A. P. Farrell, I. Forster, D. M. Gatlin, R. J. Goldburg, K. Hua, and P. D. Nichols. 2009. Feeding aquaculture in an era of finite resources. Proceedings of the National Academy of Sciences of the United States of America 106(36):15103-15110.

Nepstad, D. C., D. G. McGrath, and B. Soares Filho. 2011. Systemic conservation, REDD, and the future of the Amazon Basin. Conservation Biology 25(6):1113-1116.

Netting, R. 1993. Smallholders, Householders: Farm Families and the Ecology of Intensive, Sustainable Agriculture: Redwood City, CA: Stanford University Press.

Niles, M. T. 2013. Achieving social sustainability in animal agriculture: Challenges and opportunities to reconcile multiple sustainability goals. Pp. 193-210 in Sustainable Animal Agriculture, E. Kebreab, ed. Oxon, UK: CAB International.

Norton, B. 2012. The ways of wickedness: Analyzing messiness with messy tools. Journal of Agricultural and Environmental Ethics 25(4):447-465.

NRC (National Research Council). 2009. Emerging Technologies to Benefit Farmers in Sub-Saharan Africa and South Asia. Washington, DC: The National Academies Press.

NRC. 2010. Toward Sustainable Agricultural Systems in the 21st Century. Washington, DC: The National Academies Press.

NRC. 2011. Sustainability and the U.S. EPA. Washington, DC: The National Academies Press.

NRC. 2012. A Sustainability Challenge: Food Security for All: Report of Two Workshops. Washington, DC: The National Academies Press.

OIE (World Organisation for Animal Health). 2014. Terrestrial Animal Health Code: Glossary. Online. Available at http://www.oie.int/index.php?id=169&L=0&htmfile=glossaire.htm#terme_bien_etre_animal. Accessed September 15, 2014.

Olin, P. G. 2012. Aquaculture in the United States of America: Present status and future opportunities. Bulletin of Fisheries Research Agency 35:7-13.

O'Mara, F. P. 2011. The significance of livestock as a contributor to global greenhouse gas emissions today and in the near future. Animal Feed Science and Technology 166-167:7-15.

Oreskes, N., and E. M. Conway. 2010. Merchants of doubt: How a handful of scientists obscured the truth on issues from tobacco smoke to global warming. New York: Bloomsbury.

Otte, J., A. Costales, J. Dijkman, U. Pica-Ciamarra, T. Robinson, V. Ahuja, C. Ly, and D. Roland-Holst. 2012. Livestock Sector Development for Poverty Reduction: An Economic and Policy Perspective—Livestock's Many Virtues. Rome: FAO. Online. Available at http://www.fao.org/docrep/015/i2744e/i2744e00.pdf. Accessed September 25, 2014.

Parry, M. L., C. Rosenzweig, A. Iglesias, M. Livermore, and G. Fischer. 2004. Effects of climate change on global food production under SRES emissions and socio-economic scenarios. Global Environmental Change 14(1):53-67.

Parry, M. L., O. F. Canziani, J. P. Palutikof, P. J. van der Linden, and C. E. Hanson, eds. 2007. Climate Change 2007: Impacts, Adaptation and Vulnerability, Vol. 4. Cambridge, UK: Cambridge University Press.

PCAST (President's Council of Advisors on Science and Technology). 2012. Report to the President on Propelling Innovation in Drug Discovery, Development, and Evaluation. Online. Available at http://www.whitehouse.gov/sites/default/files/microsites/ostp/pcast-fda-final.pdf. Accessed August 17, 2014.

PCAST. 2014a. National Strategy for Combating Antibiotic-Resistant Bacteria. Washington, DC: The White House.

PCAST. 2014b. Report to the President on Combating Antibiotic Resistance. Washington, DC: Executive Office of the President.

Pelletier, N., and P. Tyedmers. 2010. Forecasting potential global environmental costs of livestock production 2000–2050. Proceedings of the National Academy of Sciences of the United States of America 107(43):18371-18374.

Pelletier, N., M. Ibarburu, and H. Xin. 2014. Comparison of the environmental footprint of the egg industry in the United States in 1960 and 2010. Poultry Science 93(2):241-255.

Peterson, H. C. 2013. Sustainability: A wicked problem. Pp. 1-9 in Sustainable Animal Agriculture, E. Kebreab, ed. Wallingford, UK: CAB International.

Reid, R. S., K. A. Galvin, and R. S. Kruska. 2008. Global significance of extensive grazing lands and pastoral societies: An introduction. Pp. 1-24 in Fragmentation in Semi-Arid and Arid Landscapes: Consequences for Human and Natural Systems, K. A. Galvin, R. S. Reid, R. H. Behnke, Jr., and N. T. Hobbs, eds. New York: Springer.

Rushen, J., A. Butterworth, and J. C. Swanson. 2011. Animal behavior and well-being symposium: Farm animal welfare assurance: Science and application. Journal of Animal Science 89(4):1219-1228.

Rushton, J., J. Pinto Ferreira, and K. D. Stärk. 2014. Antimicrobial Resistance: The Use of Antimicrobials in the Livestock Sector. OECD Food, Agriculture and Fisheries Papers 68. Online. Available at http://dx.doi.org/10.1787/5jxvl3dwk3f0-en. Accessed November 3, 2014.

Salman, M. D. 2009. The role of veterinary epidemiology in combating infectious animal diseases on a global scale: The impact of training and outreach programs. Preventive Veterinary Medicine 92(4):284-287.

Salman, M. D., J. C. New, Jr., M. Bailey, C. Brown, L. Detwiler, D. Galligan, C. Hall, M. Kennedy, G. Lonergan, L. Mann, D. Renter, M. Saeed, B. White, and S. Zika. 2008. Global Food Systems and Public Health: Production Methods and Animal Husbandry. A National Commission on Industrial Farm Animal Production Report. Pew Commission on Industrial Farm Animal Production. Online. Available at http://trace.tennessee.edu/utk_compmedpubs/32. Accessed August 22, 2014.

Sanchez, P. 2010. Tripling crop yields in tropical Africa. Nature Geoscience 3(5):299-300.

Sanchez, P., C. Palm, J. Sachs, G. Denning, R. Flor, R. Harawa, B. Jama, T. Kiflemariam, B. Konecky, and R. Kozar. 2007. The African millennium villages. Proceedings of the National Academy of Sciences of the United States of America 104(43):16775-16780.

Schillhorn, T. W. 1999. Agricultural policy and sustainable livestock development. International Journal for Parasitology 29(1):7-15.

Sen, A. 1983. Poverty and Famines: An Essay on Entitlement and Deprivation. Oxford, UK: Oxford University Press.

Seo, S. N., and R. Mendelsohn. 2008. Measuring impacts and adaptations to climate change: A structural Ricardian model of African livestock management. Agricultural Economics 38(2):151-165.

Slovic, P. 1999. Trust, emotion, sex, politics, and science: Surveying the risk-assessment battlefield. Risk Analysis 19(4):689-701.

Steinfeld, H., and P. Gerber. 2010. Livestock production and the global environment: Consume less or produce better? Proceedings of the National Academy of Sciences of the United States of America 107(43):18237-18238.

Steinfeld, H., P. Gerber, T. Wassenaar, V. Castel, M. Rosales, and C. de Haan. 2006a. Livestock's Long Shadow: Environmental Issues and Options. Rome: FAO. Online. Available at http://www.fao.org/docrep/010/a0701e/a0701e00.htm. Accessed August 15, 2014.

Steinfeld, H., T. Wassenaar, and S. Jutzi. 2006b. Livestock production systems in developing countries: Status, drivers, trends. Revue Scientifique et Technique 25(2):505-516.

Stevens, M. P., T. J. Humphrey, and D. J. Maskell. 2009. Molecular insights into farm animal and zoonotic salmonella infections. Philosophical Transactions of the Royal Society B: Biological Sciences 364(1530):2709-2723.

Sun, R. Q., R. J. Cai, Y. Q. Chen, P. S. Liang, D. K. Chen, and C. X. Song. 2012. Outbreak of porcine epidemic diarrhea in suckling piglets, China. Emerging Infectious Diseases 18(1):161-163.

Swanson, J. C., Y. Lee, P. B. Thompson, R. Bawden, and J. A. Mench. 2011. Integration: Valuing stakeholder input in setting priorities for socially sustainable egg production. Poultry Science 90(9):2110-2121.

Texas A&M AgriLife Extension. 2013. Beta-agonists for Market Steers. Online. Available at: http://deafsmith.agrilife.org/files/2013/07/2013-Beta-Agonists-General-1.pdf. Accessed November 3, 2014.

Thornton, P. K. 2010. Livestock production: Recent trends, future prospects. Philosophical Transactions of the Royal Society B: Biological Sciences 365(1554):2853-2867.

Thornton, P. K., and M. Herrero. 2010. Potential for reduced methane and carbon dioxide emissions from livestock and pasture management in the tropics. Proceedings of the National Academy of Sciences of the United States of America 107(46):19667-19672.

Thornton, P. K., J. Van de Steeg, A. Notenbaert, and M. Herrero. 2009. The impacts of climate change on livestock and livestock systems in developing countries: A review of what we know and what we need to know. Agricultural Systems 101(3):113-127.

Timmer, C. P., W. P. Falcon, and S. R. Pearson. 1983. Food Policy Analysis. Baltimore, MD: John Hopkins University.

Tucker, C., D. Brune, and E. Torrans. 2014. Partitioned pond aquaculture systems. World Aquaculture 45(2):9-17.

UK Department of Health. 2013. UK Five Year Antimicrobial Resistance Strategy 2013 to 2018. Online. Available at https://www.gov.uk/government/uploads/system/uploads/attachment_data/file/244058/20130902_UK_5_year_AMR_strategy.pdf. Accessed August 17, 2014.

Upton, M. 2004. Policy Issues in Livestock Development and Poverty Reduction. Pro-Poor Livestock Policy Initiative Policy Brief. Online. Available at http://www.fao.org/ag/againfo/programmes/en/pplpi/docarc/pb_wp10.pdf. Accessed September 10, 2014.

USAPHIS (U.S. Animal and Plant Health Inspection Service). 1997. Shedding of Salmonella by Finisher Hogs in the U.S. USDA Centers for Epidemiology and Animal Health Info Sheet N223.197. Online. Available at http://www.aphis.usda.gov/animal_health/nahms/swine/downloads/swine95/Swine95_is_salmonella.pdf. Accessed August 17, 2014.

USDA (U.S. Department of Agriculture). 2014. Porcine Epidemic Diarrhea (PED). Online. Available at http://www.aphis.usda.gov/animal_health/animal_dis_spec/swine/downloads/ped_tech_note.pdf. Accessed September 10, 2014.

Van Eenennaam, A. L., and W. M. Muir. 2011. Transgenic salmon: A final leap to the grocery shelf? Nature Biotechnology 29(8):706-710.

Veissier, I., A. Butterworth, B. Bock, and E. Roe. 2008. European approaches to ensure good animal welfare. Applied Animal Behaviour Science 113(4):279-297.

Waite, R., M. Beveridge, R. Brummett, S. Castine, N. Chaiyawannakarn, S. Kaushik, R. Mungkung, S. Nawapakpilai, and M. Phillips. 2014. Increasing Productivity of Aquaculture. Working Paper, Installment 6 of Creating a Sustainable Food Future. Washington, DC: World Resources Institute.

Welch, L. 2014. Murkowski blasts genetically modified salmon. Alaska Dispatch News: May 30 Online. Available at http://www.adn.com/article/20140531/laine-welch-murkowski-blasts-genetically-modified-salmon. Accessed August 29, 2014.

White House. 2014. Executive Order: Combating Antibiotic-Resistant Bacteria. Online. Available at http://www.whitehouse.gov/the-press-office/2014/09/18/executive-order-combating-antibiotic-resistant-bacteria. Accessed December 15, 2014.

WHO (World Health Organization). 2001. WHO Global Strategy for Containment of Antimicrobial Resistance. Online. Available at http://www.who.int/drugresistance/WHO_Global_Strategy_English.pdf. Accessed August 16, 2014.

Williams, B., and S. van't Hof. 2014. Wicked Solutions: A Systems Approach to Complex Problems. Online. Available at http://www.bobwilliams.co.nz/wicked.pdf. Accessed September 10, 2014.

Wirsenius, S. 2003. The biomass metabolism of the food system: A model based survey of the global and regional turnover of food biomass. Journal of Industrial Ecology 7(1):47-80.

WorldPoultry. 2013. Cage-free eggs grow in popularity in Costa Rica. News Release, October 15. Online. Available at http://www.worldpoultry.net/Layers/Eggs/2013/10/Cage-free-eggs-grow-in-popularity-in-Costa-Rica-1388326W. Accessed August 17, 2014.

Wright, C. K., and M. C. Wimberly. 2013. Recent land use change in the western corn belt threatens grasslands and wetlands. Proceedings of the National Academy of Sciences of the United States of America 110(10):4134-4139.

Zepeda, C., M. D. Salman, and R. Ruppanner. 2001. International trade, animal health, and veterinary epidemiology: Challenges and opportunities. Preventive Veterinary Medicine 48(4):261-272.

3

Animal Agriculture Research Needs: U.S. Perspective

INTRODUCTION

Over the last several decades, animal agriculture in the United States—including dairy, poultry, pork, and beef—has made tremendous strides in production efficiency, as described in the previous chapter. Whether this will continue is now in question. Between 1982 and 2007, more than 41 million acres of rural land (crop, pasture, range, land formerly enrolled in the Conservation Reserve Program, forest, and other rural land) in the United States has been converted to developed uses (Farmland Information Center, 2014). This represents the size of the combined areas of Illinois and New Jersey, or more than half the land that is annually planted to soybeans in the United States. If this trend of ever-increasing amounts of land taken out of rural use continues, then production efficiency and yields must increase at an even greater level to maintain the current level of food production, not to mention the future animal product consumption demands of the increasing American population (Box 3-1). These need to be increased in the face of global challenges such as climate change, increased regulations, and various public and social concerns regarding animal welfare and the use of technologies that have contributed and could contribute to production efficiency.

This chapter covers key issues for animal sciences research considerations for the United States. It provides a summary of reports by various policy and other stakeholder groups engaged in animal science research across various disciplines, and then discusses topics on the importance of the research contributions of animal sciences research to productivity, trade, funding, sustainable food security, the environment, and socioeconomic considerations. The role of animal agriculture research in U.S. welfare is described in terms of food security and food safety, health, jobs, and trade. There is also a discussion of challenges to U.S. animal agriculture, including social concerns, increasing regulations, the environment, and a decreasing agricultural land base to produce not only the same amount of food as presently produced but to also meet the increasing demand for global animal protein by the year 2050.

In keeping with its charge, the committee has focused its efforts on animal agricultural research; however, the broader context of overall agricultural R&D, for which there is more information available, provides additional reason for concern. Kennedy (2014), in a recent editorial in Science, describes the agricultural research sector as suffering from decades of neglect which must be reversed if we are to sustain a population of 9 billion people. Heisey et al. (2011) evaluated the impact of future public agricultural research spending scenarios on overall U.S. agriculture productivity growth from 2010 to 2050. A 50 percent fall in the rate of increase in U.S. agricultural productivity is predicted if agricultural funding increases continue unchanged at the current rate, which is less than the expected rate of inflation of research costs. In contrast, funding that meets the rate of research cost inflation leads to a 73 percent increase in productivity between now and 2050—and a 1 percent increase in inflation-adjusted spending is projected to lead to an 83 percent increase (Heisey et al., 2011).

Data specific to animal agricultural research are harder to find. The committee, however, believes that animal agricultural research has borne the brunt of the decades of neglect described by Kennedy (2014), particularly in private funding. Globally, private food and agricultural R&D for crops has increased 53 percent from $5,697 million in 1994 to $8,711 million in 2010, while the total for animal R&D only increased 4 percent (from $1,516 million to $1,577 million). Of note is that Fuglie et al. (2011) also reported that the relative amount of private animal as compared to crop research in 2006 was lower in the United States ($432 million animal vs. $2,392 million crops; 18 percent less) than in the rest of the world ($1,033 animal vs. $4,133 million crops; 25% less). Heisey of the U.S. Department of Agriculture (USDA) found that public funding in the United States for crop and animal research have both grown only marginally over the period 1999-2007 (K. Fuglie, USDA, personal communication, November 6, 2014).

> **BOX 3-1**
> **Animal Protein Availability in the United States**
>
> The U.S. Department of Agriculture's Economic Research Service tracks per capita food availability as a proxy for food consumption. The data show that beef availability grew substantially starting in the 1960s, reaching a peak of 88.8 pounds per capita per year in 1976. Since that time, beef availability has declined to roughly the same level as in 1909 when recordkeeping began. In contrast, chicken availability exploded in the second half of the 20th century. In 2012, 56.6 pounds of chicken were available per person per year, over 2 pounds more than beef. Pork production has fluctuated much less than beef or chicken. Fish and shellfish availability was 30 percent higher in 2012 than in 1909, but was still a relatively small portion of available animal protein compared with beef, chicken, and pork (Figure 3-1). Animal protein available from eggs and dairy products has fallen on a per capita basis, even though the supply of eggs has remained relatively constant since the 1960s and the dairy supply has more than doubled since the 1920s (Figures 3-2 and 3-3).
>
>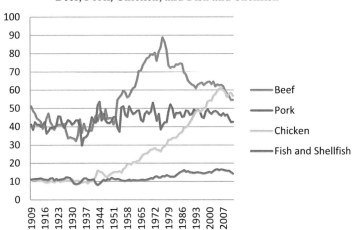
>
> **FIGURE 3-1** Beef, pork, chicken, and fish and shellfish from 1909 to 2012.
> SOURCE: USDA ERS (2014).

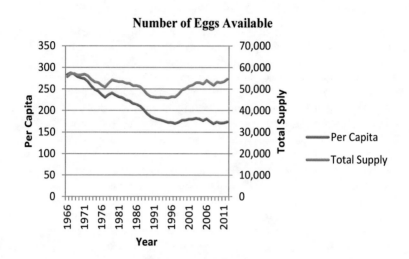

FIGURE 3-2 Number of eggs available from 1966 to 2012.
SOURCE: USDA ERS (2014)

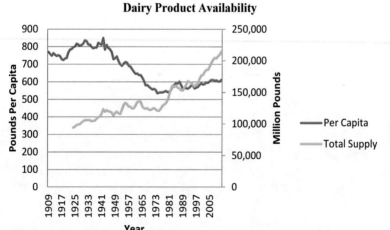

FIGURE 3-3 Dairy product availability from 1909 to 2012.
SOURCE: USDA ERS (2014).

3-1 Animal Science Research Needs

There have been multiple reports over the past decade that summarized the needs and future strategies for animal agriculture research. The USDA's Research, Education, and Economics (REE) has identified 15 key strategies for animal agriculture (USDA REE, 2012), which are included in Appendix D. In brief, these strategies focus on research related to genomics, new varieties and germplasms, sustainability of animal production and the environment, animal disease control, climate impacts, reduction of foodborne contaminants, effective management strategies, and development of animal sciences workforces. The USDA Agricultural Research Service (ARS) has proposed five main areas as FY 2015 research priorities (Appendix E), including (in ranking of budget investment) food safety, food animal protection and production, climate change, and genetic improvement and translational breeding. A stakeholder meeting sponsored by USDA ARS developed a prioritized list of research topics by species and disciplines (Appendix F).

In the 2012 Farm Animal Integrated Research (FAIR) report (Box 3-2; FASS, 2012), a collaboration of scientists, educators, producers, industry representatives, health professionals, and government representatives emphasize six crosscutting issues. These included fundamental and applied research in animal sciences, a balance of projects including large multi-institution grants and small grants, enhanced collaborations among universities and government agencies, increased public awareness of animal agricultural research, improved communication of animal science to policy makers, and an established data-mining system.[1]

A separate study (Rouquette et al., 2009) compiled recommendations from scientists at 25 land grant universities for future forage utilization research needs. Among the highest-priority needs were research on pasture systems and production efficiency, energy concerns, evaluation

---

[1] As research priorities are set and agendas are developed, it is important that comprehensive mining of historical data be conducted, to understand what is already known, to prevent unnecessary duplication, and to provide a better base on which to build future research. Data mining will reveal information that can be converted into knowledge about historical research, and these data can be used to predict future trends to be applied in research planning to support guidelines and policies (FASS, 2012).

of forage cultivars, environmental impacts, and soil fertility and nutrient management.

BOX 3-2
FAIR 2012 Report

Scientists, educators, producers, industry, health professionals, and government representatives gathered together in 2012 under the Farm Animal Integrated Research 2012 (FAIR, 2012) initiative to identify research, extension, and education priorities that will enable animal sciences to meet the key future animal agricultural challenges. Three major themes (Food Security, One Health, and Stewardship) emerged for further investment. Within Food Security, feed efficiency, energetic efficiency, and connecting "omics" to animal production and reproduction were highlighted as key research areas. The following areas were highlighted within the One Health area: new approaches to vaccine development, understanding and controlling zoonoses with an emphasis on food safety, and improving animal health through feed. Under Stewardship, the following research areas emerged: flow of nutrients and other potential pollutants from animal production systems, estimation and reduction of greenhouse gas production, and impacts of housing systems on animal well-being. FAIR 2012 emphasized six crosscutting issues:

1. Balanced portfolio where support for research spans from fundamental to applied, support for extension to help enhance adoption of the new technologies, and support for education to ensure a robust pipeline to develop new scientists, producers, and industry professionals;
2. Balanced size and scope of projects from large, multi-institutional grants to small grants;
3. Enhanced collaborations among universities, USDA, National Institutes of Health, National Science Foundation, Department of Energy, and others to leverage resources and stimulate innovation;
4. Increased public awareness through informing the consumer of the benefits and value of animal agricultural production and research;
5. Improved communication of science to policy makers such that a consistent and predictable regulatory process based on sound science is instituted; and
6. Establishment of a data-mining system.

For aquaculture research, the National Oceanic and Atmospheric Administration (NOAA) and the USDA have developed nine critical strategic goals in their 2014 National Strategic Plan for Federal Aquaculture Research (Box 3-3). These include research goals related to improving production efficiency and performance, animal well-being and nutrition, biosecurity and disease control, and the safety and nutrition of seafood for humans; developing a skilled workforce and socioeconomic and business strategies for aquaculture; and enhancing use of genetics to increase productivity and understanding of aquaculture–environment interactions.

Additionally, NOAA identified four goals for the year 2025 to expand U.S. marine aquaculture, including a comprehensive regulatory program for environmentally stable aquaculture, developing commercial aquaculture and replenishment of wild stocks, improving public understanding of aquaculture, and increased collaboration with international partners. Relevant to sustainability, Hume et al. (2011)

BOX 3-3
National Strategic Plan for Federal Aquaculture Research (2014)

This plan was developed by the Interagency Working Group in Aquaculture (IWG-A) and includes nine critical strategic goals with outcomes and milestones. The strategic goals are:

1. Advance understanding of the interactions of aquaculture and the environment;
2. Employ genetics to increase productivity and protect natural populations;
3. Counter disease in aquatic organisms and improve biosecurity;
4. Improve production efficiency and well-being;
5. Improve nutrition and develop novel feeds;
6. Increase supply of nutritious, safe, high-quality seafood and aquatic products;
7. Improve performance of production systems;
8. Create a skilled workforce and enhance technology transfer; and
9. Develop and use socioeconomic and business research to advance domestic aquaculture.

SOURCE: NSTC (2014).

identified three objectives for animal production systems: (1) maximizing the number of productive offspring per breeding male and female, (2) maximizing efficiency of converting feed and water to useful animal product, and (3) minimizing waste and losses through infectious and metabolic diseases.

As evidenced by the content of these reports, various stakeholders in the United States—from government agencies to industries to academics to producers—recognize the importance of animal sciences research as the means to ensure a safe, high-quality, plentiful, and affordable food supply to meet the future protein demand in the United States and the world. Common themes arise across these multiple reports, including climate change concerns; continued development and use of emerging technologies to improve animal production efficiency, animal health, and feed and food safety and quality; animal waste management; development of underutilized resources; and ensuring appropriate infrastructure and collaborations to achieve sufficient animal production in the future. Despite these common themes, USDA ARS only allocated 16 percent ($176 million) of its proposed FY 2015 budget to animal production research and 8 percent to product-quality/value-added research compared to 36 percent for crop research, 18 percent for food safety and nutrition, and 18 percent for environmental stewardship (Appendix J). The National Institute of Food and Agriculture (NIFA) allocated $103 million to animal science research in its FY 2012 budget. This is less than 1 percent of the value that animal products contribute to the U.S. economy. Animal agriculture annually accounts for between 60 and 70 percent of the total agriculture economy (often exceeding $100 billion/year) and plays an important role in the balance of agricultural trade (USDA ERS, 2013). Exports of pork, lamb, broiler, and dairy products equal $6 billion, $26 million, $4.2 billion, and $6.7 billion annually, respectively (USMEF, 2012a,b; Davis et al., 2013; U.S. Dairy Export Council, 2014). This chapter further describes the historical, current, and future technologies that improve animal sciences production, efficiency, sustainability, and safety in the United States. Findings and recommendations are presented for future animal science to ensure a sustainable food supply in the United States, focusing on and integrating U.S. and global perspectives. The nexus of animal science research and the economic, environmental, and social pillars of sustainability are also addressed in greater detail.

While previous reports focus on disciplines and topics important to animal sciences research, there has heretofore been less emphasis on a vision of a systems approach to research, which can incorporate the entire food production system. This is what the current report brings to bear in analysis of needs for future animal sciences research.

Recommendation 3-1

In view of the anticipated continuing increased demand for animal protein, growth in U.S. research related to animal agricultural productivity is imperative. Animal protein products contribute over $43 billion annually to the U.S. agricultural trade balance. Animal agriculture accounts for 60 to 70 percent of the total agricultural economy. In the past two decades, public funding, including formula funding and USDA ARS/NIFA funding, of animal science research has been stagnant in terms of real dollars and has declined in relation to the research inflation rate. A 50 percent decline in the rate of increase in U.S. agricultural productivity is predicted if overall agricultural funding increases in normative dollars continue at the current rate, which is less than the expected rate of inflation of research costs. If funding does meet the rate of research cost inflation, however, a 73 percent increase in overall agricultural productivity between now and 2050 is projected and a 1 percent increase in inflation-adjusted spending is projected to lead to an 83 percent increase.

Despite documenting the clear economic and scientific value of animal science research in the United States, funding to support the infrastructure and capacity is evidently insufficient to meet the needs for animal food; U.S.-based research will be needed to address sustainability issues and to help developing countries sustainably increase their own animal protein production and/or needs. Additionally, animal science research and practices in the United States are often adopted, to the extent possible, within developing countries. Thus, increases in U.S. funding will favorably impact animal production enterprises in developing countries.

With the lack of increase in public funding of animal science research, private/industry support has increased. The focus of industry funding is more toward applied areas that can be commercialized in the short term. Many of these applications are built on concepts developed from publicly funded basic research. With the increased animal protein demands, especially poultry, more publicly funded basic research is needed.

To meet current and future animal protein demand, and to sustain corresponding infrastructure and capacity, public support for animal science research (especially basic research) should be restored to at least past levels of real dollars and maintained at a rate that meets or exceeds the annual rate of research inflation. This is especially critical for those species (i.e., poultry) for which consumer demand is projected to significantly increase by 2050 and for those species with the greatest opportunity for reducing the environmental impact of animal agriculture.

Priorities for Research Support

USDA ARS spends 50 percent more on crop research and a greater percentage of its budget on food safety and nutrition and environmental stewardship than on animal production research. In addition, in the past couple of decades, public funding (USDA ARS/NIFA) of animal science research has been declining in terms of real dollars. One priority for research support in this area includes:
- USDA through its relevant agencies is encouraged to maintain high priority for funding for research, including translational research, that is commensurate with the future needs of livestock protein. USDA should maintain and enhance the current link to the livestock, poultry, and aquaculture industries in the United States with the aim of building better public–private partnerships in funding research in animal science.

RESEARCH CONSIDERATIONS FOR SUSTAINABILITY

3-2 Productivity and Production Efficiency

Productivity is a key element in achieving food security, and production efficiency relates to sustainability through its effects on economics and environmental impacts. Increasing the productivity per animal unit and land unit while concomitantly decreasing negative impacts on the environment (sustainable intensification) can ultimately produce safe, affordable, and nutritious food to help meet overall global food and protein needs. Technological advancements, genetic improvement, better nutrition, husbandry, and advances in animal health and welfare in animal production have contributed to major productivity and efficiency gains in food animals (Table 3-1; Hume et al., 2011). The mechanization of animal agricultural production systems, such as animal waste materials disposal, milking machines, milking parlors, milking robotics, poultry houses with automatic feeding, watering, and egg collection, and feed storage systems was advanced through research activities that combined both engineering and biological disciplines. The transition from draft power to fuel power in animal feed production was achieved through similar research. The preceding examples provide evidence of the combined-discipline approaches to solving issues that are directly related to the essential requirements for more efficient food production systems. The use of technological improvements to reduce negative environmental impacts associated with animal waste is the subject of Box 3-4.

From 1977 to 2012, commercial pork production in the United States increased 174 percent from slaughtering more and larger pigs than previously marketed (Mathews et al., 2013). Public and private R&D during this period led to efficiency gains that have altered the structure of the pork industry. Some of the gains in productivity are attributable to increases in the scale of production and technological innovation. A major contributing factor is the genetic improvement of animals through research, from lard-producing pigs prior to the 1980s to leaner meat-type pigs, which is evident in improvements in meat quality and weight gain per animal. Average dressed weight of barrows and gilts (young male and female hogs, respectively) has increased 25 percent since 1977, from an annual average of 163 pounds in 1977 to 203 pounds in 2011 (Mathews et al., 2013).

BOX 3-4
Technologies to Reduce Environmental Impact of Animal Wastes Associated with Maximizing Productivity

Animals require special nutrients designed to optimize their productivity and performance. However, dietary nutrients, such as nitrogen and phosphorus, when applied in excessive quantities or in a fashion that facilitates surface runoff, can exert harmful effects on the surrounding environment. Excess nitrogen can contaminate groundwater and surface-water supplies, and decrease air quality while excessive phosphorus can build up in soil after the land application of animal manure and run off into surface waters, causing eutrophication. The addition of crystalline amino acids to the diet results in a 1 percent unit decrease in crude protein concentration, which corresponds to a 10 percent decrease in nitrogen excretion. Adding the enzyme phytase into swine and poultry diets, which they lack naturally, can improve the digestibility of phosphorus by 20 to 50 percent. Recent research has found that the protein requirements of dairy animals can be met with lower crude protein diets without affecting milk production. A reduction in phosphorus concentration from 0.47 to 0.31 percent in the diet of dairy cattle results in a 50 percent reduction in phosphorus excretion. Reduced levels of dietary crude protein in dairy cattle and phase feeding in beef cattle have reduced nitrogen excretion. By using the knowledge of animal nutrient requirements gained from research, along with frequent feed nutrient analysis and animal performance data, a precision feeding strategy aims to optimize delivery of the level of nutrients required by the animals to reduce production costs and environmental pollution (Carter and Kim, 2013)

TABLE 3-1 Improvements in Food Animal Productivity over the Past 40 to 50 Years

Species	Trait	Performance		Increase (%)
		1960s	2005	
Pig	Pigs weaned per sow/year	14	21	50
	Proportion of lean meat	0.40	0.55	37
	Feed conversion ratio (FCR)	3.0	2.2	27
	kg lean meat/ton feed	85	170	100
Chicken, broiler	Days until 2 kg are reached	100	40	60
	FCR	3.0	1.7	43
Chicken, layer	Eggs/year	230	300	30
	Eggs/ton of feed	5,000	9,000	80
Dairy cow	kg milk/cow/lactation	6,000	10,000	67

SOURCE: Modified from van der Steen et al. (2005); Hume et al. (2011).

McBride and Key (2013) reviewed U.S. pig production from 1992 to 2009 and reported that technology innovation through advances in nutrition, genetics, housing, handling equipment, veterinary services, and management improved hog performance and efficiency. Adoption of artificial insemination to improve genetic potential and conception rates increased from 7 percent in 1990 to 46 percent in 2006 (USDA National

Animal Health Monitoring Service survey data). All-in and all-out housing management is another innovative practice to enhance productivity that was increased from 25 percent in 1990 to 71 percent in 2006. Subtherapeutic antibiotics for growth promotion, disease prevention, and overall animal health improvement have been used since the 1950s. Typically, an increase in feed efficiency and productivity was observed, especially in nursery pigs. With the concern of antimicrobial resistance in the human population, the use of antimicrobials in swine production has been decreasing since 2004.

With major gains in production efficiency from utilizing these systems and practices come significant reductions in environmental impacts (Hume et al., 2011). For example, modern beef production requires considerably fewer resources than the equivalent system in 1977, with 70 percent of animals, 81 percent of feedstuffs, 88 percent of the water, and only 67 percent of the land needed to produce 1 billion kg of beef (Capper, 2011b). In addition, waste outputs were also reduced, with modern beef systems producing 82 percent of the manure, 82 percent of the methane, and 88 percent of the nitrous oxide per billion kilograms of beef compared with production systems in 1977 (Capper, 2011b). The carbon footprint for the production of beef has been reduced by 16 percent as well (Capper, 2011b). The specific gains attained from increased feed efficiency in ruminants are discussed in Box 3-5.

BOX 3-5
Potential Benefits of Improved Feed Efficiency in Ruminants

Feed efficiency plays a crucial economic and environmental role in the production process for ruminant animals. Feed costs have historically accounted for 50-70 percent of beef production costs (Shike, 2013). Feed efficiency is an important determinant in the amount of greenhouse gas emissions intensities produced by livestock systems (Herrero et al., 2013). The nutrition received by a dairy cow affects the emission intensities produced by that animal and its waste (Mitloehner, 2014); more efficient nutrient use decreases costs as well as improves the surrounding environmental quality (Chesapeake Bay Foundation, 2014). Using intensive production processes to promote efficient feed use provides an opportunity to decrease environmental emissions and costs of production. Intensive production processes require particular attention to proper handling of wastes, and are far more likely to attract local public concern (NRDC, 2014).

The U.S. dairy industry realized a 59 percent increase in milk production with 64 percent fewer cows in 2007 than in 1944. As a consequence, greenhouse gas (GHG) production from the industry decreased 41 percent (Capper et al., 2009). In 1944, the average herd contained six cows that were fed a pasture-based diet with some supplemental grain (e.g., corn, soybean meal). Artificial insemination was in its infancy, and neither antibiotics nor supplemental hormones were available for animal use (Capper, 2011a).

By 2007, genetic improvements (e.g., artificial insemination, embryo transfer, sexed semen), the use of hormones for reproductive management, and improved animal management and nutrition resulted in the dairy industry requiring 21 percent of dairy animals, 23 percent of the feedstuffs, 10 percent of the land, and 35 percent of the water needed to produce an equivalent amount of milk compared to 1944. As a result, animal waste and carbon footprint per unit of milk were reduced 24 percent and 63 percent, respectively (Bauman and Capper, 2010; Capper, 2011a). Although the production volume of milk and beef per animal increased due to improvements in all fronts, genetic selection for individual feed efficiency in dairy and beef has never been widely researched, developed, and implemented as it has in the poultry and pig industries. Recently, some research has been done, but the opportunities for improvement are great.

The poultry industry has seen increases in the rate of gain and feed efficiency over the past 50 years (Havenstein et al., 2003; Havenstein, 2006). Havenstein et al. (2003) estimated that 85 to 90 percent of the improvement in broiler performance is attributed to genetics and 5 to 10 percent to nutrition and nutrition management. Pelletier et al. (2014) compared the environmental footprint of the egg industry in the United States in 1960 and 2010. Feed efficiency, feed composition, and manure management were the three primary determinants of the environmental impact of egg production. Per kilogram of eggs produced, the environmental footprint for 2010 was 65 percent lower in acidifying emissions, 71 percent lower in eutrophying emissions, 71 percent lower in GHG emissions, and 31 percent lower in cumulative energy demand compared to 1960 (Pelletier et al., 2014). Reductions in the environmental footprint were attributed to the following: (1) 27 to 30 percent to increased efficiencies of background systems, (2) 30 to 40 percent to changes in feed composition, and (3) 28 to 43 percent to increased bird performance (Pelletier et al., 2014).

In the United States, aquaculture is the fastest growing segment of agriculture, growing at a rate of 10 percent annually, and production levels have essentially doubled every 10 years. From 2007 to 2012, the number of total farms devoted to aquaculture in the United States decreased 27 percent, from 4,896 to 3,586 (USDA, 2014). Approximately 35 percent of the reduction was attributed to decreases in catfish farms as the total value of this farmed crop decreased from $461.9 to $375.9 million. The decrease in farms was primarily due to the demise of small farms that were unable to compete with larger farms, given the increased costs of operation. Most of the other categories of culture fisheries in the United States, including other food fish, sportfish, crustaceans, and mollusks, increased in value despite a reduction of total farms for most of these groups. From 1950 through 2007, production increased from 61,883 to 528,045 tonnes. From 2008 to 2012, production remained at approximately 420,000 tonnes (Olin, 2011) and meeting the increased per capita seafood consumption was due to imports amounting to $82.6 million in 2012 (NSTC, 2014).

The early aquaculture enterprise was devoted to maximizing rather than optimizing production, with marginal attention to sustainability issues. In addition, some enterprises were not successful because management practices were implemented without important information about the biology of the species under culture. In the absence of this baseline information, effective management practices to ensure consistent and efficient production were lacking. The concern for the aquaculture enterprise adversely affecting the environment led to a substantial amount of research transitioning to developing culture strategies based on sustainability concepts complemented by awareness of policies, existing infrastructure, and current and anticipated construct of markets. In comparison to its terrestrial animal production counterparts, the aquaculture enterprise is comparatively young. Accordingly, the potential of biotechnological advances to increase both productivity and efficiency is very high. The application of existing technologies is now strongly based on biological concepts, mitigation of adverse environmental effects, and management of ecosystem balance whereby in some cases aquaculture impacts are actually characterized as positive relative to ecosystem services (Millennium Ecosystem Assessment, 2005).

Biotechnological research with aquaculture species already has a strong innate foundation in achieving efficiency and sustainable intensification because among animal agricultural species, fish are the most efficient converters of feed to weight (feed conversion ratio). This efficiency is the result of their physiology and environment, whereby caloric demands for certain metabolic functions correspondingly reduces loss of consumed calories in the form of protein and fat. For example, the feed conversion ratio for salmon is 1.2 whereas this ratio is 8.7 for beef, 5.9 for pork and 1.9 for poultry (NOAA, 2014). These characteristics represent important opportunities for the expansion of animal aquaculture products as a source of protein. Greater emphasis on the production of lower-trophic-level species of fish and crustaceans would yield more efficiently produced and sustainable animal protein. In addition, Åsgård and Austreng (1995) noted that approximately 30 percent of protein, fat, and energy contained in feed is retained in the edible part of salmon, whereas 18, 13, and 2 percent is retained in the edible part of chicken, pigs, and sheep, respectively. Therefore, among animal agricultural species, fish, crustaceans, and other aquaculture products are the most sustainable end product relative to efficiencies of production and yield (i.e., protein for consumption per harvested animal).

Intensive production biotechnologies focus on increased production per unit of space. The systems derived from these technologies have focused on the elimination of any detrimental effects on the environment, particularly waste production (i.e., water pollution per unit of production). In addition, the water footprint for these systems is minimal, being based on water conservation and, at times, water reclamation. For example, the partitioned aquaculture system (PAS) is a self-confined (recirculating) aquaculture system (RAS), whereas the biofloc system is a pond-based system. The partitioned pond and biofloc pond systems have been studied and introduced to eliminate limitations of intensification that are not attainable in the earthen pond/raceway systems that have traditionally been used for farming. Both pond-based systems attempt to adhere to ecosystem management principles. For partitioned pond systems, the goal is to separate different aspects of ecosystem management, such as provision of oxygen, fish feeding, and waste treatment (Tucker et al., 2014). These systems have been the subject of research for many years and are beginning to be accepted as systems that can be cost-effective under commercial culture conditions. Biofloc technology (BFT) systems present many advantages over traditional pond culture systems through improved biosecurity, feed conversion, and water quality control, and increased efficiency in the use of water and land resources (De Schryver et al., 2008; Hargreaves, 2013).

Bioflocs are aggregates of algae, bacteria, protozoans, and particulate organic matter (feces, uneaten feed) that serve as food sources. The populations of bacteria that are part of the bioflocs remove potentially harmful concentrations of nitrogen and phosphorus, thereby ensuring that any waste products contained within the effluent from such systems will not be detrimental to the environment. With the consumption of flocs being an added source of nutrients for the farmed species, feed costs can be reduced because the proportional amount of protein in feeds can be reduced. BFT systems are an application of ecosystem management, have been in commercial use since the beginning of the 21st century, and currently are limited to the culture of shrimp and tilapia. There are many variations of biotechnology, and a greater understanding of the principles common to corresponding systems are needed before application to other aquaculture species can be realized.

RAS development and use have progressed, but this technology must continue to be improved with the goal of cost-effectiveness (Malone, 2013). The principal characteristic is a closed-loop design that allows a significant reduction of water and land use (i.e., environmental footprint). In addition, environmental conditions are controlled and not subject to seasonal variations, which allows for the optimization of growth under high density (intensive) culture conditions. An RAS also offers complete and convenient harvesting, rapid response to and effective control of disease, and the flexibility of being able to locate production facilities near large markets. Within the system, accumulation of harmful waste material due to intensive culture is avoided through a design that includes both biological (bacteria) and physical (filters) controls. This is an intensive production system that minimizes water and land resource inputs and effectively treats waste; however, there remain many potential issues facing the success of RAS systems (Box 3-6). Cost-effectiveness among RAS designs will be based on the scale and components of the system. Investment costs are a primary economic feasibility consideration (Losordo et al., 1998), and increases in investment cost per weight of production will translate into increases in the production cost of the product that will be transferred to the consumer. Future research must focus on species-specific designs and related costs that will reduce energy demands. Despite carnivorous species commanding a higher value in the marketplace, their successful RAS culture will also depend on the replacement of high-protein fish meal and fish oil–containing feeds with more sustainable and economically practical feeds. These feeds will be based on research that evaluates the effectiveness of the underdeveloped sources of protein in the form of insect, worm, or marine invertebrate meals and oils from algae and yeasts (NOAA and USDA, 2011).

Aquaculture production has the potential to increase significantly through the development and acceptance of biotechnological advances that will permit sustainable intensification via offshore cage culture. Culture in sea cages, particularly salmon, has advanced during the last two decades, and continued technological advances in the design, maintenance, and management of and information on monitoring from sea cages are necessary if the goal of increasing marine-based aquaculture production is to be realized (Vielma and Kankainen, 2013). Sea cages must be designed with ecosystem management in mind, particularly for offshore use. With these advances, commercialized cage culture could be designated or confined to certain zones where the protection of wild populations would be upheld and the possible conflicts of resource use would be minimized. Integrative aquaculture and

---

BOX 3-6
Economic Feasibility of Recirculating Aquaculture Systems (RAS)

RAS are closed-loop intensive production systems requiring small amounts of water input and generating nearly zero waste. These unique properties, yielding aquaculture products while imposing limited environmental impacts, poise RAS to meet the key needs of a sustainable aquaculture production system (Martins et al., 2010); however, this technology is facing a steep challenge in becoming economically viable. Local Ocean, a company receiving popular attention for producing fish using RAS, had been operating at a loss for its entire existence and lost its main investors in 2013 (Wright, 2014). Because of high upfront costs spent on curating an environment for aquaculture and high perceived risk, the economic viability of RAS is not necessarily attractive for investors (Wright, 2014). This poses a critical problem if sustainable systems such as RAS are to be developed and adopted in the aquaculture industry. Some analyses, however, conclude that RAS may have good profitability for niche and target markets (Federation of European Aquaculture Producers, 2013). RAS provide a valuable opportunity to pursue intensive aquaculture production with limited environmental impacts, but must first clear economic hurdles to be adopted on a larger-scale.

ecological research need to be conducted to develop a strong understanding of the tradeoffs between technological advances and ecosystems.

Not all species that are commercially cultivated will be amenable to intensive culture systems such as RAS and PAS because management of the basic biology of some species may not be applicable to such systems. Farming such species will accordingly warrant semi-intensive and, most probably, pond-based systems. Although production per unit of land is lower than that of intensive systems, the cost of feed and adverse water quality effects that lead to disease are significantly reduced, which contribute to profitability and economic sustainability. Research and corresponding biotechnological developments are needed for less intensive systems and must be guided by similar principles of sustainable management practices involving efficient resource use.

Technological developments in finding substitutes for fish meal and fish oil in formulated diets have been realized. During the past 15 years, the ratio of the weight of fish meal and fish oil in feed to the weight of fish produced has decreased from 3-4.1:1 to 1.5:1, demonstrating proportionately lower levels of or no fish meal in feeds produced for aquaculture (Tacon et al., 2011), but more research to find effective and efficient substitutes is needed to attain sustainable intensification. This need is particularly acute for U.S.-based production of carnivorous fish and crustacean species. Because of sources and the market for these materials, the United States has little control over prices and quantities sold. As of 2012, 35 percent of the fish meal used in aquaculture feeds was derived from wastes of fish processing rather than whole fish from captured fisheries (FAO, 2014b). An understanding of the nutrient value of fish meal derived from the combined processing waste of capture (natural fisheries) and culture (aquaculture) fisheries relative to the traditional "trash fish" source is essential. Identified nutrient deficiencies could be managed through appropriate, cost-effective enhancements.

In 2007, NOAA and USDA began a consultation program with a variety of stakeholder groups to synthesize information and provide future recommendations for the development and availability of effective and alternative (i.e., not fish meal– or fish oil–based) feeds for aquaculture. These efforts were designed to identify and prioritize research that would lead to results considered essential in realizing sustainable intensification. The cost of feeds increased threefold from 2002 to 2012, and demand and vacillations in availability, mostly originating from South America, of this nonrenewable resource was a strong impetus. The study produced a group of 20 recommendations (Appendix L; NOAA and USDA, 2011).

As the preceding examples demonstrate, researchers in animal science disciplines have been instrumental in developing new technological applications for producing efficient and safe food. Through these advancements in animal production and food processing, the food industry (including crop production) has better optimized the efficiency of production; however, there are public concerns associated with increases in production efficiency made by animal agriculture, ranging from the environmental impacts of concentrated animal feeding operations to the welfare of animals in confinement systems (Box 3-7). The stagnation in public funding of animal science research has hindered researchers' abilities to sufficiently address public concerns about some aspects of U.S. production systems that have led to increased productivity, including aspects that negatively affect animal welfare and environmental sustainability. Although productivity gains have occurred in the past several decades because of publicly funded animal science research, the current stagnation in funding may result in a slowing of productivity gains right at a time when the increase in global demand for animal protein is accelerating.

BOX 3-7
Biological Limits to Productivity?

As described in this chapter, animal science research on topics such as genetics, nutrition, and physiology have led to striking increases in animal productivity over time, usually with concomitant decreases in the environmental impacts of animal protein production. These increases in productivity, however, have sometimes come at the cost of negative effects on animal health, leading some academics to suggest that we may be approaching the biological limits of productivity (Grandin and Johnson, 2009). For example, there have been increasing problems with musculoskeletal disorders in poultry and dairy cattle, leading to high rates of lameness. It is estimated that 30 percent of dairy cattle in the United States and Canada experience problems with lameness (Hoard's Dairyman), and a similar percentage of broiler chickens have gait disorders (Hocking, 2014). The high egg-laying rate of commercial laying hens is associated with the development of osteoporosis, since bonecalcium is depleted to be used for eggshell formation; consequently, nearly 80 percent of hens in some production sys-

> tems experience bone breaks at some point during the production cycle (Lay et al., 2011). Although the causes of musculoskeletal disorders are multifactorial, a major contributor is genetic selection and management of animals for productivity. Rauw et al. (1998) documented a large number of studies that indicated an association between the selection for production traits and undesirable metabolic, reproductive, and health traits in cattle, swine, and poultry. This finding draws attention to the need for a balanced approach to genetic improvement in livestock. Future research focused on increasing productivity will need to take into account the potential for these kinds of undesirable effects, along with strategies to mitigate or avoid them.

Academic institutions have only limited input in setting overall future directions for animal agriculture research. Seemingly unrelated to any vision of a future of animal agriculture, several animal science departments including dairy and poultry sciences, for instance, were either eliminated or combined with other departments in well-known U.S. agricultural universities (Roberts et al, 2009). A reenvisioning and reinvigoration of the U.S. animal science research enterprise is needed to meet future animal protein demands sustainably.

Recommendation 3-2

Regarding productivity and production efficiency, the committee finds that increasing production efficiency while reducing the environmental footprint and cost per unit of animal protein product is essential to achieving a sustainable, affordable, and secure animal protein supply. Technological improvements have led to system/structural changes in animal production industries whereby more efficient food production and less regional, national, and global environmental impact have been realized.

> **Support of technology development and adoption should continue by both public and private sectors. Three criteria of sustainability—(1) reducing the environmental footprint, (2) reducing the financial cost per unit of animal protein produced, and (3) enhancing societal determinants of sustainable global animal agriculture acceptability—should be used to guide funding decisions about animal agriculture research and technological development to increase production efficiency.**

Other Research Priorities

Technological advancements in the animal protein production system, including genetics, breeding, reproduction, nutrition, animal health and welfare, management, food and feed safety, and food product quality, have been critical to improvements in the production of more environmentally friendly and sustainable animal protein products. One research priority in this area includes:

- Research in sustainable intensification should continue to focus on land, energy, water, and nitrogen utilization. The relevant U.S. government agencies should build a professional interagency network with the aim of maintaining sustainable animal production systems.

3-3 Genetics, Genomics, and Reproduction

The genetics of animal species has far-ranging effects in animal agriculture, including reproductive performance, farm economics, productivity, and environmental impacts. Johnson and Ruttan (1997) suggested that the adoption of breeding technologies has been the most significant factor contributing to farm animal productivity, including livestock and poultry, since the 1940s. Adoption of advanced breeding technologies has resulted in positive impacts on farm profits and milk produced per cow, but has negatively impacted cost of production (Khanal and Gillespie, 2013). Animal breeders have effectively manipulated the genomes of food animal species by using the natural variation that exists within a species. Traditional breeding was done in the past in the absence of molecular knowledge of the genes acting on a quantitative locus, and although much progress has been made, efficiency decreases when traits such as fertility, longevity, feed efficiency and disease resistance are difficult to measure and have a low heritability (Eggen, 2012). Selection for these traits must be accomplished using genomic selection, which will require high-quality phenotypic data from animals.

The application of genomics to the design and implementation of animal and poultry breeding programs is now being actively implemented, and the costs of genomic selection tools have greatly declined in recent years. Combined selection for growth, body composition, and feed efficiency continues to deliver 2-3 percent improvement per year in the efficiency of meat production (Gous, 2010), and milk yields will continue to increase by 110 kg per cow per year

(Eggen, 2012). Commercial breeding goals in poultry and pigs has widened considerably since the 1970s, with the number of genetic traits now typically between 30 and 40 (Neeteson-van Nieuwenhoven et al., 2013). Continued movement away from single-production trait selection (e.g., milk yield) toward more indices that take into account animal reproductive performance, animal behavior, animal welfare, form and function, adaptability to environmental change, and longevity will be critical for advancing the sustainability of animal agriculture. Currently, the genome, transcriptome, epigenome, and the metagenome of many species are being investigated by high-throughput sequencing methods. Benefits are already being reported in the improvement of beef quality (e.g., tenderness, marbling) and muscle development or growth (Box 3-8, Hocquette et al., 2007).

Epigenetics is a newer field in animal science investigating the changes in gene expression, not from alterations of DNA sequence, but rather changes in chromatin structure (Funston and Summers, 2013). These changes reflect broad environmental influences that add methyl groups or other structural elements to DNA beyond those involved in basic inheritance. As such, epigenetics can be considered to reflect the effects of nurture on chromosomal activities, which is a way to understand the cumulative exposure to chemical and nonchemical stressors in humans (Olden et al., 2014). Although research has investigated fetal and neonatal programming due to nutritional differences (Soberon et al., 2012; Funston and Summers, 2013), further research is needed to understand the underlying mechanisms of epigenetics and advances in statistics, bioinformatics, and computational biology (Gonzalez-Recio, 2011). Linking researchers with these basic skills to other animal scientists will help bridge gaps between the "omics" and the whole animal. Additionally, longer-term, multigenerational production systems research (particularly for extensive animal production systems) is required to reveal potential epigenetic effects, while simultaneously evaluating the environmental, social, and economic sustainability of different production systems.

> **BOX 3-8**
> **Applying Innovations in Genomic Selection to Animal Breeding Programs**
>
> Technological developments in genomics are being directly applied in livestock breeding programs, facilitating greater certainty and efficacy in promoting beneficial genetic traits in raised livestock. As a result, genomic selection is being increasingly integrated into breeding paradigms, creating opportunities for efficiency gains and sustainability improvements related to meat and dairy farming. Improvements in genetic sequencing technology have facilitated the discovery of many single-nucleotide polymorphisms (SNPs) in livestock and poultry. These SNPs have allowed for predictive equations to estimate more accurately a calf's genomic estimated breeding value (GEBV), which in turn has allowed breeders to better select males and females with describable genetic qualities and improved breeding efficiency (Schefers and Weigel, 2012). Being able to genetically screen cattle prior to sexual maturity has allowed breeders to improve the genetic makeup of livestock at a faster rate. Genetic improvement programs based on genomics are currently only used by a small number of commercial dairy producers and almost no commercial beef producers. In fact, only 5 to 6 percent of the beef herds in the United States use artificial insemination. Increasing the adoption of these technologies would accelerate the dissemination of improved genetics.

Emphasis on recent funding has been placed on genomic information and tools, whereas fundamental research on the biology of birds has been neglected (Fulton, 2012). Poultry genetic diversity is needed to better understand gene function and the effects of variation on traits. The tremendous loss of farm animal diversity on a commercial level in the United States needs to be reversed. Microbial genomics is an area that has the potential to transform the breeding industry because adjustments of the microflora could reduce environmental impacts and improve sustainability. As geneticists continue to make improvements in broiler and layer performance, there will be more emphasis on bird welfare and the ability to cope with widely different environments, especially heat tolerance and dissipation (Gous, 2010).

Green (2009) reviewed the role of animal breeders and the challenges they will face in the genomics era. He suggested that animal breeders will need to lead the way in the integration of genomic and phenotypic data. There will be a need to better understand how the interactions of genes, proteins, mechanisms, and the external environment produce the phenotype of an animal. To accomplish this, animal breeders will need to work with colleagues in physiology, nutrition, muscle biology, growth and development, lactation, immunology, microbiology and economics. The challenge is articulated by Green (2009) in the following quote:

> The reduction in number of academic animal breeding programs around the world, and particularly in North America, has resulted in a deficit of human talent and resources ready to take on these challenges. Not only are quantitative geneticists lacking, those few who are being produced in the remaining programs are entering a highly competitive job market where few are remaining in academia, with many of the trained animal breeding scientists being pulled into the plant and biomedical arenas. The need to rebuild infrastructure for developing scientists and expertise in the animal breeding area is critical and must be addressed through new strategies and models. More direct investment of private industry into the funding of these programs is beginning to occur and will need to increase in the near term. Increased reach of federal research programs into the education sector is also needed, principally through the increased availability of research funding in USDA-ARS for graduate student and postdoctoral training. The time may have come for U.S. programs to break down past barriers that have prevented consortia-type funding so that large-scale, common sense interdependent partnerships can be instituted between federal and state agencies, universities, the private sector, and industry commodity organizations. This model has certainly had success in other areas of the world, and the area of animal breeding is one that could benefit greatly from such approaches. Finally, because of the rapid escalation in the development of genome- and phenome-based technologies, the need for public education and outreach efforts has never been greater (Green, 2009).

It has been estimated that less than 10 percent of all current aquaculture production can be attributed to improvement in stocks (Gjedrem et al., 2012). Hence, aquaculture is far behind other animal genetic breeding programs. Genetic selection will generally have an end product of increased production and increased efficiency relative to the amount of feed, reduction in space needed, and the amount of labor required. The potential for high gains in growth in aquatic species has been well documented, but breeding programs remain quite limited. For example, research efforts with European seabass have recently been directed toward selection for high growth rates on diets containing plant-based rather than feedstuffs of fish origin (Le Boucher et al., 2013). Efficiency of use of resources must be the ultimate determining factor in breeding programs, such that breeding for higher production must be weighed against a prioritized array of values. Olesen et al. (2000) argued that in the quest for sustainable production systems, breeding goals include consideration of environmental and social concerns, such that market and nonmarket trait values should be dually examined. Combining marker-assisted selection with classic selection programs will speed up the process of producing those stocks that would have a positive impact on aquaculture production relative to meeting the protein consumption needs of the world. The use of genetic markers (i.e., transgenic fish) has not been positively received by the public, and a strong and sustained educational effort must be made to remove misconceptions of consequences of use of these fish for aquaculture enterprise.

Major advances in genetic editing technologies have been made possible with the development of the TALEs (transcription activator-like effectors) and CRISPRs (clustered, regularly interspaced, short palindromic repeats). These emerging technologies are promising tools for precise targeting of genes for agricultural and biomedical applications (Montague et al., 2014; Tan et al., 2013). In the future, there should be a focused effort on the factors that contribute to faster genetic gain across all species, which include: (1) a greater accuracy of predicted genetic merit for young animals; (2) a shorter generation interval; and (3) an increased intensity of selection because breeders can use genomic testing to screen a larger group of potentially elite animals (Schefers and Weigel, 2012). Genomic breeding strategies and transgenic approaches for making farm animals more feed efficient will be needed (Niemann et al., 2011). It is anticipated that genetically modified animals will play an important role in shaping the future of feed-efficient and thus sustainable animal production.

3-3.1 Advancements in Reproduction and Transgenesis

The earliest biotechnology procedure used to improve the reproduction and genetics of food animals was artificial insemination (AI), followed by embryo transfer and in vitro fertilization (Hernandez Gifford and Gifford, 2013). With the advent of AI came the opportunity to devise ways to control the estrous cycle for timed AI (Moore and Thatcher, 2006). More recent tools include super ovulation, in vitro production of embryos, cloning, sexed semen, and transgenics (Moore and Thatcher, 2006; Hernandez Gifford and Gifford, 2013). The benefit of cloning is in the increased accuracy in evaluation of bull dams and the acceleration of the rate of genetic progress. In addition, cloning provides a tool for the producer to propagate highly efficient and productive animals that require fewer inputs and a more positive environmental footprint. The transfer of new genetic material into animals via recombinant DNA protocols results in a transgenic animal. This technology has been available for over two decades and has numerous applications in food animal production including the potential to improve productivity, carcass composition, growth rate, milk production, disease resistance, enhanced fertility, and production of animals with reduced environmental impact (Table 3-2; Hernandez Gifford and Gifford, 2013). Despite its potential, the adoption of cloning and transgenics in animals for food use has not occurred because of societal concerns.

Artificial insemination is an example of a technology that was rejected or approached with much hesitation when first introduced and yet became a conventional breeding system in food animals, including specific lines of poultry production. During the last four decades, research has improved the collection and screening process for specific diseases, storage, and delivery of semen to females. This improvement has made the application of artificial insemination a practical and efficient way for breeding across the world with several food animal species. The improvement, in conjunction with several advancements in the research of embryology, including in vitro maturation, fertilization, and culture, has led to efficiency gains in the embryo transfer of food animal species. Embryo transfer is currently considered a way to improve breeding and avoid inbreeding issues that can affect production and cause diseases to spread. Embryo transfer in animals pioneered the application of this procedure in humans. Research finding recommendations applied to breeding procedures in specific food animal species, including sexed semen, cloning, and gene transfer, have contributed to food production efficiency and protein conversion. Looking ahead, application of reproductive biotechnologies such as cloning and transgenic animals needs to be considered as ways to more rapidly affect genetic change to produce the next generation of superior animals that will benefit the environment, producers, and consumers (Hernandez Gifford and Gifford, 2013).

TABLE 3-2 Examples of Successful Transgenic Food Animals for Agricultural Production

Year	Transgenic Trait	Gene	Species
1999	Increased growth rate, less body fat	Growth hormone (GH)	Pig
1999	Increased growth rate, less body fat	Insulin-like growth factor (IGF-1)	Pig
2004	Increased level of polyunsaturated fatty acid in pork	Desaturase (from spinach)	Pig
2006	Increased level of polyunsaturated fatty acids in pork	Desaturase (from *C. elegans*)	Pig
2001	Phosphate metabolism	Phytase	Pig
2001	Milk composition (lactose increase)	α-lactalbumin	Pig
1992	Influenza resistance	Mx protein	Pig
1991	Enhanced disease resistance	IgA	Pig, sheep
1996	Wool growth	Insulin-like growth factor (IGF-1)	Sheep
1994	Visna virus resistance	Visna virus envelope	Sheep
2001	Ovine prion locus	Prion protein (PrP)	Sheep
2004	Milk fat composition	Stearoyl desaturase	Goat
2003	Milk composition (increase of whey proteins)	B-casein κ-casein	Cattle
1994	Milk composition (increase of lactoferrin)	Human lactoferrin	Cattle
2005	*Staphylococcus aureus* mastitis resistance	Lysostaphin	Cattle

SOURCE: Adapted from Niemann and Kues (2007).

Control of avian influenza by genetic modification benefits the poultry industry as well as the consumer. Lyall et al. (2011) demonstrated suppression of avian influenza transmission in genetically modified chickens in which transgenes were introduced without affecting other genetic properties of the bird. Research and education efforts should focus on understanding the barriers to consumer acceptance of such technologies to realize their full sustainability potential. Other areas for research utilizing reproductive technologies include improved semen technologies and cryopreservation (Rodríguez-Martínez and Peña Vega, 2013), improved fertility (Niemann et al., 2011), and development of technologies that will result in greater success for early pregnancy (Spencer, 2013). Early pregnancy translates into higher production and greater economic efficiency and sustainability of food animal production enterprises. Systems approaches that address nutrition, disease mitigation and prevention, environmental effects, and better adaptation through genetic selection will be important.

Recommendation 3-3

Further development and adoption of breeding technologies and genetics, which have been the major contributors to past increases in animal productivity, efficiency, product quality, environmental, and economic advancements, are needed to meet future demand.

Research should be conducted to understand societal concerns regarding the adoption of these technologies and the most effective methods to respectfully engage and communicate with the public.

Other Research Priorities

Reproductive technologies, such as artificial insemination and embryo transfer, continue to result in more rapid genetic improvement, especially in ruminants. In addition, epigenetics, which is the study of the impact of relatively subtle alterations of the genetic code through such processes as methylation of DNA bases, is an exciting new research area that shows promise of providing insights into factors related to post-embryo processes that affect animal growth and health. Improved understanding of genetic x environment interactions can further our understanding of production efficiency while considering other aspects of sustainability such as animal welfare. Research priorities for this area include:

- Research in understanding gene–environment interactions, epigenetics, genomics, nutrigenomics, for example, using biotechnological tools and genetic editing should be conducted to meet the goal of production efficiency while also considering other meaningful components of sustainability such as animal welfare.
- Research leading to advances in the integration of genomic and phenotypic information, genomic breeding strategies, obtaining faster genetic gain, greater accuracy of predicted genetic merit, shorter generation intervals, and increased intensity of selection should continue, particularly in aquaculture.
- Resources should be devoted to the new field of epigenetics, with its potential to provide insight into predicting and integrating the effects of multiple growth promoters and stressors into animal agriculture.
- Animal science departments should integrate computational biology and bioinformatics expertise into traditional animal science disciplines in order to capitalize on the potential of genomics, epigenetics, and metagenomics to improve animal husbandry, productivity, and sustainability.
- Reproductive technologies, such as embryo transfer and cryopreservation, should be further developed and utilized to accelerate the rate of genetic gain. Further emphasis is needed on semen characterization, storage, and quality.

3-4 Advancements in Nutrition

Animal science research has contributed significantly to our current understanding of nutrition concepts and principles in energetics (Johnson, 2007), carbohydrates, and lipids (Nafikov and Beitz, 2007), proteins (Bergen, 2007), and body composition and growth (Mitchell, 2007). The first half of the 20th century may be thought of as the qualitative era of nutritional research where most of the essential nutrients and functions were discovered, whereas the second half of the century was a quantitative period, as nutrient requirements, nutrient–nutrient interactions and pharmacologic aspects of nutrients were the focus of animal nutrition research (Baker, 2008). Translation of research into preparing and feeding diets that closely met an animal's requirements led to increased productivity and efficiency.

Many advances in understanding animal nutrition have been the result of the public funding of animal science research. The committee refers the reader to the following published reviews on nutritional advancements and innovations that are too numerous to mention in detail in this report: (1) innovations in indirect methods of estimating dry matter intake and digestibility (Burns, 2008); (2) innovations in energy determination (Burns, 2008); (3) innovations in forage nutritive value (Burns, 2008); (4) innovations in the evaluation of harvested forage (Burns, 2008); (5) innovations related to grazing animals (Burns, 2008); (6) landmark discoveries in swine nutrition (Cromwell, 2009); (7) major advances in fundamental dairy cattle nutrition (Drackley et al., 2006); and (8) major advances in applied dairy cattle nutrition (Eastridge, 2006).

With a major focus on genetic improvement, knowing nutrient requirements of animals at their various stages of life is critical to achieving maximum productivity and reproduction and optimal health and preventing performance and health problems. For example, intense breeding resulted in broilers growing faster than their immature skeletal system could support, resulting in leg and health problems (McCarthy, 2013). Energy and nutrient requirements (amino acids, lipids, carbohydrates, water, minerals, and vitamins) of various food animal species are established from the results of animal nutrition research. The National Research Council (NRC) of the National Academies commissions species-specific scientific committees about every 10 to 15 years to review the scientific literature and establish nutrient requirement guidelines. These requirements are used as the gold standard for formulating diets, especially for studies that will be submitted to U.S. regulatory agencies. In addition, in legal cases these requirements are used as the comparator to assure that diets have met or exceeded the requirements of the species of interest.

The NRC publishes requirements for beef cattle, dairy cattle, swine, poultry, small ruminants (goats, sheep, cervids, new world camelids), horses, rabbits, fish and shrimp, dogs, cats, and laboratory animals (NRC, 2014), which are used in many parts of the world. Published NRC requirements that are relevant to the genetic capabilities of animals in 2050 will be critical. Applegate and Angel (2014) reviewed the history of the NRC and nutrient requirement publications and advocate for keeping these documents current. Based on their perception of the historical poultry nutrient requirement guidelines, the definition of requirement may be changing from a requirement to prevent a nutrient deficiency to a requirement to optimize growth or egg production response per unit of nutrient intake. To keep the nutritional requirements of animals current with the genetic advancements, species-specific research needs to continue in many areas, including energy, protein and amino acids, lipids, carbohydrates, water, minerals, vitamins, models for estimating nutrient requirements, alternative feedstuffs, the effects of environment on requirements, nonnutritive feed additives, feed contaminants, feed processing, feed intake, nutrient excretion and the environment, and growth- and lactation-enhancing products and their effects on nutrient requirements.

Current versions of NRC publications on animal nutrient requirements include chapters that articulate the current research needs as viewed by the committees that review the latest data pertaining to nutrient requirements. For example, gaps in knowledge of swine nutrition include methods of nutrient requirement assessment, nutrient utilization and feed intake, energy, amino acids, minerals, lipids, vitamins, and feed ingredient composition (NRC, 2012, ch. 15). Critical research needs for fish and shrimp include (1) establishment of essential amino acid requirements; (2) an understanding of whether the main difference in qualitative essential fatty acid requirements among species is dependent on environmental habitat or whether feeding habit is equally or more important; (3) a more definitive understanding of the role of carbohydrate sources on metabolism and energy partitioning; (4) micronutrient requirements; (5) delivery of nutrients in different types of manufactured feed and more rigorous methodology to assess bioavailability of essential amino acids in feed ingredients; (6) alternative lipid sources; (7) nutrient interactions among amino acids, fatty acids, soluble carbohydrates, and lipids; (8) delivery of water-soluble nutrients in sinking and larval feeds; (9) application of modern molecular techniques to determine the effects of nutrients on cells and organelles; and (10) gene-specific regulation of physiological processes (NRC, 2011, ch. 17).

Nutritional advancements in food animals are being made to enhance animal reproduction (e.g., long-chain polyunsaturated fatty acids, selenium, and vitamin E), improve the quality and nutritional value of animal products for the consumer (e.g., dietary vitamin E to extend meat shelf-life, enhanced omega-3 fatty acids in eggs, fresh grass to increase conjugated linoleic acid in dairy products), and improve animal health (e.g., copper and gut health, selenium and vitamin E for improved oxidative stability). The crucial need to study and understand the role of gut microflora is analyzed in Box 3-9. Major accomplishments in monogastric nutrition have been made with development of energy systems, amino acids and available amino acids, lipids, carbohydrates, minerals, gut metabolism, and gut microbiome work. Research in improving the utilization of fiber in feed ingredients for swine, poultry, and aquaculture is needed. For ruminants, major advancements have been in energy metabolism with the development of the net energy system, protein and amino acid requirements, rumen digestion, technology for protecting nutrients (e.g., amino acids, lipids) from microbial degradation in the reticulo-rumen, improved forage digestibility, and lipid and mineral metabolism.

BOX 3-9
Gut Microflora and the Microbiome

Relatively little is known about the function or the beneficial attributes of many microbes that live in the gastrointestinal (GI) tracts of animals; however, it is clear from studies of many species that the coevolution of animals and microbes led to many symbiotic and commensal relationships between microbes and the GI tract (Hooper and Gordon, 2001; Ley et al., 2008). A better understanding of these relationships could create innovative advances in livestock production to the benefit of the environment, human and animal health, and production efficiency. Selecting animals for breeding stock that have better nutrient retention or altering the microflora in the gut has the potential to increase the efficiency of nutrient use and to reduce waste production and greenhouse gas emissions from livestock. For example, the microbes in the foregut of ruminants could be altered to metabolize nutrients more efficiently, reduce nutrient excretion, and mitigate methane emissions. One method proposed for altering the gut microflora is to increase the amount of acetogenic bacteria in the gut, which decreases the amount of hydrogen available to reduce carbon dioxide to methane (Jeyanathan et al., 2014). Another method is to select breeding stock that produces less methane and uses nitrogen more efficiently, which could decrease ammonia and nitrous oxide emissions and presumably decrease greenhouse gas emissions.

Subtherapeutic levels of antibiotics are administered to livestock, in part, to increase weight gain. In a feeding study comparing pigs fed performance-enhancing antibiotics with those not fed antibiotics, Looft et al. (2012) found that the pigs fed antibiotics had an increase in certain bacterial populations and an increase in energy production and conversion that were related to microbial functional genes. Subtherapeutic use of medically important antibiotics in livestock production is currently under scrutiny because of its possible linkage to increased antibiotic

> resistance in humans (CDC, 2013) and the relationship between subtherapeutic antibiotic use and overcrowded housing conditions for livestock (FAO, 2013). It may be that probiotics could be used in the place of antibiotics. Probiotics—live microorganisms suspected of conferring health benefits to the host—are already used as additives to the feed of many kinds of livestock for their ability to increase production efficiency and improve food safety by outcompeting pathogenic microbes in the gut (Callaway et al., 2003). For example, the direct feeding of probiotics to cattle has been shown to decrease fecal shedding of E. coli O157:H7 (Brashears et al., 2003); however, the mechanisms of action for many of these beneficial microbial feeds are poorly understood (Gaggìa et al., 2010).

The existence of successful feeds for the culture of larval fish and crustaceans, particularly marine and estuarine species, limits the ability to increase production. In contrast to live food diets that require much labor and can be subject to variable nutrient composition, formulated diets offer control over nutrient content and substantially eliminate labor. A number of technologies have been used in attempts to deliver nutrients effectively to larval forms; however, success, defined as the exclusive use of these feeds, has been very limited. Technological advances are essential to achieve cost-effective preparation and content of these feeds whereby they will be readily consumed and well digested and assimilated to deliver required nutrients for growth (NRC, 2011).

All of these advancements have led to improved productivity and production efficiency with less environmental waste in the form of nitrogen, phosphorus, and GHG emissions per unit of output; however, not all agricultural systems take advantage of these nutritional advancements that contribute to environmental sustainability. For example, in the U.S. National Organic Program, participants cannot use synthetic amino acids such as lysine, methionine, and tryptophan, which are used to balance the dietary amino acids to better meet the animal's requirements. Because of this restriction, organic diets may cost more and animal productivity may be reduced with less efficient use of those nutrients, resulting in more nitrogen being excreted into the environment. Many nutritional advancements are not allowed to be used in organic agriculture according to the U.S. National Organic program (list of approved ingredients can be found in the *Code of Federal Regulations* [7 CFR §§ 205.600-205.607]). Only synthetic methionine is approved for poultry up to specified maximum levels. Additionally, the U.S. National

Organic Program has requirements for the production of feedstuffs, such as the prohibition of synthetic fertilizers and pesticides for 3 years before the feed can be certified organic, which can limit feed availability and increase costs for organic animal production systems (Mainville et al., 2009). Organic crop yields tend to be lower than conventional crop yields; however, the difference in yields is dependent on the system and location (Seufert et al., 2012; Rosegrant et al., 2014). Cows on organic farms produced 43 percent less milk per day than conventional nongrazing cows and 25 percent less than conventional grazing herds (Stiglbauer et al., 2013). Future research on improving crop productivity and feed quality while addressing environmental sustainability concerns and accommodating socioeconomic and cultural needs, regardless of the production system considered (i.e., organic or conventional), will benefit the nutrition of food animals.

Recommendation 3-4

The committee notes that understanding the nutritional requirements of the genetically or ontogenetically changing animal is crucial for optimal productivity, efficiency, and health. Research devoted to an understanding of amino acid, energy, fiber, mineral, and vitamin nutrition has led to technological innovations such as production of individual amino acids to help provide a diet that more closely resembles the animal's requirements, resulting in improved efficiency, animal health, and environmental gains, as well as lower costs; however, much more can be realized with additional knowledge gained from research.

> **Research should continue to develop a better understanding of nutrient metabolism and utilization in the animal and the effects of those nutrients on gene expression. A systems-based holistic approach needs to be utilized that involves ingredient preparation, understanding of ingredient digestion, nutrient metabolism and utilization through the body, hormonal controls, and regulators of nutrient utilization. Of particular importance is basic and applied research in keeping the knowledge of nutrient requirements of animals current.**

Other Research Priorities

Organic animal agriculture is growing in the United States. For example, organic livestock increased from ~56,000 head in 2000 to

~492,000 in 2011, and poultry numbers have steadily increased from 3.1 million to 37 million. Although in certain circumstances organic animal agriculture can contribute to sustainability at the local level, such as by decreasing energy needs for transportation and by lessening the concentration of waste in a single location, there is no evidence to suggest that organic animal agriculture in its present form can be scaled up to make a substantial contribution to current or future needs for animal protein. Research priorities for this area include:

- For organic agriculture to contribute significantly to meeting global protein demand, research directed toward quantifying the potential for organic agriculture to be sustainably scaled up (equal or better environmental footprint per unit of animal protein produced, equal to or more affordable supply of animal protein, equal to or greater efficiency in animal protein production compared to conventional) to meet more than local needs is essential.
- Very high priority should be given to research into how best to enhance informed and respectful engagement between the scientific community and the public on issues related to the potential role of organic agriculture in achieving sustainable local and global impacts on food security.

3-5 Animal Models

3-5.1 Agricultural Animal Models Benefiting Animals

Research using animal models together with characterized diets has been beneficial in advancing animal nutrition and has contributed to what is known about nutrient–nutrient interactions, bioavailability of nutrients and nutrient precursors, and tolerance levels for excessive intakes of nutrients (Baker, 2008). In the last 50 years, efforts were made to understand the biology of animals and poultry in quantitative terms. Baldwin and Sainz (1995) and Dumas et al. (2008) provide a historical perspective to the mathematical models used in animal nutrition as well as a discussion on the mechanistic dynamic growth and lactation models. Many of these mathematical models with sensitivity analyses helped animal science researchers to focus on areas where knowledge gaps needed to be filled and to have major impacts on the biological system as a whole. Other animal models were developed to accurately predict nutrient requirements of animals in different stages of development, including maintenance, growth, lactation, and reproduction. Energy and protein feeding systems exemplified in the NRC nutrient requirement publications describe the nutrient requirements of more than 20 species of domestic animals and have evolved considerably over the past 40 years. Although systems have remained static, factorial, and largely empirical in nature, mechanistic concepts have been incorporated into the statistical models used in analyses of input-output data and applied in the revised systems (Baldwin and Sainz, 1995). The NRC nutrient requirement publications in the last 15 years for dairy, beef cattle, and swine all contain animal models that use environmental, management, and feed inputs to predict animal requirements and performance. In the future, agricultural animal models will need to become more dynamic and mechanistic, and more integrated into biological systems to reflect and accurately predict whole-animal performance for both research and field applications. They will also need to include consequences for and from environmental change, especially regarding performance as it relates to biogeochemical cycling and climate change. Although the committee believes that comprehensive system models that include soil, crop, animal, land use, for example, would be useful in studying such environmental tradeoffs, the scope of the work focused on animal science research and thus animal models is specifically highlighted in this section. McNamara and Shields (2013) and Lantier (2014) suggest that animal models be used as tools for understanding emerging or reemerging infectious diseases. Biomathematical system models will be needed for integrating nutrition, growth, reproduction, and lactation to predict animal performance. Such models will enhance animal science research in exploring and predicting the effects of different feeds, feeding systems, management, environment, and health.

3-5.2 Agricultural Animal Models Benefiting Humans

Agricultural animal models have been used in biomedical research to study a wide range of physiological and disease factors important to humans (Ireland et al., 2008). Seventeen Nobel Prize winners have used farm animals such as cattle, pigs, sheep, goats, horses, and chickens as research models (Roberts et al., 2009), yet their value is generally underappreciated. "There are numerous examples of compelling domestic species models relevant to diverse areas of biomedical research, including comparative physiology and genomics, cloning, artificial insemination, 'biopharming' to produce high-value pharmaceuticals in milk, osteoporosis, diabetes-induced accelerated atherosclerosis, asthma, sepsis, alcoholism, and melanoma" (Roberts et al., 2009). Agricultural animals should continue to be used in biomedical research.

Research Priorities

Mathematical animal model systems have been helpful in identifying key areas to research as well as providing a better understanding of the function of biological systems and predicting nutrient requirements and performance. They have also been helpful as models for studying physiological processes and diseases that may have important implications for other animals and humans. Such efforts also are critical to improving our understanding of complex biogeochemical cycling affecting climate and environment, including geographical variability. Research priorities for this area include:

Funding of research to encourage the development of animal models that are more dynamic, mechanistic, with more comprehensive integrated biological systems (i.e., growth, maintenance, reproduction, lactation) is needed to accurately predict whole-animal performance for and provide better assessments of current and potential future practices in animal agriculture.

Funding agencies should increase their emphasis on biomathematical modeling systems research at all levels (e.g., tissue, whole-animal, and production systems levels) and building and strengthening connections between modelers and experimental animal scientists.

3-6 Feed Technology and Processing

Feed technology and processing are important in enhancing productive efficiency and production in food animals. Particle size is important in rate and extent of digestion. Particle size of the diet can be too small and cause health problems such as stomach ulcers in pigs and acidosis in cattle, or too large in size and cause decreased digestibility and reduced feed intake. Heating has become important in deactivating antinutrients in feeds such as soybeans, enhancing starch digestibility through gelatinization in the steam flaking of corn and milo, and pelleting and extrusion processes; however, too much heat can be detrimental to certain amino acids, vitamins, for example (Institute for Food Technology, 2010). Feed processing is important in dust control and achieving higher feed intakes, maintaining a uniform composition, and reducing waste. Feed technology is important in aquaculture to deliver a floating, slowly sinking or sinking feed depending on the feeding behavior of the intended species. Better methods are needed to deliver dietary nutrients that are acceptable to the animal as well as contain the required nutrient density without suffering quality. This has involved grinding, pelleting, extrusion, and coating technologies as well as others. In ruminants, processing of forages and delivering feeds as total mixed rations, which reduces the animal's ability to sort out feed ingredients so that each mouthful contains the same consistent nutrient profile, has proven to improve production by reducing the variability of nutrient intake.

For a review of issues and challenges in feed technology pertaining to particle size, pelleting, and feed uniformity, see Behnke (1996) and Neves et al. (2014) and for a review on heat treatment and extrusion in animal feeds, see the proceedings of the 2nd Workshop, "Extrusion Technology in Feed and Food Processing" (Institute for Food Technology, 2010).

Research Priorities

Feed technology and processing is important to improving feed acceptability to the animal, feed efficiency, and animal health, and reducing feed wastage. One research priority in this area includes:
- Research in feed technology and processing should continue to make further step changes in the most efficient utilization of current and alternative ingredients in animals with minimal wastage (e.g., research areas of potential significance include improvements in water stability for aquaculture, enzyme stability in pelleting feed).

3-6.1 Alternative Feedstuffs

The feed industry and food animal producers are constantly looking for alternative feed ingredients that are more economical but maintain the right digestible nutrient profile, no contaminants, and compatibility with other currently used ingredients, and are organoleptically acceptable to the animal. Typically, co-products produced in making human food are used and include bakery byproducts, wheat midds, wheat bran, corn gluten feed, corn gluten meal, brewers' grains, fish meal, meat and bone meal, poultry byproducts, peanut meals, peanut hulls, rice mill byproduct, rice hulls, soybean meal, canola meal, almond hulls, citrus pulp, sugar beet pulp, sugar cane molasses, yeast, animal fat, whey, and blood meal. Other co-products are produced in nonfood industries such as biofuels (wet or dried distillers' grains, glycerin), cotton production (cottonseed meal, cottonseed hulls), fermentation waste products, and algae. Feed ingredients that do not compete with human food sources and can be converted to value-added nutritious animal products contribute to food security and are more favorable in terms of carbon footprint and environment.

Fish meal and fish oil, for example, are the main protein and oil sources used in the formulation of aquaculture diets, especially for carnivorous fish such as salmon. In 2007, principally in response to the use of unsustainable ingredients (feedstuffs) such as fish meal and fish oil in feeds for aquaculture, NOAA and USDA initiated the study of alternative feedstuffs and feeds for aquaculture. There is a finite supply of fish meal and fish oil, and with increased demand for salmon, alternative less expensive protein and oil sources are being advocated and evaluated (Naylor et al., 2009). Soybean meal can partially replace fish meal, and rapeseed oil can partially replace fish oil. However, there are limitations because alternative ingredients do not have the correct nutrient profile to provide 100 percent replacement of currently used feeds. For example, fish oil contains high levels of long-chain polyunsaturated fatty acids (eicosapentaenoic acid [EPA] and docosahexaenoic acid [DHA]) that vegetable oils do not have. Thus, a minimum amount of fish oil must remain in the diet or an alternative source of EPA and DHA needs to be provided. Recently, insect meal has been being evaluated as a protein source. Use of insect meal derived from the larvae of insects feeding on seafood waste is an attractive area of future research (van Huis et al., 2013). Research has been devoted to an evaluation of the use of fish offal (processing waste) for the mass culture of insects that in turn can be prepared as a meal for use as an ingredient in formulated feeds. This is a promising area of research that offers the potential impact of ultimately eliminating the dependence on fish meal and fish oil as ingredients in feeds for carnivorous species. The insect meal would serve as a good source of amino acids and fatty acids that characterize fish meal and fish oil. For rainbow trout, Oncorhynchus mykiss, soldier fly prepupae served as an effective substitute for 25 percent of fish meal and 38 percent of fish oil ingredients. These larvae can be grown on fish offal (waste), and noteworthy increases in total tissue lipid (30 percent) and n-3 polyunsaturated fatty acids (3 percent) were observed in just 24 hours when compared to tissue of larvae that were fed on manure for an equivalent period of time (St. Hilaire et al., 2007).

Continued efforts need to be directed toward the use of mixtures of processed animal protein ingredients and plant-derived meals as partial or complete substitutes. Cost and benefits of these alternative feedstuffs can only be assessed through an understanding of the content and availability of nutrients in the feedstuffs, as well as the environmental footprint associated with their preparation and production. Research in improving the nutrient capture from co-products by livestock, poultry, and aquaculture is critically needed, especially the improvement in nutrient utilization of fiber by poultry, swine, and aquaculture. Research should also be directed to the development of technology of mass culture of microorganisms (e.g., algae, yeasts) using food or industrial waste streams as the substrate, and to the use of oilseeds that contain important required nutrients such as long-chain polyunsaturated fatty acids that are found in fish meal and fish oil and can potentially serve as ingredients of formulated feed (Byrne, 2014). Wastes (e.g., protein and oils) derived from the production of biofuels and byproducts from the processing of terrestrial animals also hold potential as alternative feedstuffs. Typically, the industry that is generating the new product funds the animal science research needed to commercialize the product. For example, Texas A&M AgriLife is evaluating algal residue for animal feed after the biofuel has been extracted (Texas A&M University, 2012). The environmental impacts of producing an alternative feedstuff will also need to be assessed (NOAA and USDA, 2011).

Recommendation 3-6.1

Potential waste products from the production of human food, biofuel, or industrial production streams can and are being converted to economical, high-value animal protcin products. Alternative feed ingredients are important in completely or partially replacing high-value and unsustainable ingredients, particularly fish meal and fish oil, or ingredients that may otherwise compete directly with human consumption.

> **Research should continue to identify alternative feed ingredients that are inedible to humans and will notably reduce the cost of animal protein production while improving the environmental footprint. These investigations should include assessment of the possible impact of changes in the protein product on the health of the animal and the eventual human consumer, as well as the environment.**

3-6.2 Feed Additives, Growth Promotants, and Milk Production Enhancers

Feed additives and growth promotants are used to enhance productivity and improve feed efficiency, often resulting in decreased environmental impacts per unit of output (Capper and Hayes, 2012). In the U.S. National Organic Program for animal production, use of most of these additives and growth promotants is not allowed. Feed additives may improve feed intake, alter the gut microbiome in a positive way, enhance nutrient absorption, decrease antinutritives, increase digestibility of fiber, improve mineral utilization such as phytate phosphorus, improve carbohydrate metabolism, decrease methane production and improve health. Examples of feed additives include exogenous enzymes, yeast, ionophores, organic acids, probiotics, and prebiotics. Ionophores, such as monensin, lasalocid, lailomycin, salinomycin, and narsin are rumen modifiers that improve feed efficiency while potentially reducing methane emissions in ruminants (Appuhamy et al., 2013).

Another area of feed-based research that has the potential of increasing production efficiency of aquaculture organisms is the use of probiotics and prebiotics in feeds. These live microbial or nondigestible compounds, such as oligosaccharides, are feed additives that have shown promise by increasing feed efficiency and/or immune response or resistance to pathogenic infection (NRC, 2011). Both additives have been shown to alter the resident bacteria within the gastrointestinal tract of fish and crustacean species. Although there are limited reports about the benefits of probiotics and prebiotics in aquaculture organisms, future efforts to increase production efficiency from their use are expected. Laboratory research that identifies positive responses to probiotics and prebiotics ideally should be complemented by testing under production settings.

Exogenous enzymes have improved the digestibility of feedstuffs and reduced waste (Meale et al., 2014). There is a need for a better understanding of the enzymatic machinery involved in cell-wall degradation through a better understanding of the microbial microbiome. The use of dietary enzymes as a means to increase the availability of nutrients to fish and crustacean species has had limited application, but experimental results have shown positive effects on the digestibility of protein and carbohydrates (Buchanan et al., 1997; Glencross et al., 2003). The lack of stability of these enzymes when exposed to high temperature as part of the aquafeed manufacturing process is still a challenge (NRC, 2011). A dietary supplement of phytase to increase the digestibility of phytic acid, a source of phosphorus contained in plant feedstuffs, has been successful in releasing phosphorus for uptake (Gatlin and Li, 2008). This procedure also has a positive environmental effect because less undigested phosphorus is then released into the environment as a potential pollutant. Biotechnological advances are needed whereby enzymes can be added to diets without surrendering their specific metabolic capabilities.

Food and Drug Administration–approved growth promoters and milk production enhancers have been key technologies resulting in improved production efficiencies in the U.S. cattle industry for over 50 years (Table 3-3). For example, researchers recently reviewed the results of 26 peer-reviewed journals or reviews by regulatory agencies on the efficacy and safety of sometribove zinc suspension (rbST-Zn), a form of recombinant bovine somatotropin, in lactating dairy cows (St-Pierre et al., 2014). The results of the meta-analysis indicated that rbST-Zn administration to dairy cows effectively increases milk production with no adverse effects on cow health and well-being. The findings are consistent with anecdotal reports regarding the use of recombinant bovine somatotropin in more than 35 million U.S. dairy cows over 20 years (St-Pierre et al., 2014). The development through research, adoption, and application of these feed additives, growth promotants, and milk production enhancers has contributed to food security and sustainability in several ways: (1) production costs have been reduced through reductions in the amount of feed required per unit gain; (2) the land necessary to produce equivalent amounts of food for consumers has been reduced; (3) the production of GHGs has been limited by reducing the number of animals required to produce equivalent amounts of beef or milk; and (4) cost savings have been extended to consumers by providing a year-round affordable supply of beef and milk at reduced prices (Johnson et al., 2013).

Research Priorities

FDA-approved feed additives, growth promoters, and enhancers of milk production have been successfully used for decades, resulting in improved efficiency and economics and reduced environmental footprint. In aquaculture production systems, for example, feed additives called prebiotics or probiotics have shown promise for increasing growth/feed efficiency and/or immune response or resistance to pathogenic infection. Research priorities for this area include:

TABLE 3-3 Chronological Sequence of FDA Approval of Growth and Lactation Promotants Used in the U.S. Beef and Dairy Cattle Industries

Growth Promotant	Year of FDA Approval
Oral diethylstilbestrol (DES)	1954
DES implant	1957
Estradiol benzoate/progesterone (steers)	1956
Estradiol benzoate/testerone propionate (heifers)	1958
Oral melengestrol acetate (heifers)	1968
Zeranol (36 mg) implants (steers)	1969
Oral DES removed from the market	1972
DES implants removed from the market	1973
Silastic estradiol implant (cattle)	1982
Estradiol benzoate/progesterone (calves)	1984
Trenbolone acetate (TBA) implants (cattle)	1987
Estradiol (17-β)/TBA implants (steers)	1991
Estradiol (17-β)/TBA implants (heifers)	1994
Zeranol (72 mg) implants (cattle)	1995
Estradiol (17-β)/TBA implants (stocker cattle)	1996
Ractopamine hydrochloride (cattle)	2003
Zilpaterol l hydrochloride (cattle)	2006
Increase Milk Production Efficiency Agents	**Year of FDA Approval**
Bovine somatotropin (lactating dairy cows)	1993
Monensin sodium (dry and lactating dairy cows)	2004

SOURCE: Adapted from Johnson et al. (2013).

The committee supports the initiation and/or the continuation of research devoted to the development of products that enhance growth and lactation of animals.

Research needs to be conducted to identify prebiotics and probiotics that have documented efficacy in the improvement of growth and immunocompetence.

Social science research needs to be conducted to understand the social ramifications and public concerns about the use of these technologies in the production of animal protein and how to better engage and respectfully communicate the substance of the technology and its benefits and risks to the public, to food security, and to a sustainable environment.

3-7 Animal Health

It is important from a food security and food safety standpoint that food animals are healthy, animal diseases are not transmitted to humans, and animal products are safe and affordable for human consumption.

Animal health is an essential component of sustainable animal production systems. Highly contagious animal diseases can be a major disruption to global food security. Since the creation of the World Trade Organization, animal health issues have been a major hurdle for trading animal protein globally, and animal health issues are becoming a trade component for the United States and other countries. USDA ARS and NIFA have given top priority based on budget allocation to addressing animal health issues in the United States. In 2010 and 2011, USDA ARS and NIFA held stakeholder workshops that determined the high-priority health issues by species in which funding should be directed (see Appendixes D and E).

The progression in understanding animal disease ecology has led to better approaches to managing diseases that have serious negative impacts on food animal populations. During the last four to five decades, vaccination for highly contagious animal diseases has improved both in its delivery system and in vaccine production. The majority of these diseases with high economic consequences can be controlled or even eradicated from specific regions if disease intervention measures are taken. These intervention measures, including vaccination, teat dipping, dipping and housing management were researched by veterinary and animal scientists prior to being recommended to animal agriculture industries. Animal agriculture and society in general have gained major benefits from research on animal diseases, with the ability to improve animal welfare as well as reduce the risk of transmitting these diseases to humans. Advancements have resulted in less human labor so that producers can handle a greater number of animals at improved efficiencies, increasing output per animal as well as per farm. Innovative research has also lead to better animal health, such as the use of an antibacterial delivery system for aquaculture (Box 3-10).

With the potential future elimination of subtherapeutic use of medically important antibiotics, research is needed to discover substitutes that will provide the same or greater benefits in improved feed efficiency, disease prevention, and overall animal health. Confronting disease issues will be essential for private investment in aquaculture operations in the future. More research is needed to address the systems-specific production losses due to disease. Many of the disease problems arise from poor management practices that compromise the value and nutritional value of the animal product. Intensification that is sustainable will require responses to disease other than the use of medically important antibiotics. Effective responses to disease lie in proactive investigations of the physiological relationships between host and pathogen. To meet these challenges, a specific animal model needs to be developed whereby an understanding of this basic physiology of host and pathogen can be achieved. A model organism could be used to establish the basis for understanding this relationship.

BOX 3-10
The Adoption of a Novel Antibacterial Delivery Method in Aquaculture

Preventing disease in aquaculture populations is critical to the industry's financial success and the well-being of raised fish, because losses due to disease pose a serious challenge. For example, the Chilean salmon farming industry lost $2 billion, 30,000 jobs and 350,000-400,000 tons of fish from an infectious salmon anemia outbreak in 2007 (Kobayashi and Brummett, 2014), and climate change will likely lead to further outbreaks due to sea surface temperature changes (Handisyde et al., 2006). The collection of farmed fish is usually achieved through the processes of netting, grading, or pond draining (Humane Society of the United States, 2014). The frequent use of nets to sort and collect aquaculture yields presents a unique opportunity for the delivery of antibacterial agents to protect farmed fish and marine animals. The application of active substances on netting has already been tested and used for antifouling purposes on fishing equipment (Bazes et al., 2006) and as such, provides a possible medium for antibacterial coating to protect aquacultures from diseases and parasites of concern. The potential for protective coatings on nets is enormous in terms of preventing disease and minimizing aquaculture losses; however, there are substantial risks, such as the danger of creating antimicrobial-resistant bacteria in humans and animals (Park et al., 2012). If managed effectively and applied appropriately, antibacterial coatings for netting could improve aquaculture yields and better protect raised species.

In aquaculture, the zebrafish is an attractive model to gain the necessary understanding of the physiological relationships of the host and the pathogen whereby appropriate management procedures, whether preventive or therapeutic, can be implemented. The development of a standardized diet for zebrafish culture will be essential to these efforts because nutrition must be controlled. Otherwise, responses may vary depending on the diet fed to the model. Immersion vaccines were the first type of vaccines developed for fish, and vaccination methodology has progressed to injection of antigens into the body cavity (Sommerset et al., 2005). Most of the vaccines are produced by animal health companies, and an array of them are used in the aquaculture industry with a focus on salmonid species (i.e., salmon and trout) but delivery through feed has not achieved success because of breakdown and resulting lack of effectiveness (Sommerset et al., 2005). Vaccines in fish species are most effective against bacterial rather than viral diseases (Sommerset et al., 2005). Killed virus–based vaccines are not effective unless delivered at high dosages and by injection, conditions that compromise cost-effectiveness. Live-virus vaccines have yet to be developed or used because of concerns about environmental safety (Sommerset et al., 2005). Research devoted to the possible development of vaccines for invertebrates, specifically cultured crustaceans, has yielded inconclusive results (Sommerset et al., 2005); however, Rowley and Pope (2012) offer evidence that vaccination through an immune priming mechanism may have application, but will probably be species and pathogen dependent.

Vaccine development and delivery in aquaculture appear to be frustrated by the limited knowledge of the immune systems and the effect on other organisms and the consumer. Research devoted to understanding the innate immune response of aquaculture species should lead to the identification of stimulants that will promote immunocompetence. Appropriate standards related to the assessment of the health status of an array of animal and aquaculture species need to be developed. The identification of DNA probes that would determine whether or not live or processed food animals are pathogen-free could lead to attractive, field-based "health tests" that could be used by veterinarians, field technicians, or farmers. Research is needed on scientific, managerial, and educational tools and practices to enhance identification of and response to an animal disease outbreak. Continuing research is needed on technological tools that can rapidly identify diseases, defects, or contamination in animal products.

Recommendation 3-7

The subtherapeutic use of medically important antibiotics in animal production is being phased out and may be eliminated in the United States. This potential elimination of subtherapeutic use of medically important antibiotics presents a major challenge.

There is a need to explore alternatives to the use of medically important subtherapeutic antibiotics while providing the same or greater benefits in improved feed efficiency, disease prevention, and overall animal health.

Research Priorities

Animals must be healthy to achieve optimum production and efficiency. Animal diseases can result in market and trade interruptions, thus impacting food security, and may pose a significant human health threat. Advancements in animal health such as through vaccines and other preventive measures have resulted in less human labor, so that producers can handle a greater number of animals at improved efficiencies, increasing output per animal as well as per farm. Development of vaccines, however, is often frustrated by a limited knowledge of immune systems. Development must include cost analyses relative to production, delivery, and effectiveness, with particular emphasis on the development of successful vaccines using live viruses. Research priorities for this area include:

Funding agencies should support research to address the entire production system's specific production loss due to disease. The research should integrate various components in the production system and not restrict it to animal health aspects. An example is PED virus, for which production, animal movement, and feed ingredients must be considered (see Box 2-1).

Scientific, husbandry, and educational tools and practices to enhance identification of and rapid response to an animal disease outbreak must be a focus of public-sector research. Continuing research is needed on technological tools, including biological models that can rapidly identify diseases, defects, or contamination in animal products.

Research should continue to achieve an understanding of the innate immune response of species, include aquatic species, leading to the identification of stimulants that will promote immunocompetence and the rapid development and production of efficacious vaccines or other preventive measures.

Research should be conducted to address threats of new and emerging zoonotic diseases that may be exacerbated by climate change and intensification (e.g., tickborne diseases).

3-8 Animal Welfare and Behavior

Animal welfare research was considered a high priority for many of the stakeholders participating in the 2011 USDA NIFA animal health workshop (Appendix F). The scientific study of animal welfare is still relatively new compared to other disciplines and fields within the animal sciences. The empirical evaluation of animal welfare is a complex undertaking, and there is no single definitive indicator (Mason and Mendl, 1993). Such evaluation involves investigation of biological functioning (e.g., health and physiological normality), the consequences of an animal's ability or inability to perform highly motivated behaviors, and their subjective states (e.g., pain, fear, pleasure) (Fraser, 2008a,b). There is an ongoing need to develop and validate new experimental methodologies for welfare assessment, particularly indicators of animals' subjective states (Dawkins, 2006).

Animal welfare research is, of necessity, multidisciplinary, although for historical reasons discussed in Chapter 2 the development of the field was stimulated by concerns about animal production systems that restricted animal movement, and thus applied animal behavioral scientists (ethologists) initially played a major role (Fraser et al., 2013). Note, however, that ethology is an important discipline in the animal sciences in its own right, independent of its importance to animal welfare. An understanding of animal behavior is important to improve animal management and productivity in many production systems (Price, 2008). For example, animal behavior research has been influential in improving reproductive performance through understanding factors affecting expression of mating behavior and success. It has also contributed to better management methods for animals kept in social groups, methods for improving maternal care of offspring, and better handling techniques for food animals.

Ethology is also still a key discipline within animal welfare science; however, many other disciplines now contribute to the increasing body of animal welfare research and research application (Fraser, 2008a,b; Mellor et al., 2009; Appleby et al., 2011; Fraser et al., 2013) including physiology (e.g., stress, environment, neurophysiology), veterinary medicine (e.g., preventive, pathology, epidemiology), agricultural engineering/environmental design, comparative psychology, nutrition, genetics, microbiology and social sciences (e.g., the study of human–animal interactions).

Despite being a relatively new field, there is now a significant research base for animal welfare (Fraser et al., 2013); the output of scientific publications has increased by 10-15 percent annually during the last two decades and nearly half of the approximately 8,500 animal welfare papers in the ISI database were published in just the last 4 years (Walker et al., 2014). There have been many important theoretical and applied research gains. Theoretical advances include the development of testing and assessment methodologies, such as assessing animal preferences, feelings, and motivational states (Widowski, 2010). Research has also increased knowledge regarding animal pain and its alleviation (Stafford and Mellor, 2011), as well as the factors that lead animals to develop detrimental abnormal behaviors that necessitate the use of painful management practices such as beak trimming and tail-docking (Mason and Rushen, 2008). From an applied perspective, animal welfare researchers have contributed to both knowledge and new technology development for improving animal handling, transport, and slaughter (Grandin, 2010) and alternative housing and management systems (Appleby et al., 2011).

Many of these advances have come about because of a sustained commitment within the European Union to funding animal welfare research during the last few decades. The United States has lagged behind in this effort, and the number of researchers and the knowledge base relative to public concern are now imbalanced (Johnson, 2009). There is a need to build capacity and increase funding for research in animal welfare, with a focus on U.S. production systems and management and in the context of the broader sustainability considerations affecting U.S. agriculture. There is also a need for additional research on animal welfare that incorporates consideration of public values and attitudes and addresses public concerns about topics such as animal feelings and the ability of animals to perform their natural behaviors.

While some animal welfare research will be commodity specific, there are also critical overarching research areas (FASS, 2012; Walker et al., 2014), some of which overlap with identified animal health research needs. These include development of (1) alternatives/refinements for painful management procedures such as beak trimming and dehorning; (2) alternatives/refinements of euthanasia and slaughter methods to reduce pain and distress; (3) improvements in handling and transportation to reduce injury and distress; (4) new or modified production systems that provide animals with more behavioral opportunities while maintaining good animal health and production; (5) genetic and management methods to reduce the incidence of musculoskeletal disorders (e.g., lameness and osteoporosis); (6) methods for dealing with nutritionally based problems, including nutritional deficiencies/imbalances and feed restriction; (7) new preventive and prophylactic strategies to maintain good animal health in alternative management systems, such as organic and antibiotic-free, that prohibit the use of current disease prevention or treatment methods; (8) outcome-based (i.e., health and behavior) welfare assessment criteria that can be measured noninvasively for on-farm assessment; and (9) management methods for young/neonatal animals to improve later stress-competence and adaptability. These research priorities should be elaborated on to encompass the areas of emphasis identified by the OIE (World Organization for Animal Health) as part of the general principles for animal welfare (Box 2-3). The need for more translational research is also becoming apparent due to the extent to which animal welfare research is being incorporated into codes of practice, industry standards, and regulation (Appleby, 2003; Mench, 2008). This is increasing the focus on on-farm studies and the use of epidemiological methods and modeling to assess the environmental and management factors that are risks for poor welfare outcomes in commercial flocks and herds (Dawkins et al., 2004; Butterworth et al., 2011).

Although the research priorities identified above focus mainly on food animals, there is also growing emphasis on and an increasing body of knowledge about the welfare of farmed fish (Branson, 2008). Active areas of research and continuing research need to include the effects of stressors (e.g., stocking density, water quality, conspecific aggression, transport, slaughter) on fish behavior and physiology and the development of handling and slaughter methods that reduce pain, injury, and distress (Conte, 2004; Huntingford et al., 2006; Ashley, 2007). Cognizance and application of management practices that reduce stress during rearing will correspondingly minimize the incidence of disease. One such method could be the development of feeds that have the target of reducing stress through the identification of the optimal amount and balance of nutrients in feeds. Simpson and Raubenheimer (1995) offered a geometric multidimensional framework strategy to identify optimal amounts and balance of nutrients for insects with a growth target in mind. This approach may have application to vertebrates and would most probably compromise some growth in the interest of the welfare of the organism during rearing.

Recommendation 3-8

Rising concern about animal welfare is a force shaping the future direction of animal agriculture production. Animal welfare research, underemphasized in the United States compared to Europe, has become a high-priority topic. Research capacity in the United States is not commensurate with respect to the level of stakeholder interest in this topic.

There is a need to build capacity and direct funding toward the high-priority animal welfare research areas identified by the committee. This research should be focused on current and emerging housing systems and management and production practices for food animals in the United States. The Foundation for Food and Agricultural Research, USDA AFRI, and USDA ARS should carry out an animal welfare research prioritization process that incorporates relevant stakeholders and focuses on identifying key commodity-specific, system-specific and basic research needs, as well as mechanisms for building capacity for this area of research.

3-9 Feed Safety

With the need to produce more protein for the projected human population in 2050, a variety of traditional and alternative animal feed commodities (feedstuffs) will be incorporated into feeds. In the United States, individual states regulate the contents of commercial animal feeds while the FDA regulates the content and manufacture of animal feed at the federal level. In particular, efforts are directed to eliminate the presence of unwanted contaminants that adversely affect the health of farmed species as well as the human consumer by passage through the food chain (Sapkota et al., 2007). These contaminants include polychlorinated biphenyls and other persistent organic pollutants (POPs), excessive levels of heavy metals (lead, mercury, cadmium, and chromium) or mineral salts (arsenic, selenium, and fluorine) (Reis et al., 2010), and pesticides. These contaminants are contained in either the feedstuffs or forage while others arise from external contamination during storage. For example, in 1997, the FDA found animal feeds contaminated with dioxin, which resulted in elevated levels of dioxin in chickens, eggs, and catfish due to bioaccumulation in fatty tissue

(Hayward et al., 1999). Dioxin is a polychlorinated aromatic hydrocarbon POP, which originated from a mined product termed "ball clay" that was used as an anticaking agent in the soybean meal feedstuff. Another example is elevated levels of selenium, which although it is a required micronutrient, acute or chronic consumption of forage or feed containing excessively high levels can result in an array of clinical symptoms that include lethargy, reduced feed consumption and growth, and a variety of pathologies (Zain, 2010).

As an importer of animal protein products, the United States must be involved in the development of international standards, guidelines, and recommendations that protect the health of farmed animals and consumers. For example, over 80 percent of all seafood consumed in the United States is imported; international standards are in the best interests of the United States. To ensure feed safety for both farmed animals and consumers, outreach, education and quality assurance programs need to be developed for feed manufacturing facilities. In developing countries, outreach and education will be critical to the successful transition from extensive to semi-intensive farming and will require complete or supplemental manufactured feeds as part of farming practices. The Codex Alimentarius Commission is responsible for codes of practice designed for the prevention and reduction of food and feed contamination (WHO and FAO, 2012) and should serve as the foundation for the development of international standards, guidelines, and recommendations to protect the health of farmed animals and consumers. Development of rigorous policies, regulations, and successful outreach programs combined with improved analytical methods are necessary to effectively detect contaminants both in the United States and internationally. Assurance of high-quality feeds is an important component of sustainable intensification by mitigating or eliminating the incidence of any potential adverse effect on the health and productivity of farmed animals and the health of consumers. This knowledge should be complemented by educational efforts to increase an understanding of the many poisonous plants that can be consumed by certain species of farmed agricultural animals, the conditions that might lead to consumption, and the resulting adverse effects on health.

Feed safety challenges include contaminants of microbial, chemical, and biological origin; as well as consideration of the safety of feed additives. Feed additives include technological additives such as preservatives, colors, antioxidants, emulsifiers, stabilizing agents, acidity regulators, and silage additives; as well as sensory and nutritional additives, digestibility enhancers, medications, and gut flora stabilizers (EFSA, 2014). Typically, before companies commercialize these additives, they are evaluated for safety and efficacy. The FDA has established guidance documents for evaluation of food additives in diets fed to food animals (FDA, 2014b).

Inadvertent presence of feed contaminants can cause significant animal health risks, which may or may not translate to human health risks. FDA (2014a) regulates animal feed safety and has provided documents relevant to bovine spongiform encephalopathy (BSE), medicated feed in which the medications are additives that enhance nutritional concentrations in diets or are prophylactics against disease, and contaminants, including dioxin in anticaking agents and mycotoxins. In the context of sustainability, it is important to consider how future climate scenarios may affect the safety of animal feed. Perhaps the greatest concern in this regard is the risk of increased levels of mycotoxins in animal feed crops. Mycotoxins are secondary metabolites of foodborne fungi that have toxic effects on the animals and humans that consume them. For animal feed in industrial production systems, the mycotoxins of greatest concern are found in feeds based on corn, peanuts, and small cereals such as wheat, barley, and oats. The three most important groups of mycotoxins are aflatoxins, fumonisins, and deoxynivalenol (DON, "vomitoxin") (Table 3-4).

The FDA has set multiple action levels for total allowable aflatoxin in feeds for different groups of animals (FDA, 1994). Likewise, it has industry guidelines for allowable fumonisins and DON in various animal feeds. Two recent concerns have emerged regarding levels of mycotoxins in animal feed. The first is the use in feed of distillers' dried grains with solubles (DDGS), which are co-products of corn-based ethanol production. The ethanol production process concentrates the original mycotoxins in the corn up to three times in the DDGS, which are then sold for animal feed in the United States. Although animal feed is never composed entirely of DDGS, even adding it in the proportions suggested by animal nutritionists could lead to higher mycotoxin exposures that cause observable animal effects, such as reduced productivity (Wu and Munkvold, 2008). Additionally, in the near future, there is reason to believe that increased climate variability associated with climate change trends may result in higher preharvest levels of certain mycotoxins in crops, posing both economic and health risks (Wu et al., 2011). As the leading producer of corn in the world market, the United States must be well posed to address the potential impacts of climate change on corn for animal feed in terms of economic losses from excessively high mycotoxin levels at grain elevators and in terms of animal health losses.

Implications for the future of U.S. food security from mycotoxins in animal feed are twofold. First, there is the direct toxicological concern of whether mycotoxins are bioavailable in animal products that humans consume. Although most mycotoxins do not bioaccumulate in meat, dairy animals that consume aflatoxin metabolize it to aflatoxin M1, which is then present in milk and other dairy products. Second, higher mycotoxin exposures in animal feed translate to reduced growth and productivity, which means economic loss to producers as well as reduced supplies of meat and animal products for human consumption. Hence, reducing mycotoxin levels in animal feed is critical for sustainable food animal industries to ensure future food security.

TABLE 3-4 Adverse Animal Health Effects Associated with Three Classes of Mycotoxins

Mycotoxin	Fungi That Produce Toxins	Crops Contaminated by Toxins	Animal Health Effects
Aflatoxins	*Aspergillus flavus, A. parasiticus*	Corn, peanuts, tree nuts (almonds, pistachios, hazelnuts, etc.), copra, cottonseed, spices	Liver tumors Immune suppression, Reduced weight gain and productivity Lower eggshell quality in poultry
Fumonisins	*Fusarium verticillioides, F. proliferatum, A. niger*	Corn	Equine leukoencephalomalacia Porcine pulmonary edema Reduced weight gain and productivity
Deoxynivalenol (DON)	*F. graminearum, F. culmorum*	Corn, wheat, barley, oats	Gastrointestinal disorders Immune dysfunction Reduced weight gain and productivity

SOURCE: Wu et al. (2011).

Research Priorities

As an importer and exporter of animal protein products, the United States is affected by the development of international standards, guidelines, and recommendations that protect the health of the farmed species and the consumers through regulation of contaminants in feeds and feedstuffs. One research priority for this area includes:
- Research through U.S. agencies in the identification of feed and feedstuff contaminants should continue so that rigorous policies of regulation can continue to be updated to protect the health of farmed species and consumers and provide information to feed manufacturers through outreach programs.

3-10 Food Safety and Quality

3-10.1 Food Technology

Supplying affordable, safe, and desirable animal products as a source of nutrients to the human diet is important. Americans annually purchase about $100 billion of animal products (e.g., meat and meat products, poultry products, fish, shellfish, and dairy products, and nonfood products, such as wool, mohair, cashmere, and leather) at the farm gate and retail levels (USDA NIFA, 2009). The quality and safety of animal products prior to harvest are influenced by genetics, nutrition, and management systems, while after harvest they are affected by handling, processing, storage, and marketing practices (USDA NIFA, 2009). The threat from foodborne illness due to animal products has been reduced during the last few decades, particularly because of the improvement in the hygienic process of food (Morris, 2011). The hygienic process of animal origin food was mainly derived through a series of recommendations from several research initiatives conducted via experimental and observational studies of foodborne agents (e.g., bacterial, viral, and toxins). The practice of Hazard Analysis of Critical Control Points (HACCP) was initiated through research and is currently becoming the standard for food safety in the United States and abroad as many food-producing systems are adapting HACCP in their operations. One obstacle in overcoming foodborne illnesses is public fear and resistance to food safety techniques such as meat irradiation (Box 3-11).

> **BOX 3-11**
> **Irradiating Meat to Combat Foodborne Illness**
>
> The Centers for Disease Control (CDC) estimates that one in six Americans (or 48 million people) become sick from foodborne illnesses each year (CDC, 2014). Developing an effective means of ensuring that meat products are safe for consumption is paramount. Irradiating meat products is one means of killing potential pathogens and sanitizing meat for consumers. The FDA approved the use of irradiation in 1963 for wheat and flour, and has since expanded its allowed use to include fruit, meat, vegetables, and spices (USDA FSIS, 2013). Academic studies in the following years confirmed the safety of irradiation as a food safety practice, also finding that the practice did not harm food properties such as nutritional value (Stevenson, 1994). Irradiation has the positive benefit of damaging the DNA of living organisms when used in high energy rays, which kills potential foodborne pathogens (Tauxe, 2001). The USDA Food Safety and Inspection Service (FSIS) noted in its final rule on meat and poultry irradiation that "irradiation of meat and poultry does not increase human exposure to radiation since the energy used is not strong enough to cause food to become radioactive," adding that during testing scientists used higher levels of radiation than those approved for use and discovered no negative health effects from consuming the irradiated meat (USDA, 1999). But despite the scientific acceptance of irradiated meat's safety, there are still difficulties in selling its safety to the public. As was documented in a survey by the Foodborne Diseases Active Surveillance Network, only half of surveyed adults were willing to purchase irradiated ground beef or chicken, and only a fourth of respondents were willing to pay a "premium" for these meats which (at the time of that survey) cost more to produce than nonirradiated products (Frenzen et al., 2000). Despite scientific consensus of the public health benefits from meat irradiation, the public acceptance of this practice faces an uphill battle.

A major goal of food safety is to determine a way to measure the impact of interventions at different phases of the production chain. Research is needed to examine individual interventions and those used in combination, and the economic considerations of such interventions. The continued identification and evaluation of risk factors is needed to understand the transmission and persistence of foodborne organisms, and the development and implementation of interventions, mitigation, prevention, and/or control strategies are needed (Oliver et al., 2009). Research should target more rapid, sensitive, and specific diagnostic tests for many foodborne pathogens.

With the growing popularity of "organic" and "natural" food products and the number of recalls of products from these production systems, an increased emphasis on food safety is required, especially in microbial pathogen control. Research is needed to determine better and faster pathogen detection methods for laboratory and field application (Sofos, 2009). The concept of food quality has been evolving within the animal product industry to where there is (1) hygienic quality (i.e., absence of heavy metals, pesticides, mycotoxins, feed additives, drugs, pathogens, and microbial contaminants); (2) compositional quality (e.g., water, protein, fat, and ash); (3) nutritional quality (i.e., biological value of protein and fatty acid composition); (4) sensorial quality (e.g., taste, color, flavor, odor, tenderness, and juiciness); and (5) technological quality (i.e., suitability for processing, storage, and distribution) (Nardone and Valfrè, 1999). Hocquette et al. (2005) provided other definitions of quality; however, most experts make a distinction between intrinsic and extrinsic quality attributes. Intrinsic quality refers to safety and health aspects, sensory properties and shelf-life, chemical and nutritional attributes, and reliability and convenience of food. Extrinsic quality refers to production system characteristics and marketing variables such as traceability. The most important intrinsic quality traits include color, texture, flavor, juiciness, and healthiness, and the most important extrinsic trait is traceability (Hocquette et al., 2005). Traceability needs ongoing validation because it is a vital component of operational standards and certification programs. The implementation of traceability for aquaculture products is a work in progress that is confronted with many challenges (Schroder, 2008). Traceability systems are particularly pertinent to aquaculture products that are taking on increasing significance in global trade and are highly perishable. Product supply chains are monitored with the ultimate goal of guaranteeing an accurate determination of product sources to protect consumer health and prevent fraud (e.g., misrepresentation of species). A practical guide for traceability in the U.S. seafood industry has been produced (Petersen and Green, 2006).

There have been many advancements and innovations in animal product quality in the past 30 years (Aumaître, 1999; Nardone and Valfrè, 1999; Higgs, 2000; Hocquette et al., 2005, 2007; Goff and Griffiths, 2006; Henning et al., 2006; Sofos, 2009). Nardone and Valfrè

(1999) evaluated the effects of production methods and systems such as genetics, nutrition, management, and animal health on poultry meat, cattle meat, eggs, and milk quality. Genetics provides the greatest opportunity for improvement in quality, followed by nutrition and management. An example of the application of functional genomics on beef quality such as tenderness and marbling is provided by Hocquette et al. (2007). Research has also determined how quality attributes such as color, texture, tenderness, flavor, and juiciness can be altered (Hocquette et al., 2005). In the dairy industry, fluid milk processing has made technological advancements in extending the shelf life of milk by high-heat treatment, microfiltration, and aseptic packaging (Goff and Griffiths, 2006). Membrane filtration technology has been an important advancement in dairy processing in the past few decades for all dairy products, and cheese production has benefited from increased knowledge of the genetics of microorganisms used during their production (Henning et al., 2006; Johnson and Lucey, 2006). Red meat production has benefited from selective animal breeding and the increased trimming of fat, which has resulted in a decrease in the total fat and saturated fat content of red meat (Higgs, 2000).

In the future, animal product research in the United States needs to focus on the quality of animal food products. Genomics and proteomics are likely to be the most important factors in affecting animal product quality (Nardone and Valfrè, 1999; Hocquette et al., 2005; Johnson and Lucey, 2006). Animal nutrition research needs to further assess the sanitary and nutritional characteristics of meat (i.e., fatty acid composition such as conjugated linoleic acid and omega-3s, polyunsaturated fatty acids, vitamins and trace minerals, antioxidants, and chemical and technological characteristics of fat). The effect of different management systems on meat, milk, and egg quality will need further research with a focus on hygienic and organoleptic qualities Fundamental research is important to understand the biological and physical mechanisms, and to develop new techniques to improve quality. Sensory quality and methods using a systems approach to improve the nutritional value of fish, meat, milk, and eggs and associated products should remain a research priority (Hocquette et al., 2005).

Research Priorities

The quality and safety of animal products prior to harvest are influenced most by genetics, nutrition, and management. Genetics provides the greatest opportunity for further improvement in animal product quality; although technological advances have played a major

role in milk and dairy food processing, genetics and nutrition have been the major contributors to beef, pork, and chicken quality. Traceability is an important component of the assurance of food quality and needs ongoing validation because it is also a vital component of operational standards and certification programs. Research priorities for this area include:
- Studies to identify and evaluate risk factors to understand the transmission and persistence of foodborne organisms should be initiated and conducted by relevant public institutions including universities. These studies should be aimed toward the development and implementation of interventions or mitigations across all the phases of the production chain. The validity of prevention or control strategies should continue to be assessed.
- More rapid, sensitive, and specific foodborne pathogen diagnostic assays need to be developed for laboratory and field application.
- Research should be conducted to understand the biological and physical mechanisms relevant to improving nutritional, functional, and organoleptic qualities of animal products.

3-10.2 Food Losses and Food Waste

Food is lost or wasted throughout the food chain (Figure 3-4). Food losses refer to qualitative (i.e., reduced nutrient value, undesirable taste, texture, color, and smell) and quantitative (i.e., weight and volume) reductions in the amount and value of food. Food losses represent the unconsumed edible portion of food available for human consumption. Food waste is a subset of food loss and generally refers to the deliberate discarding of food because of human action or inaction.

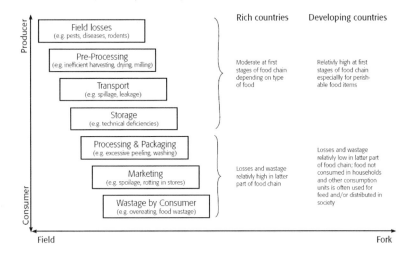

FIGURE 3-4 Areas of food losses and wastage. Illustration by Britt-Louise Andersson, Stockholm International Water Institute.
SOURCE: Lundqvist et al. (2008). Reprinted with permission from the Stockholm International Water Institute.

Food loss and waste are important for the following reasons (Buzby and Hyman, 2012): (1) food is needed to feed the growing population and those already in food insecure areas; (2) food waste represents a significant amount of money and resources; and (3) there are negative externalities (i.e., GHG emissions from cattle production, air pollution and soil erosion, and disposal of uneaten food) throughout the production of food that affect society and the environment. Annual global food loss and waste by quantity is estimated to be 30 percent of cereals; 40-50 percent of root crops, fruits, and vegetables; 20 percent of oilseed, meat, and dairy products; and 35 percent of fish (FAO, 2014a). One-third or 1.3 billion tons of the food produced for human consumption is lost or wasted globally (Gustavsson et al., 2011). In 2008, the estimated total value of food loss at the retail and consumer levels in the United States as purchased retail prices was $165.6 billion or 30 percent of the total value (Table 3-5; Buzby and Hyman, 2012). In terms of the value of food loss, meat, poultry, and fish accounted for 41 percent and dairy products for 14 percent of the total loss in food supply value (Buzby and Hyman, 2012).

TABLE 3-5 Estimated Total Value of Animal Product Loss at 2008 Retail and Consumer Levels in the United States

Animal Product	Food Supply, Million $	Losses from Food Supply					
		Retail Level		Consumer Level		Total Retail and Consumer Level	
		Million $	%	$ Million	%	Million $	%
Dairy products	97,622	9,023	9	14,679	15	23,703	24
Fluid milk	23,655	2,844	12	4,164	18	7,008	30
Other dairy products	73,957	6,180	8	10,515	14	16,695	23
Meat, poultry, fish	176,284	8,453	5	59,844	34	68,297	39
Meat	83,127	3,747	5	27,911	34	31,658	38
Poultry	69.100	2,694	4	25,810	37	28,504	41
Fish	24,058	2,012	8	6,124	25	8,135	34
Eggs	12,826	1,154	9	1,751	14	2,905	23

SOURCE: Adapted from Buzby and Hyman (2012).

The total food loss at the combined retail and consumer level in the United States was estimated to be 28 percent dairy products, 38 percent meat, 41 percent poultry, 33 percent fish and seafood, and 23 percent eggs (Buzby and Hyman, 2012). The major contributors to food loss at this level include spoilage, expired sell-by dates and confusion over label dates (NRDC, 2014), all of which pertain to shelf life. In the United Kingdom, 60.9 percent of the food wasted could be attributed to storage and management, which is directly linked to shelf life with smelling and tasting bad and past due dates as the major reasons for disposal (WRAP, 2008). One extra day of shelf life could result in savings of £2.2 billion due to less food waste in the United Kingdom (Robinson, 2014).

One major research focus for food loss in animal products is shelf life. Shelf life can be defined as the time it takes a food product to deteriorate to an unacceptable degree under specific storage, processing, and packaging conditions. Time to deteriorate depends on the product composition, storage conditions (e.g., temperature, atmosphere), processing conditions, distribution conditions, initial quality, and packaging. Modes of food deterioration include microbial, insects and rodents, chemical (i.e., oxidation, flavor deterioration, color change or loss, vitamin loss, enzymatic), and physical (i.e., moisture or textural change) (Hotchkiss, 2006). For aquaculture, technologies that can be effectively applied post-harvest are needed to extend shelf life.

Improving animal product shelf life would reduce wastage and open or expand export markets, and may make niche-oriented, lower-volume products more economical.

A lot of research has gone into extending the shelf life of dairy products. Dejmek (2013) provides a historical overview of milk shelf life including heat treatment, pasteurization, ohmic heating, and microwave and ultrahigh-temperature (UHT) sterilization. Post-pasteurization shelf life is affected by contamination of milk by psychrotrophic bacteria, presence of heat-stable enzymes (e.g., proteases and lipases), or presence of thermoduric psychrotrophs (e.g., bacterial spores that survive pasteurization) (Tong, 2009). With pasteurization and good storage temperatures, a 17- to 21-day shelf life can be attained. For extended shelf life, ultrahigh-temperature processes are used to sterilize all milk contact surfaces and disinfect packaging to prevent environmental contamination at filling. Shelf life for UHT milk may last 30 to 90 days in refrigerated storage. Bacterial growth is usually not an issue, but physical and chemical changes due to heating and prolonged storage can be major contributors to food loss (Tong, 2009). Research has found that storing milk at low refrigerated temperatures will reduce the rate of these changes affecting milk quality. Nonthermal technologies are being evaluated, such as pulsed electric field, high-pressure processing, membrane filtration, controlled-atmosphere packaging, and natural microbial inhibitors.

Extending the shelf life of eggs is another area of investigation. Eggs cooled under current methods lose AA grade (highest grade for eggs) in 6 weeks. If eggs could maintain AA grade for 8 weeks, then they could be shipped anywhere in the world. Scientists at Purdue University have discovered a rapid cooling technology that provides 12 weeks of AA grade rating (Wallheimer, 2012). Research efforts continue in the extension of meat shelf life. Major contributors to meat waste include off-color in the retail case and rancidity caused by oxidation, microbial contamination, and microbial load. Consumers in the United States prefer red meat to be red in appearance. The red color is the result of myoglobin in the meat binding to oxygen and iron (ferrous form) and converting it into oxymyoglobin. Over time, the oxymyoglobin is converted to metmyoglobin, due to oxidation, resulting in the meat changing color to brown, which is unappealing to consumers. In an effort to extend the shelf life and keep the appearance of the meat red longer, cattle feeders can feed a high level of Vitamin E (antioxidant) to animals. The Vitamin E deposited in the muscle tissue by the animal results in an antioxidant effect, thereby slowing the color change in meat. Currently, cattle producers have not widely adopted the practice because the economic benefit to the retailer has not been passed back to the producers who must increase production costs to use Vitamin E.

Microbial load on or in meat can reduce shelf life and potentially cause human safety concerns, resulting in massive recalls and waste. Irradiation technology was developed to mitigate this problem. Pasteurization technologies such as irradiation, freezing, flash pasteurization, and ultrahigh-pressure exposure have been successfully used to inactivate pathogenic enteric viruses such as Vibrio in mollusks (Richards et al., 2010). New products and concepts to address the microbial load and contamination, including edible antimicrobial films being placed on the surface of meats to extend shelf life, are being explored (Morsy et al., 2014). Packaging technologies are also being researched to help extend meat shelf life. The goal of packaging is to reduce the rate of quality loss and extend shelf life. Recent technologies to extend shelf life include modified-atmosphere packaging (MAP), vacuum-skin containers, higher-barrier packaging, direct addition of carbon dioxide to products, broader use of irradiation, high pressure, ohmic heating, and pulsed-light treatment (Hotchkiss, 2006; Spinner, 2014). MAP is a packaging technology advancement that includes directly adding carbon dioxide (CO_2 pad technology) to the package to displace oxygen and ethylene, which prevents the growth of aerobic bacteria. In February 2011, FDA granted approval for a CO_2 pad technology to be used for meat, poultry, seafood, and fruit and vegetables, which is now in commercial use (JS Food Brokers, 2014). In addition, antimicrobial packaging is being investigated where the packaging material contains antimicrobial or bioactive materials, selective and adjusting barriers, indicating and sensing materials, and/or flavor-maintenance and -enhancing materials (Hotchkiss, 2006).

Research efforts need to continue to reduce animal product waste in the United States. Dejmek (2013) outlined the following areas to be further explored for dairy products: (1) sterile extraction of milk and the maintenance of carbon dioxide/oxygen status of raw milk; (2) a more complete understanding of the biology of spore germination, which would lead to new methods of reliably inducing germination and thus removing the need for sporicidal treatments; (3) affinity-based separations of pathogens and product spoiling enzymes; (4) a better understanding of the casein micelle to make it possible to manipulate the micelle size on the technical scale and thereby improve the efficiency of bacteria separating methods; and (5) improvement of centrifugation efficiency by detailed computational fluid dynamics simulations of centrifuge flow patterns, which are now feasible. Research to extend fish, meat, milk, and egg shelf life and mitigate and prevent microbial contamination at the point of animal product harvest and packaging should be pursued. Research is also needed into how to better communicate the benefits and safety of technologies such as irradiation, which have been met in the past by consumer resistance.

3-11 Environment and Climate Change

Sustainability encompasses economic, environmental, and social considerations; however, sustainability is often assumed to address only environmental issues. At its core, animal agriculture is about converting natural resources of lower human value (e.g., forages and grains) to food and fiber products of higher value to humans (e.g., meat, milk, eggs, and wool). Although agriculture is essential for human civilization, it can also contribute to environmental degradation and change, natural resource depletion, and biodiversity loss. Animal agricultural productivity is also impacted by environmental change and climate variability. Incorporating sustainability into animal agriculture is challenging because it requires balancing increased global demand for animal products, mitigating environmental impacts, and addressing social concerns about animal welfare, food safety, and labor issues while also keeping food production economical for producers and consumers.

3-11.1 Implications of Historical U.S. Animal Agricultural Trends

In many cases, the changes in production efficiency that have been made across species in animal agriculture as previously discussed translate into reduced environmental emissions and resource use per unit of output (Table 3-6). Increased daily body-weight gains, daily milk yields, reproductive performance (e.g., eggs/hen per year), and shortened times to slaughter have all had a net effect of decreasing the amount of resources required for animal maintenance relative to animal production on an individual animal and a herd/flock basis (Capper and Bauman, 2013).

TABLE 3-6 Published Historical Comparisons of U.S. Egg, Dairy, and Beef Industries' Environmental Impacts and Resource Use

Reference	Industry	Years Compared	Change Relative to Historical Year
Pelletier et al. (2014)	Egg	1960 vs. 2010	65% lower acidifying emissions, 71% lower eutrophying emissions, 71% lower GHG emissions, and 31% lower cumulative energy demand per kilogram of eggs produced
Capper et al. (2009)	Dairy	1944 vs. 2007	79% fewer animals, 77% fewer feedstuffs, 65% less water, 90% less land, and 63% lower carbon footprint per 1 billion kg of milk produced
Capper (2011b)	Beef	1977 vs. 2007	30% fewer animals, 19% fewer feedstuffs, 12% less water, 33% less land, and 16% lower carbon footprint per 1 billion kg of beef produced

Concomitantly with increased production efficiency, there has also been tremendous intensification and geographical concentration of U.S. terrestrial animal operations over the last several decades. Across all major U.S. animal agricultural industries, there has been a consistent trend toward decreasing the number of operations and increasing animal numbers per operation; however, total U.S. animal populations for the ruminant industries have greatly declined from historic highs (Figure 3-5).

The concentration and confinement of animals in one geographical location have increased public concern about how nutrients (particularly manure nutrients) are managed and lost to the environment and about human and animal health and welfare (Fraser et al., 2001; Donham et al., 2007). One notable exception to the concentration trend is the cow-calf and stocker/backgrounder phases of the U.S. beef industry. More cow-calf production now occurs from larger herds; however, there are still over 700,000 cow-calf producers in the United States with an average herd size of 40 beef cows (USDA Census of Agriculture, 2012b).

Additionally, both of these phases of the beef production chain rely heavily on extensive grazing of pasture and rangeland; therefore, there are inherent limits to the degree of consolidation that can occur in these sectors when extensive production practices are used.

Although improvements in production efficiency have largely reduced environmental impacts and resource use across animal industries per unit of output in the United States, there are nutrient concentration or resource supply challenges that are affecting and are affected by animal production. One example is water use in the western United States. California is currently experiencing a severe drought, which is highlighting the challenges of allocating water to municipalities, agriculture, and ecosystems within the state (Howitt et al., 2014). Another example is the High Plains Aquifer (or Ogallala), which extends from western Texas and eastern New Mexico to South Dakota and has dropped precipitously in certain areas, especially in the aquifer's southern portion (USGS, 2014). The area's economy is highly dependent on agriculture, and the region produces significant amounts of dairy, beef, and pork. The long-term viability of those industries in their current state may be impossible if the drop in the aquifer continues at its current rate (Steward et al., 2013). Water for animal feedstuff production contributes to 98 percent of the water footprint of animal products (Hoekstra, 2012). Research into improved water use, improved cropping systems (e.g., using lower-water-use crops and cover cropping), and better understanding of the social components of water use in the region (e.g., public policy) will be critical for the continued existence of animal agriculture in the region.

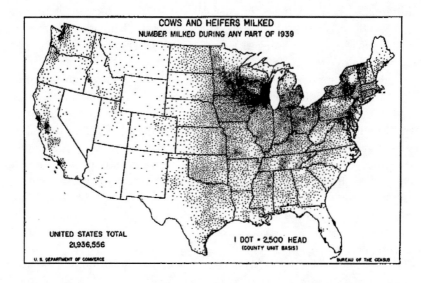

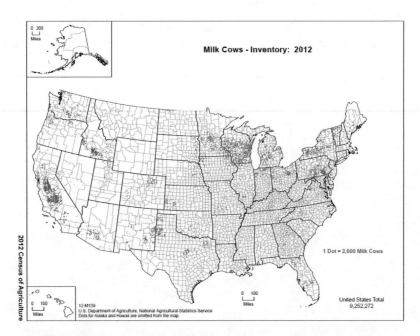

FIGURE 3-5 Distribution of 21.9 million dairy cows and heifers in 1939 (Top) and distribution of 9.3 million dairy cows in 2012 (Bottom).
SOURCES: USDA ESMIS (2012); USDA Census of Agriculture (2012a).

Clearly, these challenges to resource use and allocation extend far beyond the traditional boundaries of animal science research; however, incorporation of other sciences into a transdisciplinary approach will be necessary to manage such sustainability challenges related to animal production. As a result of increasing public concern and environmental regulatory measures, the past few decades have resulted in an increased focus on research, extension, and education of the environmental management of terrestrial animal agriculture. Efforts have been focused primarily on the impacts of animal agriculture on water quality; however, there has been an increasing interest in the impacts of animal agriculture on air quality and climate change.

Research Priorities

Animal agriculture is a major consumer of surface and groundwater resources, with 98 percent of the water footprint of animal agriculture due to the production of feedstuffs for food animals. One research priority in this area includes:
- Research to increase the water-use efficiency of animal agricultural systems (including feed production) is necessary (e.g., better irrigation, water recycling, identifying animals that are more water-use efficient); however, such research should also consider the complicated issues of water resource allocation in areas of stressed freshwater availability.

3-11.2 Climate Change and Variability

Adaptation and building resilience to climate variability and change in animal agricultural systems will be a major challenge in the coming decades. Climate change will likely affect feed-grain production, availability, and price; pasture and forage crop production and quality; animal health, growth, and reproduction; and disease and pest distributions (Walthall et al., 2012). Increasing concentrations of carbon dioxide and certain impacts of climate change may also have beneficial implications for animal agriculture. Increasing atmospheric carbon dioxide concentrations may have a positive effect on the yields of plant species important for animal agriculture, particularly those with C3 photosynthetic pathways (e.g., wheat, barley, and soybeans; Wheeler and Reynolds, 2013). However, plants that use the C4 photosynthetic pathway have much less of a yield gain than C3 plants (Wheeler and Reynolds, 2013). For both C3 and C4 plants, water-use efficiency is improved in conditions of higher carbon dioxide concentration (Drake et al., 1997).

Benefits of increasing carbon dioxide concentration on plant growth may be mediated by other environmental changes due to climate change, such as soil water availability (Polley et al., 2013). Increased variability in temperatures and precipitation will likely lead to decreased feed quality and increased environmental and nutritional stress for some animal agricultural species (Craine et al., 2010; Nardone et al., 2010; Wheeler and Reynolds, 2013). For example, an investigation of 76 years of growth records for Hereford beef cattle in the Northern Great Plains revealed that calf growth was greater in longer, cooler growing seasons, suggesting that increased temperatures decrease calf growth in the region (MacNeil and Vermeire, 2012). Heat stress already has a significant impact on the dairy, beef, swine, and poultry industries in the United States, with estimated annual economic costs at $1.69 billion to $2.36 billion due to decreased growth, reproductive performance, and lactation performance (St-Pierre et al., 2003). The prospect of more frequent heatwaves and higher average temperatures in the coming decades emphasizes the need for creating animal agricultural systems that are more adaptable and resilient to increased thermal stress.

As with terrestrial animal agriculture, climate change and variability will likely have both positive and negative effects on aquaculture. The United States currently imports 85 percent of its fish and shellfish; therefore, the impacts of climate change in major aquaculture regions such as Southeast Asia can have a significant impact on the ability to sustainably meet U.S. demand (Hatfield et al., 2014). Adaption and mitigation of the detrimental effects will be in response to changes in sea level and sea temperatures. Increases in sea level will eliminate or require the movement of existing coastal ponds, and increases in sea temperature will also affect growth of species cultured in cages, possibly increasing the incidence of disease. Oceans are absorbing approximately 25 percent of the CO_2 emitted into the atmosphere, which is reducing oceanic pH (acidification), and ultimately affecting the physiology of species such that their presence and/or productivity would be reduced (Walsh et al., 2014). Temperature changes in the water can affect the direction of currents and changes in the wind magnitude and direction. These changes may ultimately compel an adaptive response founded in the need to relocate offshore operations. The prospects of expansion of the aquaculture enterprise along the coasts of the marine environment will frustrate continued development. Policy decisions will have to be directed toward actions whereby mitigation and adaptation can be successfully implemented without compromising production. Because 65 percent of current aquaculture production is inland, mitigation activities for these freshwater production activities may be on a lower scale of magnitude than marine production activities (De Silva and Soto, 2009). One major concern is the lack of sufficient water resources; however, other concerns are water stratification in ponds or changes in disease-causing microbial flora.

Heffernan et al. (2012) offered the following recommendations to address the effects of climate change on animal infectious diseases. There is a need for frameworks and/or approaches that support collective learning and approaches within this emerging field. There is a need to enhance our abilities in predictive decision making. Critical areas that were identified as gaps include: (1) a greater understanding of the role of extra-climate factors on animal health, such as management issues and socioeconomic factors, and how these interplay with climate change impacts; (2) a greater cross-fertilization across topics/disciplines and methods within and between the field of animal health and allied subjects; (3) a stronger evidence base via the increased collection of empirical data in order to inform both scenario planning and future predictions regarding animal health and climate change; and (4) the need for new and improved methods to both elucidate uncertainty and explicate the direct and indirect causal relationships between climate change and animal infections.

Research will be needed to address the effects of climate change on crop productivity (Wheeler and Reynolds, 2013) and crop nutrient content. Bloom et al. (2014) reported a decrease of 8 percent in protein concentration in wheat, barley, rice, and potatoes with higher atmospheric carbon dioxide levels. Climate change effects on infectious diseases (Heffernan et al., 2012; Sundström et al., 2014) and animal production (Sundström et al., 2014) also need more research. Walthall et al. (2012) suggested research into technologies that improve management of agricultural products through automation of processes and tools, sensor development, and enhancement of information technologies.

The impacts of climate change and variability outlined above will have negative consequences for the efficiency of animal production and, thus, the environmental impact per unit of production. Consequently, adaptation to climate change and the mitigation of GHG emissions (and other environmental emissions) are intertwined, and future research should increase the understanding of the impact of animal agriculture on the environment and the impact of the environment on animal agriculture rather than researching these relationships in isolation.

Recommendation 3-11.2

While there is uncertainty regarding the degree and geographical variability, climate change will nonetheless impact animal agriculture in ways as diverse as affecting feed quality and quantity and causing environmental stress in agricultural animals. Animal agriculture affects and is affected by these changes, in some cases significantly, and must adapt to them in order to provide the quantity and affordability of animal protein expected by society. This adaptation, in turn, has important implications for sustainable production. The committee finds that adaptive strategies will be a critical component of promoting the resilience of U.S. animal agriculture in confronting climate change and variability.

> **Research needs to be devoted to the development of geographically appropriate climate change adaptive strategies and their effect on GHG emissions and pollutants involving biogeochemical cycling, such as that of carbon and nitrogen, from animal agriculture because adaptation and mitigation are often interrelated and should not be independently considered. Additional empirical research quantifying GHG emissions sources from animal agriculture should be conducted to fill current knowledge gaps, improve the accuracy of emissions inventories, and be useful for improving and developing mathematical models predicting GHG emissions from animal agriculture.**

Other Research Priorities

The committee supports previous recommendations by others that animal science research should increase the understanding of the impacts of climate change and variability on animal health and disease, animal productivity and welfare, and crop yields, quality, and availability. One research priority in this area includes:

- Research should develop viable climate change adaptation strategies for animal agriculture systems, including the development of information technologies to enhance animal agriculturalist decision making at the farm and operation levels.

3-11.3 Greenhouse Gas Emissions

In animal agricultural production, the GHGs that have received the most attention are nitrous oxide and methane. Carbon dioxide emissions result from the production of animal protein via the burning of fossil fuels throughout the supply chain and from soil in cropping systems. Respiratory carbon dioxide emissions from animals typically are not considered a net source of GHG emissions because the respiratory carbon dioxide is assumed to be offset by the carbon dioxide sequestered by the plants that animals consume (Steinfeld et al., 2006). Researchers, however, have evaluated GHG emissions from animal agricultural systems by accounting for carbon dioxide sequestered by plants and also respired by animals (Stackhouse-Lawson et al., 2012). Other GHG emissions from the production of animal protein include high-global-warming-potential refrigerants that can leak to the environment from the transportation and retail sectors of the supply chain (Thoma et al., 2013).

Nitrous oxide emissions from animal production typically result from soils applied with manure (i.e., deposited by grazing animals, deposited in open dry-lot facilities, or applied to pasture or cropland as a fertilizer source) or synthetic nitrogen fertilizers due to denitrification processes carried out by anaerobic bacteria (Robertson et al., 2013). Nitrous oxide emissions can also occur from manure storage systems, although total emissions from manure storage systems tend to be much lower than nitrous oxide emissions from manure-amended soils (EPA, 2014). This is due to the required nitrification transformation of manure ammonium to nitrate often being inhibited by the anaerobic nature of many manure storage systems in intensive U.S. animal operations (Montes et al., 2013). Although there is current and past research quantifying and investigating nitrous oxide emissions mitigation strategies, further collaborative efforts across crop and soil science, engineering and animal science disciplines will be needed to make significant improvements in emissions reductions. A particular emphasis should be placed on knowledge transfer of cost-effective mitigation strategies to animal agricultural producers. Furthermore, there is currently very little research on nitrous oxide emissions and mitigation opportunities in aquaculture systems, which is the fastest growing segment of animal agriculture; therefore, better understanding nitrous oxide emissions from aquaculture systems will be critical in coming decades (Hu et al., 2012).

Methane emissions can occur in anaerobic environments, such as manure storage systems and the foregut of ruminants. In 2012, enteric fermentation was estimated to account for 24.9 percent of U.S. methane emissions and 23 percent of GHG emissions from agriculture in CO_2 equivalents (EPA, 2014). Emissions of methane from manure management were estimated to be 9.3 percent of total U.S. methane emissions and 8.6 percent of GHG emissions from agriculture (EPA, 2014). The vast majority of methane emissions from enteric fermentation result from ruminant species (e.g., cattle, goats, and sheep); methane emissions from the digestive tracts of nonruminant herbivores and omnivorous species (e.g., pigs, poultry, and horses) are relatively minor (Crutzen et al., 1986). Methane emissions from ruminants generally increase with increasing feed intake and are affected by the type of carbohydrate and lipid content of the diet, feed processing, feed additives, or other strategies that can alter rumen microbial populations (Johnson and Johnson, 1995). Under strict anaerobic conditions, prokaryotic archaea primarily reduce carbon dioxide with hydrogen gas to generate energy for their own life processes and, as a consequence, produce methane, most of which is eructated out of the ruminant animal's mouth (Ellis et al., 2008). Mitigation options for enteric methane emissions typically focus on inhibiting the archaea directly or reducing the amount of hydrogen gas available for methane production (i.e., methanogenesis). Mitigation options include the use of compounds that directly inhibit methanogenesis, alternative hydrogen gas sinks, ionophores, plant bioactive compounds, exogenous enzymes, direct-fed microbials, defaunation, manipulation of rumen archaea and bacteria, dietary lipids, forage quality, forage-to-concentrate ratio of the diet, feed processing, and precision feeding (Hristov et al., 2013). For many of the above-listed enteric methane emission techniques, more data from live animal experiments and long-term experiments are needed to determine their long-term efficacy (Hristov et al., 2013); however, collecting such data can be expensive due to the equipment and labor requirements involved. For U.S. ruminant systems, there is a dearth of data from grazing ruminant emissions because of the methodological challenges of measuring enteric methane emissions while allowing animals to graze. Additionally, increased consideration should be given to the cost-effectiveness of enteric methane mitigation strategies and the potential tradeoffs and unintended consequences of mitigating enteric methane emissions on other aspects of animal agricultural sustainability.

Methane from manure is typically emitted from liquid or slurry manure storage systems, with very little methane produced once manure is applied to land (Montes et al., 2013). Methane mitigation options from manure storage typically focus on either preventing anaerobic conditions in storage systems, or, if anaerobic conditions exist, capturing or transforming methane through a flare to oxidize methane to carbon dioxide (Montes et al., 2013). Recent developments in new manure management and treatment systems include anaerobic digestion, which has the potential to decrease odor and pathogen loads. Anaerobic digestion also reduces nitrous oxide emissions from the land application of digested manure and is a source of energy from the biogas produced (Holm-Nielsen et al., 2009; Montes et al., 2013). Currently, installing and operating anaerobic digestion systems on animal operations is fairly cost-prohibitive, which has limited its adoption by animal agricultural operations despite the potential to improve environmental sustainability. A recently released report, *Climate Action Plan—Strategy to Reduce Methane Emissions* (White House, 2014) highlighted the importance of manure management and treatment technologies, such as anaerobic digestion, in mitigating methane emissions and strengthening the nation's bioeconomy. Additionally, the report announced the partnership of the U.S. Environmental Protection Agency, Department of Energy, and USDA with the Innovation Center for U.S. Dairy to form a "Biogas Roadmap" to outline strategies to increase the adoption of anaerobic digesters. Furthering interdisciplinary research and engagement with stakeholders and policy makers is necessary to improve the economic viability of anaerobic digesters and increase their adoption.

Another incentive to develop cost-effective anaerobic digestive systems for animal agriculture is to "close the loop" and use food waste and other organic matter as feedstock for the anaerobic processes (Holm-Nielsen et al., 2009). With such systems, products previously considered waste with no human value and an environmental burden can be converted into valuable products and energy. Food waste at the retail, consumer, and food service portions of the animal agricultural supply chain are estimated to range from 16 percent of edible food for red meat, poultry, fish, and seafood to 32 percent of edible food for dairy products (Kantor et al., 1997). Waste-to-worth type of energy production systems can make animal production more sustainable and potentially avoid "food vs. fuel" dilemmas that occur with other biofuels, such as corn-derived ethanol.

Research Priorities

There are uncertainties associated with current GHG emission predictions; however, ruminant (cattle) meat production systems tend to have higher GHG emission intensities compared to monogastric (swine, poultry, fish) production systems. Dual-purpose ruminant systems (e.g., milk production) tend to have lower GHG emission intensities than ruminant meat production systems. GHG emission mitigation strategies can offer considerable synergies with other desirable outcomes (e.g., reducing methane emissions from ruminants can increase feed energy conversion efficiency, and anaerobic digestion of animal manure can lead to the generation of renewable energy). However, research testing GHG emissions mitigation strategies has often been conducted without consideration of economic and social sustainability concerns. Research priorities for this area include:

- Research investigating GHG emission mitigation strategies should consider the impacts of such mitigation strategies on other environmental pollutants (e.g., ammonia emissions), as well as the economic and social viability of such mitigation strategies. Whole farm systems and whole-food supply-chain systems-based evaluations of the effects of mitigation strategies are needed.
- Research should build on the potential value of anaerobic digestion of animal manure and other organic wastes as well as other approaches that help to close energy and nutrient loops in the food system.

3-11.4 Air Quality and Nuisance Issues

The concentration of animal production has shifted how animal manure and litter are managed. While animals in extensive systems will deposit their urine and feces while grazing or foraging, manure or litter in confinement systems must be collected and usually stored in some way. While anaerobically stored manure can be a source of methane emissions, fresh animal manure and litter in animal housing systems can be a source of ammonia emissions when applied to land. Ammonia emissions occur when manure is exposed to air and can be affected by pH, temperature, and wind speed (Robertson et al., 2013). Past research has investigated optimizing nitrogen use in animal diets, and improved manure management techniques that can reduce ammonia emissions. It has been estimated by the EPA (2014) that 54 percent of ammonia emissions come from animal waste. Emissions of ammonia from animal agricultural operations can contribute to human respiratory health problems, acidification of environments, and eutrophication of water bodies (Aneja et al., 2001; Pinder et al., 2007). Although these emissions are currently unregulated, U.S. regulation in the next 40 years is likely;

therefore, further research, extension, and education efforts into increasing nitrogen-use efficiency and ammonia emission mitigation techniques from animal agricultural systems should be a priority to ensure science-based policy approaches and to provide viable mitigation options to animal agriculturalists.

Additionally, animal agriculture can contribute to the formation of tropospheric ozone through the emissions of volatile organic compounds and methane. Tropospheric ozone can both directly and through the formation of secondary aerosol particles lead to negative human health outcomes and decrease the net primary productivity of ecosystems and agricultural systems (Lippmann, 1991; Ainsworth et al., 2012; Shindell et al., 2012). Methane emissions from animal agriculture are typically not thought of as a local driver of ozone formation within airsheds, but rather contribute to the continued increase in global background tropospheric ozone concentrations (Fiore et al., 2002). As a consequence, research that identifies viable methane emission mitigation strategies can reduce animal agriculture's contributions to both climate change and reduced air quality. Recent research has found that most volatile organic compounds are emitted from fermented feed sources rather than fresh and stored manure sources (Howard et al., 2010a,b). Mitigation strategies that reduce emissions from fermented feed also have the potential to reduce feed losses, which is beneficial to animal agricultural producers (Place and Mitloehner, 2014). More research is needed to better characterize emission sources from animal agriculture that lead to reduced air quality, to develop economically viable mitigation strategies for animal agricultural producers, and to understand the relationships and tradeoffs of mitigation strategies with other aspects of sustainability.

Odor is a continual issue with animal agriculture in the United States, particularly in areas where suburban encroachment results in people previously unexposed to animal agriculture living in close proximity to animal farms. Research on characterizing and mitigating odor from animal operations has been ongoing for the past few decades; however, an improved understanding of the compounds that cause odor from animal operations and measurement technologies is required (Rappert and Müller, 2005). Additionally, pests, such as flies and rodents, can increase due to animal waste (e.g., feces and urine) and feedstuffs on animal operations, which can be another point of conflict with nonagricultural neighbors close to animal agricultural operations. Additionally, these pests can spread disease (CDC, 2014). With the increased trend toward intensification both in the United States and globally, it is likely that there will be more conflicts within communities between the general public and animal agricultural producers. Research can help with these conflicts both by providing more technical knowledge of odor mitigation and improving the communication between these two groups.

3-11.5 Nutrient Management

Most animal manure is applied to cropland, which can be beneficial as a source of nutrients and organic matter for soil, but can also be a source of groundwater and surface-water pollution. Animal agriculture can be a significant source of groundwater nitrates in areas of concentrated animal agriculture (Harter et al., 2012). Animal agriculture is a non-point source of nitrogen and phosphorus that contributes to algal blooms and subsequent dead zones in water bodies such as the Chesapeake Bay and Gulf of Mexico (Diaz and Rosenberg, 2008). As food animal operations have grown larger, the EPA and corresponding state agencies have developed regulations for animal feeding operations (AFOs) and concentrated animal feeding operations (CAFOs). These regulations have historically focused on the impact of terrestrial animal operations on water quality; however, some states, such as California, have moved toward permitting facilities for air emissions as well. In response, animal science and affiliated departments (e.g., crop and soil sciences) at land-grant universities have conducted research to address manure nutrient management and to reduce nutrient excretion from animal operations to the environment.

A criticism of CAFOs is that animal and crop systems have become decoupled, leading to the concentration of manure nutrients within relatively small geographical areas (Naylor et al., 2005). Previous integrated crop and animal agricultural production system research has demonstrated benefits of such systems, including increased biodiversity and better water quality and soil carbon sequestration, and nutrient cycling when compared to crop systems that do not integrate food animal production (Tomich et al., 2011; Barsotti et al., 2013; George et al., 2013). Examples of integrated crop-animal production systems include grazing ruminants on grain corn residue in the Corn Belt of the United States, dual-purpose wheat pasture systems in the southern Great Plains, agroforestry and silvopasture systems in the southeastern and western United States, and sod-based crop rotations (Sulc and Franzluebbers, 2014), as well as integrated terrestrial and aquatic systems (Godfray et al., 2010). Despite these examples, separate and specialized crop and animal production systems have become conventional in the United States (Sulc and Franzluebbers, 2014). Further basic and applied research on integrated crop and animal systems, including research into understanding the barriers and limitations of their adoption by agricultural producers would improve our understanding of ways to mitigate the undesirable environmental effects of the concentration of animal operations.

Research Priorities

Nutrients are necessary for crop production systems, and integrated animal-crop systems can enhance nutrient cycling; however, concentrated manure nutrients, if improperly managed, can also lead to environmental pollution caused by nutrient runoff and leaching. One research priority in this area includes:

- Continue and enhance research on the nutrient management of animal agricultural systems. Research should be directed to assessment of current and novel integrated crop-animal production systems, as well as the potential barriers and limitations to their adoption by agricultural producers.

3-11.6 Systems Evaluation: Environmental Metrics and Life-Cycle Assessment

A key challenge with sustainability is measurement and determining key indicators or metrics of sustainability in animal agriculture. When considering environmental impacts and, in particular, GHG emissions, recent research has focused on quantifying and mitigating environmental emissions per unit of output (e.g., per kilogram of product or protein) rather than per animal or per unit of land. The rationale for expressing environmental impacts per unit of output (known as the "functional unit" in life-cycle assessment [LCA]) is to report impacts on the basis of the animal production system's value to humans (de Vries and de Boer, 2010). Expressing impacts per unit of land can be problematic due to variations in soil types and climate, while expressing impacts per animal creates difficulties for comparisons across animal breeds and species. The efficiency of calorie production per hectare has been analyzed and animal production has been shown to be less efficient than crop production, with beef and other ruminant production being the least efficient (Cassidy et al., 2013; Eshel et al., 2014). The aquaculture enterprise is recognized as having the least adverse environmental impact of any animal production system. When compared to other animal protein production systems (beef, chicken, or pork), aquaculture is the most efficient and environmentally friendly as manifested in feed conversion, protein efficiency, nitrogen emissions, phosphorus emissions and land use (Bouman et al., 2013). For example, average nitrogen and phosphorus emissions for fish are 2.5 to 3.0 times lower than emissions for beef and pork at equivalent levels of protein production. However, others have suggested that when comparing ruminant efficiency to other species, the conversion of human edible or digestible energy and protein should be considered, rather than simply kilograms of feed or calories to produce a kilogram of beef or milk (Oltjen and Beckett, 1996; Wilkinson, 2011; Eshel et al., 2014). The proper functional unit of any one environmental analysis of animal production systems will depend on the goals of the analysis and should take into account that there are often multiple outputs of value from animal production systems (e.g., meat, milk, leather, and manure, which can have fertilizer value, are all outputs from dairy production). Transparency and clearly delineating assumptions are crucial for any environmental assessment of an animal agricultural production system. When evaluating environmental impacts from animal agricultural production, LCA is often used (Box 3-12). Different species and production systems have been evaluated in recent LCAs, including aquaculture (Cao et al., 2013), poultry (Pelletier, 2008; Pelletier et al., 2014), swine (Pelletier et al., 2010a; Thoma et al., 2011), beef (Pelletier et al., 2010b; Capper, 2011a; Lupo et al., 2013) and dairy (Capper et al., 2009; Thoma et al., 2013).

BOX 3-12
Life-Cycle Assessment and Application to Animal Agricultural Production

LCA has been often applied in studying and evaluating the environmental impact related to the production of meat and other agricultural products. A technique that is intended to measure a product's environmental impact throughout all stages of its life (cradle to grave) (Curran et al., 2005), LCA is often employed in analyses of production processes. There are four main components to any LCA analysis: (1) goal defining and scoping, where the objective and limitations/boundaries of the analysis are established; (2) inventory analysis, where the use of inputs and environmental releases is recorded; (3) impact assessment; and (4) interpretation (SAIC, 2006). These steps account for the environmental impact generated within the defined scope of the

> system for a specific quantity of a manufactured product. LCA has been applied in studying the environmental impact of pork, lamb, and beef, as well as related products such as dairy and leather (Peters et al., 2010). Although LCAs can provide valuable insights into the environmental intensity of meat production, they are also very much subject to the assumptions made in terms of the scope of study. Additionally, it has been noted that LCA is "a continuously evolving science, and the studies produced are time-point, market, and region specific. It is therefore difficult to compare across studies with any degree of certainty" (Capper and Hayes, 2012). Additionally, many LCAs use prediction models to estimate the environmental emissions from animal agriculture that are empirical and derived from site-specific data. Increasing the use of processed-based simulation models to derive estimates of emissions, such as methane and nitrous oxide, could reduce the error associated with LCA results (Rotz and Veith, 2013). While there is definite room for improvement in terms of standardization and comparing the results from LCA studies, it still remains a valuable tool used in the environmental study of production processes.

Adapted from its original use in evaluating industrial processes, LCA is essentially an environmental accounting system that sums impacts of interest across the entire production chain, from cradle to grave. For animal agriculture, this can include impacts from crop production, animals and their manure, processing, and transportation. Few LCA of animal agricultural products have conducted full cradle-to-grave assessments, but have instead focused on impacts generated at the farm gate with varying scopes and geographical locations assessed. Variation in scope and scale across LCA is a key challenge for making comparisons across published estimates. In an attempt to harmonize LCA methodological differences, the Livestock Environmental Assessment and Performance Partnership (LEAP), a collaborative effort of stakeholders from governments, nongovernmental organizations, and the private sector released draft guidelines regarding LCA for feed supply chains, small ruminants, and poultry production in early 2014 with plans to release guidelines for large ruminants in the near future.

Life-cycle assessment allows for evaluation of the impacts from the full production chain of a given animal production system, which can better account for the potential unintended consequences of changing one aspect of the production chain on the other phases of production. Additionally, consequence-related LCA techniques could be used to understand the environmental and economic tradeoffs associated with alternative uses of the biomass fed to animals (e.g., forages, byproducts, and grains), such as the production of biofuels or human food.

Quantifying uncertainty in LCAs is yet another challenge. Enteric methane emissions, for example, are not easily measured; therefore, LCAs rely on prediction equations to estimate methane emissions produced by ruminants via enteric fermentation. However, the accuracy of such models is fairly poor (Moraes et al., 2014), and there is a severe lack of empirical methane emissions data from grazing cattle. Remedying this uncertainty in enteric methane emissions should be a priority if there is interest in better understanding the formation of enteric methane emissions and mitigating those emissions. Because many of the environmental emissions from terrestrial animal agriculture and aquaculture are dependent on biogeochemical processes and can be altered by factors such as weather, microbial populations, and animal diet, there is a great need for empirical data quantifying emissions, testing mitigation strategies, and further developing mathematical models that can better represent observed data. A further challenge to LCAs is how to incorporate the social aspects of sustainability. While both environmental and economic impacts lend themselves to quantitative metrics, such a strategy with social considerations is difficult or impossible due to their subjective, qualitative nature.

Research Priorities

Incorporation of social sustainability concerns into LCA methodology is challenging. Currently, LCA researchers use different scopes, scales, and environmental emission prediction methods that can make comparisons across the published literature difficult or inappropriate. (e.g., environmental impacts are commonly expressed per unit of production such as per kilogram of meat but are also expressed on the basis of land area or feed or calorie conversion). Research priorities for this area include:

- LCA, an effective methodology to evaluate whole-animal food production systems, should be improved by further efforts, such as LEAP, to harmonize methodologies and increase the transparency of methods and assumptions.
- Researchers in animal science and related disciplines should work collaboratively with LCA researchers to conduct empirical research that improves the methodologies and reduces the uncertainties of LCA. Research should provide empirical data to quantify emissions,

test mitigation strategies, and further develop and evaluate mathematical models that can better represent observed data and incorporate social sustainability.

3-12 Socioeconomics

Given the growing effects of social, cultural, and ethical concerns on the direction and effectiveness of research and technology transfer in animal agriculture (see Chapter 2), it is becoming increasingly important to employ transdisciplinary approaches that include both the natural and social sciences. One obvious area where social science and animal science research need to be better aligned is with respect to the economic consequences of research applications. From the point of view of commercial animal producers, market forces, government actions, and technical factors that affect production are essentially an economics problem. For example, diseases that reduce productivity also reduce production and therefore profitability. Diseases also impose costs on producers from treating animals and from the related disruptions to local and international markets and trade. Consequently, nearly all aspects of animal agriculture from breed selection to pasture and herd management have critical economic dimensions. If the net economic effect of all production activities is not positive, then production eventually ceases as producers search for more profitable land uses and production alternatives. At the same time, changes in economic forces impact the growth potential of animal agriculture and its ability to meet the growing demand for protein (see discussion on economics in Chapter 2). Some of the salient economic issues for the growth of U.S. animal agriculture relate to the expansion of biofuel production, consumer preferences, government intervention in markets, and market organization and comparative advantage.

The rapid expansion of biofuel production in the United States along with other economic factors, such as the rapid expansion of demand for food across the world and particularly in China due to income growth, have contributed to much higher feed prices which, in general, have negatively affected U.S. animal production and the profitability of animal agriculture (Taheripour et al., 2013). Changing consumer preferences also have effects on the economics of animal agriculture. For example, changing food preferences related to health concerns not only affect red meat consumption but also alter the relative profitability of the production of different animal types. The power of this economic force is pressuring the ruminant animal industry to innovate and develop new technology and more efficient systems to compete with nonruminant systems, which are more vertically integrated, and to deliver products to consumers that better satisfy their desires for less fat in their diet.

Government intervention in the animal agricultural industry, such as grazing fees on public lands, animal waste regulations, and sanitary controls related to animal product trade, can provide positive social benefits; however, it also can distort market prices and costs, reduce market efficiency and ultimately limit the achievable profit from animal production, which constrains the ability of producers to adopt new technology. The way in which markets are organized (market structure) can also affect the profitability of food animal production. In the United States, cattle, sheep, and hog producers have historically been faced with relatively few buyers (packers). Consequently, food animal producers have expressed concerns over the years that the concentration of the packing industry provides packers with oligopoly-type market power, allowing them to pay low prices for slaughter animals and to sell meat products at higher prices than would occur with a more competitive packing industry. Such a market organization has the potential to limit the profitability of animal production while reducing consumer demand for animal products because of higher prices. On the other hand, government subsidies to animal producers, such as disaster assistance payments, enhance the profitability of animal agriculture and therefore promote the adoption of new technology and growth in food animal production.

Comparative advantage affects animal agriculture because unless a country has a comparative advantage in the production of food animals, the country will tend to import animal products rather than produce them.[2] Thus, countries with plentiful land, which has few alternative uses, will have an advantage in extensive food animal production. For example, U.S. sheep production continues to decline and lamb imports continue to increase, not because U.S. producers are inefficient or high-cost producers but because, among other factors, the opportunity cost of continuing to use land traditionally dedicated to sheep production in the United States continues to rise as the demand for that land for urban expansion and for producing increasingly high-value corn and other crops and cattle continues to rise dramatically. The opportunity cost of

---

[2] A country has a comparative advantage in the production and export of livestock products if that country can produce those products at a relatively lower cost in terms of resource use (e.g., land and labor).

land used for sheep production in Australia, on the other hand, is extremely low. Thus, Australia has a comparative advantage in the production and export of lamb because of relatively low opportunity cost of that production in Australia compared to the United States. Technology development and diffusion have helped maintain the low relative cost of U.S. cattle production and thus helped the country maintain a competitive industry despite growing beef imports from Australia and elsewhere with extensive land resources with low opportunity costs.

Potential contributions of social scientists to successful animal science research and technology transfer extend beyond the analysis of economic implications. The expertise of the committee members did not allow these potential contributions to be outlined in detail in this report, nor for specific recommendations to be made about the highest-priority social sciences research topics for the U.S. animal science research enterprise. In general, social science and bioethics research is critical to understanding the attitudes and values of both consumers and producers of animal products, and how those attitudes and values affect the acceptability of new technologies or practices (Croney and Anthony, 2010). Although some information is available regarding public attitudes about agriculture in the United States, particularly agricultural biotechnology (Food Insight, 2014), a more comprehensive approach toward obtaining this kind of information has been taken in Europe. The European Commission regularly conducts Eurobarometer surveys (EC, 2014) to assess public attitudes toward agriculturally related topics such as agricultural biotechnology, agricultural trade policy, food safety and security, and rural development. The European Union also recently funded a large-scale project to assess consumer and farmer attitudes to animal welfare throughout Europe with an associated animal sciences research program on this topic (Blokhuis et al., 2013). The successful development of a U.S. research agenda to increase sustainable intensification of animal food production nationally and globally would be facilitated by obtaining information about what the public knows about topics such as global food security and the application of technology to animal production, as well as their views about the ethics of raising and consuming animals, the environmental impacts of animal production, and the role of animal agriculture in society.

Karami and Keshavarz (2010) identify some other ways that social science research contributes to the achievement of agricultural sustainability. These include understanding gender impacts; developing different paradigms for interpreting and achieving sustainability; exploring the cultural, economic, demographic, and attitudinal variables that explain the adoption of sustainable practices by farmers; and developing predictive models for technology and practice adoption. The authors also highlight the role of social scientists in informing decision makers about social impacts of their decisions. Social scientists and ethicists can also help to inform integration and communication strategies more generally, which is important both for structuring the animal science research agenda and for providing scientifically based information about animal agriculture within the framework of diverse stakeholder values (Swanson et al., 2011). Very few animal science departments in the United States have social science or bioethics faculty to carry out this type of research. Enhancing expertise in this area is also critical to properly prepare undergraduate, graduate, and veterinary students to address the societal issues underlying growing public concerns relative to food animal production and, ultimately, to serve the key stakeholders in animal agriculture, including all sectors of food animal production as well as retailers, consumers, and policy makers.

Recommendation 3-12

Although socioeconomic research is critical to the successful adoption of new technologies in animal agriculture, insufficient attention has been directed to such research. Few animal science departments in the United States have social sciences or bioethics faculty in their departments who can carry out this kind of research.

Socioeconomic and animal science research should be integrated so that researchers, administrators, and decision makers can be guided and informed in conducting and funding effective, efficient, and productive research and technology transfer.

3-13 Communication and Public Understanding

Although many people probably have opinions about food production practices related to their personal consumption of animal products, there is little evidence that the public is aware of the magnitude of the potential global food security crisis. The popular media report that there will be significant population growth by 2050, but often focus more on the environmental, health, and social sustainability aspects of modern production practices than on the relationships between those practices and global food security needs. Animal agriculture is often the focus of significant criticism. Fraser (2001) has characterized what he refers to as the "New Perception" of animal agriculture in the growing popular literature—that it is detrimental to animal welfare, mainly controlled by large corporations, motivated by profit, causes increased world hunger, produces unhealthy food, and is harmful to the environment. He notes that this "New Perception" has raised important ethical issues about animal agriculture that cannot be dealt with simply by making counterclaims, and that the level of communication and discussion about these issues so far has been "simplistic" and failed to create a "climate of dialogue and consensus building."

Croney et al. (2012) point out that although it is often assumed that reframing these kinds of issues or the language used to describe them are effective communication methods, this generally only has short-term effects on peoples' views of issues. In addition, simply providing factual information does not necessarily inform peoples' opinions—surveys have shown, for example, that although knowledge about biotechnology increases with educational level, acceptance of biotechnology does not (Priest, 2000). For animal biotechnology, or any other new technology or research application, to be accepted by the public, effective and responsible communication among scientific, community, industry, and government stakeholders to create a climate of dialogue is critical (Van Eenennaam, 2006).

Numerous researchers have begun the process of better understanding current communication gaps and methods to bridge them. For example, the University of Illinois maintains an extensive collection of international research publications in its Agricultural Communications Documentation Center (ACDC). Areas of research included in the ACDC include history, current status, trends, and outlook for agricultural communications and the impact of communications (1) on farmers, farm families, and farming; (2) on agri-marketing systems and processes; (3) on the food industry; (4) on food systems; (5) within agricultural organizations; (6) on relationships between agriculture and society; (7) on risk and issue management related to agriculture; (8) on nutrition and health; (9) on consumer decisions related to food and nutrition; (10) on rural communities and their development; (11) on agricultural policies; (12) on natural resources, environment, and sustainability; (13) on the renewable energy mission of society; (14) on agricultural trade and international affairs; and (15) on social movements related to agriculture.

ACDC has identified a number of priority areas that need research attention. Specifically, they suggest that communication research should be focused into three primary areas that explore how: (1) communication can enhance decision making within the agriculture sector; (2) communication can aid the public in effectively participating in decision making related to agriculture; and (3) to build competitive societal knowledge and intellectual capabilities. One important element of effective communication is trust in the person or organization providing that communication (Siegrist et al., 2010), since trust affects whether or not the information is perceived as being truthful, accurate, and respectful of values.

Collaborative efforts between agricultural schools and other university components can be helpful in achieving better understanding of communication issues. For example, schools of public health have departments with research expertise in understanding the role of communication in behavior, as well as departments focused on environmental health and public policy, which also overlap with issues of concern to animal agriculture research.

Recommendation 3-13

The committee recognizes a broad communication gap related to animal agriculture research and objectives between the animal science community and the consumer. This gap must be bridged if animal protein needs of 2050 are to be fulfilled.

> **There is a need to establish a strong focus on communications research as related to animal science research and animal agriculture, with the goals of enhancing knowledge dissemination, respectful stakeholder participation and engagement, and informed decision-making.**

3-14 Integration and Systems Approaches

The above sections have discussed changes and potential knowledge gaps that should be addressed within disciplines and research areas; however, to address the sustainability of animal agriculture food systems, animal science disciplines must integrate across traditional boundaries and conduct collaborative, transdisciplinary research, extension, and education efforts. Sustainability encompasses environmental, economic, and social concerns. To best serve public needs, animal science researchers must continue basic research to reduce scientific uncertainties within their respective disciplines. Researchers also must expand beyond traditional disciplines to increase understanding of the tradeoffs between areas of sustainability, and better understand public perceptions of animal agricultural systems as well as the value judgments made by researchers, animal agriculturalists, and consumers regarding sustainability. A systems approach that holistically integrates across disciplines is necessary.

3-14.1 Integrating Research, Extension, and Education Efforts

As previously discussed, animal agriculture in the United States has become increasingly intensified and concentrated, leading to concerns of nutrient pollution. Researchers, particularly those at land-grant universities, have responded to the increasing concentration of animal operations by creating research, extension, and education programs in nutrient management and environmental quality relating to livestock and poultry operations. For example, the Nutrient Management Spear Program (NMSP) in the Department of Animal Science at Cornell University has been developed to conduct applied research, extension, and education. The program combines animal science, manure management, and crop and soil science disciplines to advance holistic nutrient management of dairy farms in New York State. One tool that the program has developed is the whole-farm nutrient balance calculator, which has been used by both dairy farmers and students, and can be used as an important component of adaptive management[3] approaches that improve environmental quality and economic viability over time (Figure 3-6; Soberon et al., 2013). Additionally, a capstone course in whole-farm nutrient management is taught to animal science and crop science students. The course covers a wide range of topics including CAFO regulations, crop nutrient requirements, and the nitrogen cycle, with the final portion of the class allowing students to create a comprehensive nutrient management plan for a cooperating case farm (Albrecht et al., 2006). Similar courses are offered at University of Wisconsin (Managing the Environmental Impacts of Livestock Operations) and Penn State University (Nutrient Management in Agricultural Systems). Although the NMSP offers a great example of combining research, extension, and education efforts to advance the environmental sustainability of food

---

[3] From Rist et al. (2013), adaptive management is natural resource management conducted in a manner that purposely and explicitly increases knowledge and reduces uncertainty (Holling, 1978; Walters, 1986).

animal production, similar programs do not exist at all land-grant universities. Many have a single waste management or nutrient management specialist housed on campus, who may not be associated with the Department of Animal Science but rather with the Department of Agricultural or Biosystems Engineering. Although this should not eliminate the ability to collaborate with animal scientists, the separation of animal waste specialists from departments of animal science can sometimes create an obstacle to integrated, collaborative research. This "silo effect" of separating disciplines, along with often limited funding support, can stifle the creation of transdisciplinary environmental management or sustainability programs for animal agricultural production that contain research, extension, and education elements.

One useful tool in eliminating regional barriers and enhancing knowledge transfer in extension programing is the eXtension website (eXtension, 2014b). In the area of nutrient management, there is the Livestock and Poultry Environmental Learning Center, which has fact sheets, tools, and webinars on topics such as air quality, environmental planning, manure treatment, pathogens, and climate change (eXtension, 2014a). Additionally, there is a separate extension project funded by USDA NIFA, called "Animal Agriculture and Climate Change," which created a free online course covering topics such as the science behind climate change, GHG emission sources, GHG emission mitigation and adaptation to climate change (USDA NIFA, 2013). In addition to the extension-focused "Animal Agriculture and Climate Change" project,

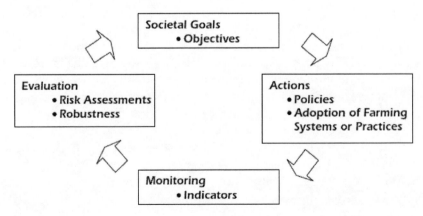

FIGURE 3-6 Adaptive management for sustainability of agricultural farming systems.
SOURCE: NRC (2010).

there are other ongoing USDA NIFA-funded integrated research projects (integrated with extension and/or education components) that are related to either adaptation to climate change and/or mitigation of GHG emissions. Some of these projects include "Climate Change Mitigation and Adaptation in Dairy Production Systems of the Great Lakes region" (Project No. WIS01693) and "Resilience and Vulnerability of Beef Cattle Production in the Southern Great Plains Under Changing Climate, Land Use and Markets" (Project No. OKL02857). USDA also supports the Sustainable Agriculture Research and Education (SARE) program, which has provided funding for over 5,000 projects to producers, extension educators, researchers, nonprofits, and communities since the late 1980s (SARE, 2012). Funded projects have included on-farm renewable energy, no-till and conservation agriculture, nutrient management, crop and food animal diversity, and systems research (SARE, 2012).

While these examples provide an indication of potentially successful strategies, there is a need to further improve understanding and encourage wider application of outcome-based solutions for environmental management. Such an effort requires integrated research, extension, and education activities that are unlikely to be funded solely by a single producer group or private company because of resource limitations and, in many cases, a lack of clear commercial outcomes for research. Public funding and strategic public–private partnerships will be critical if progress in mitigating environmental impacts is desired. An example of a broad public–private partnership sustainability research effort, which goes beyond focusing just on environmental impacts, is the Coalition for a Sustainable Egg Supply (Box 3-13). Land-grant universities are uniquely qualified to conduct integrated research, as they have done for 100 years, particularly field-to-fork research that encompasses all phases of animal production to the consumer; however, the decline in public funding (both state and federal) in recent decades has led to an erosion of land-grant universities' abilities to meet societal demand for integrated research.

BOX 3-13
Coalition for a Sustainable Egg Supply

The Coalition for a Sustainable Egg Supply (CSES) provides a model for how public–private partnerships can be effective in initiating and supporting multidisciplinary research on the sustainability of animal agriculture. As discussed in Box 2-4, the egg industry in the United States is in a state of transition due to stakeholder pressure to transition from conventional cage housing for hens to alternative housing because of concerns about hen welfare. In 2006, the American Egg Board provided funding for teams of social and natural scientists to conduct literature reviews to determine the knowledge base and information gaps regarding the effects of different hen housing systems on environmental, economic, and social sustainability (Swanson et al., 2011). These reviews identified not only significant knowledge gaps but concluded that most research had been conducted in the European Union on an experimental rather than a commercial scale. A need was identified for commercial-scale studies conducted under U.S. conditions, which stimulated the formation of the CSES.

The CSES is a multistakeholder group involving academic institutions, retailers, food distributors, egg producers, scientific and veterinary organizations, and nongovernmental organizations (http://www2.sustainableeggcoalition.org/). Building on the findings of the American Egg Board project, the CSES provided approximately $6.6 million in funding to researchers at academic institutions (Michigan State University, University of California, and Iowa State University) and USDA ARS to conduct a multiyear commercial-scale study of three housing alternatives for egg-laying hens in the United States The Coalition researchers worked to better understand the impact of these laying-hen housing systems on a sustainable supply of eggs by conducting integrated multidisciplinary research on the environmental, worker health and safety, hen health and welfare, economic, and food safety and quality aspects of those systems. In conjunction, the Center for Food Integrity (http://www.foodintegrity.org/), which facilitates the CSES, is conducting research to better understand consumers' attitudes toward egg production and how those attitudes are affected when consumers are informed about the research results. This project will identify the sustainability tradeoffs between systems in order to allow stakeholders to make informed decisions going forward.

3-14.2 Integrating Research Disciplines to Understand Sustainability Tradeoffs

While there are often synergies across the three pillars of sustainability (environmental, economic, and social), there are also tradeoffs that have been largely neglected by past animal science research. For example, Capper et al. (2008) compared the environmental impact of three dairy production systems used to produce equivalent amounts of milk. The three systems included a conventional system, conventional system plus the adoption of recombinant bovine somatotropin (rbST), and an organic system. Capper et al. (2008) found that the adoption of rbST technology resulted in 8 percent fewer animals, 5 percent less land area, 5-6 percent reduced animal waste, and 6 percent reduction in GHG production compared to the conventional system. By contrast, the organic system resulted in a more negative environmental impact with a 25 percent increase in animal numbers, 20 percent increase in land area, and 13 percent increase in GHG emissions compared to the conventional system (Capper et al., 2008). Although the adoption of technologies such as rbST have economic and environmental sustainability benefits due to improving the efficiency of resource conversion into dairy products, the social sustainability implications of technologies used in animal production are less clear. In the United States, there has been a major shift away from selling fluid milk from cows treated with rbST in the past decade. Indeed, it is nearly impossible to find fluid milk sold without the label "from cows not treated with rbST (or rBGH)." The caveat to such labels is the "Food and Drug Administration (FDA) has determined there is no significant difference between milk from rbST treated cows and non-rbST treated cows." Despite the assurance by the FDA that milk from rbST treated cows is safe, consumer resistance to rbST has been driven by concerns of IGF-1 levels in milk, the possibility of increased health risks to children exposed to milk from rbST treated cows, and potential impacts on dairy cattle welfare (Collier and Bauman, 2014). While these concerns have been determined to be unfounded (Collier and Bauman, 2014), consumer resistance remains, translating into retailer pressure for dairy farmers to stop using rbST (Olynk et al., 2012). It is unclear if consumers are aware of the potential environmental benefits of using rbST and if that would have any impact on the reluctance of many consumers to buy dairy products from cattle treated with rbST.

A more recent example of the tradeoff challenges for biotechnology is the use of the beta-agonist zilpaterol hydrochloride in finishing beef cattle. In the literature, both live animal experiments and a study using modeling techniques have found that use of growth-promoting technologies, such as beta-agonists, can decrease environmental impacts and resource use per unit of beef (Cooprider et al., 2011; Capper and Hayes, 2012; Stackhouse-Lawson et al., 2013). However, in August 2013, because of concerns over zilpaterol hydrochloride's impact on animal welfare, the company Tyson announced that it would stop harvesting cattle that had been fed beta-agonists in their slaughterhouses. The manufacturer of zilpaterol, Merck Animal Health, eventually removed the product from the market while investigating the animal welfare concerns. Lonergan et al. (2014) analyzed the effect of beta-agonists (both zilpaterol and ractopomine hydrochloride [Elanco Animal Health]) on cattle mortality. Though the mortality rates for the thousands of cattle in the datasets analyzed were very low, there was a significant increase in death loss for cattle treated with beta-agonists, particularly in summer months (Lonergan et al., 2014). Although the mechanisms or potential interactions that cause increased mortality and potential lameness (the primary welfare concern that led Tyson to its decision) still need to be elucidated, the case of zilpaterol hydrochloride illustrates the need for cooperation across disciplines within animal and veterinary sciences, as well as with social sciences to determine acceptability of biotechnologies such as beta-agonists. Consumers are likely largely unaware of the array of technologies used in animal agriculture, the reason for use, and the potential costs and benefits. The tradeoffs surrounding the social, environmental, and economic facets of sustainability have gone largely ignored and unquantified in the animal sciences, pointing to a need for further collaboration, training, and research in the future.

Beyond biotechnology, improved animal husbandry practices and housing systems that can affect animal longevity, fertility, and productivity can also affect environmental impacts per unit of output. Place and Mitloehner (2014) and Tucker et al. (2013) highlighted the synergy between some, but not all, animal welfare concerns and environmental quality. An example of a synergistic relationship between animal health and welfare and environmental impact is mastitis in the dairy industry. Reducing the incidence of mastitis (inflammation of the mammary gland that can be painful to the cow) has been shown to reduce dairy production's contribution to climate change, eutrophication, and acidification by decreasing the losses of human-consumable milk (Hospido and Sonesson, 2005).

Currently, there is limited research linking the disciplines of animal welfare, animal health, genetics, and the environment. The data gap on the connections and interactions of these disciplines represents an opportunity for animal science research to increase our collective knowledge and improve tradeoff analyses between environmental quality and other aspects of sustainable animal agriculture.

Research Priorities

There are several examples of integrated research, extension, and education efforts; however, relatively few were considered to be focused on sustainable animal agriculture and food systems. In some instances, the use of biotechnologies can simultaneously improve production efficiency and reduce environmental impacts per unit of output; however, the social acceptability and impacts of these biotechnologies on animal welfare should also be considered in research. Research priorities for this area include:

- A coordinated approach to engage with the public, policy makers, animal agriculturalists, and animal scientists is required to better understand, respond to, and communicate the tradeoffs of biotechnological solutions across the economic, social, and environmental components of sustainability.
- To tackle the issues of sustainability in animal agriculture, public funding should be guided by a twofold objective for support—projects that take a transdisciplinary approach and integrate research with extension and/or education efforts, and projects that continue to increase fundamental knowledge within the disciplines of animal science.

REFERENCES

Ainsworth, E. A., C. R. Yendrek, S. Sitch, W. J. Collins, and L. D. Emberson. 2012. The effects of tropospheric ozone on net primary productivity and implications for climate change. Annual Review of Plant Biology 63:637-661.

Albrecht, E., F. Teuscher, K. Ender, and J. Wegner. 2006. Growth- and breed-related changes in marbling characteristics in cattle. Journal of Animal Science 84:1067-1075.

Aneja, V. P., P. A. Roelle, G. C. Murray, J. Southerland, J. W. Erisman, D. Fowler, W. A. H. Asman, and N. Patni. 2001. Atmospheric nitrogen compounds, II: Emissions, transport, transformation, deposition and assessment. Atmospheric Environment 35(11):1903-1911.

Appleby, M. C. 2003. The European Union ban on conventional cages for laying hens: History and prospects. Journal of Applied Animal Welfare Science 6(2):103-121.

Appleby, M. C., J. A. Mench, I. A. S. Olsson, and B. O. Hughes, eds. 2011. Animal Welfare, 2nd Ed. Cambridge, UK: CAB International.

Applegate, T. J., and R. Angel. 2014. Nutrient requirements of poultry publication: History and need for an update. Journal of Applied Poultry Research 23(3):567-575.

Appuhamy, J. A., A. B. Strathe, S. Jayasundara, C. Wagner-Riddle, J. Dijkstra, J. France, and E. Kebreab. 2013. Anti-methanogenic effects of monensin in dairy and beef cattle: A meta-analysis. Journal of Dairy Science 96(8):5161-5173.

Åsgård, T., and E. Austreng. 1995. Optimal utilization of marine proteins and lipids for human interest. Pp. 79-87 in Sustainable Fish Farming, H. Reinertsen and H. Haaland, eds. Rotterdam, The Netherlands: A. A. Balkema.

Ashley, P. J. 2007. Fish welfare: Current issues in aquaculture. Applied Animal Behaviour Science 104(3-4):199-235.

Aumaître, A. 1999. Quality and safety of animal products. Livestock Production Science 59(2-3):113-124.

Baker, D. H. 2008. Animal models in nutrition research. Journal of Nutrition 138(2):391-396.

Baldwin, R. L., and R. D. Sainz. 1995. Energy partitioning and modeling in animal nutrition. Annual Review of Nutrition 15:191-211.

Barsotti, J. L., U. M. Sainju, A. W. Lenssen, C. Montagne, and P. G. Hatfield. 2013. Crop yields and soil organic matter responses to sheep grazing in U.S. northern Great Plains. Soil and Tillage Research 134:133-141.

Bauman, D. E., and J. L. Capper. 2010. Efficiency of Dairy Production and Its Carbon Footprint. Online. Available at http://dairy.ifas.ufl.edu/rns/2010/11-Bauman.pdf. Accessed September 4, 2014.

Bazes, A., A. Silkina, D. Defer, C. Bernède-Bauduin, E. Quéméner, J. P. Braud, and N. Bourgougnon. 2006. Active substances from Ceramium botryocarpum used as antifouling products in aquaculture. Aquaculture 258(1-3):664-674.

Behnke, K. C. 1996. Feed Manufacturing Technology: Current Issues and Challenges. Online. Available at http://www.ker.com/library/advances/204.pdf. Accessed August 18, 2014.

Bergen, W. G. 2007. Contributions of research with farm animals to protein metabolism concepts: A historical perspective. Journal of Nutrition 137(3):706-710.

Blokhuis, H., M. Miele, I. Veissier, and B. Jones, eds. 2013. Improving Farm Animal Welfare (e-book). Science and Society Working Together: The Welfare Quality Approach. Online. Available at http://www.wageningenacademic.com/welfarequality-e. Accessed August 18, 2014.

Bloom, A. J., M. Burger, B. A. Kimball, and P. J. Pinter, Jr. 2014. Nitrate assimilation is inhibited by elevated CO_2 in field-grown wheat. Nature Climate Change 4(6):477-480.

Bouman, A. F., A. H. W. Beusen, C. C. Overbeek, D. P. Bureau, M. Pawlowski, and P. M. Gilbert. 2013. Hindcasts and future projections of global inland and coastal nitrogen and phosphorus loads due to finfish aquaculture. Reviews in Fisheries Science 21(2):112-156.

Branson, E. J., ed. 2008. Fish Welfare. Hoboken, NJ: Wiley-Blackwell.

Brashears, M. M., D. Jaroni, and J. Trimble. 2003. Isolation, selection, and characterization of lactic acid bacteria for a competitive exclusion product to reduce shedding of Escherichia coli O157:H7 in cattle. Journal of Food Protection 66(3):355-363.

Buchanan, D. V., M. L. Hanson, and R. M. Hooton. 1997. Status of Oregon's Bull Trout. Portland, OR: Oregon Department of Fish and Wildlife.

Burns, J. C. 2008. ASAS centennial paper: Utilization of pasture and forages by ruminants: A historical perspective. Journal of Animal Science 86(12):3647-3663.

Butterworth, A., J. A. Mench, and N. Wielebnowski. 2011. Practical strategies to assess (and improve) welfare. Pp. 200-214 in Animal Welfare, 2nd Ed., M. C. Appleby, J. A. Mench, I. A. S. Olsson, and B. O. Hughes, eds. Cambridge, UK: CAB International.

Buzby, J. C., and J. Hyman. 2012. Total and per capita value of food loss in the United States. Food Policy 37(5):561-570.

Byrne, J. 2014. "Bacteria beats insects and algae hands down," Nutrinsic CEO talks up new feed protein source from factory wastewater. Feed Navigator.com. Online. Available at http://www.feednavigator.com/Suppliers/Bacteria-beats-insects-and-algae-hands-down-Nutrinsic-CEO-talks-up-new-feed-protein-source-from-factory-wastewater/?utm_source=newsletter_daily&utm_medium=email&utm_campaign=01-Aug-2014&c=FmRQxaYHW8H090MvpDzohbG3gBhofps2. Accessed August 18, 2014.

Callaway, T. R., R. O. Elder, J. E. Keen, R. C. Anderson, and D. J. Nisbet. 2003. Forage feeding to reduce preharvest Escherichia coli population in cattle: A review. Journal of Dairy Science 86(3):852-860.

Cao, L., J. S. Diana, and G. A. Keoleian. 2013. Role of life cycle assessment in sustainable aquaculture. Reviews in Aquaculture 5(2):61-71.

Capper, J. L. 2011a. Replacing rose-tinted spectacles with a high-powered microscope: The historical versus modern carbon footprint of animal agriculture. Animal Frontiers 1(1):26-32.

Capper, J. L. 2011b. The environmental impact of beef production in the United States: 1977 compared with 2007. Journal of Animal Science 89(12):4249-4261.

Capper, J. L., and D. E. Bauman. 2013. The role of productivity in improving the environmental sustainability of ruminant production systems. Annual Review of Animal Biosciences 1:469-489.

Capper, J. L., and D. J. Hayes. 2012. The environmental and economic impact of removing growth-enhancing technologies from U.S. beef production. Journal of Animal Science 90(10):3527-3537.

Capper, J. L., E. Castañeda-Gutiérrez, R. A. Cady, and D. E. Bauman. 2008. The environmental impact of recombinant bovine somatotropin (rbST) use in dairy production. Proceedings of the National Academy of Sciences of the United States of America 105(28):9668-9673.

Capper, J. L., R. A. Cady, and D. E. Bauman. 2009. The environmental impact of dairy production: 1944 compared with 2007. Journal of Animal Science 87(6):2160-2167.

Carter, S. D., and H. Kim. 2013. Technologies to reduce environmental impact of animal wastes associated with feeding for maximum productivity. Animal Frontiers 3(3):42-47.

Cassidy, E. S., P. C. West, J. S. Gerber, and J. A. Foley. 2013. Redefining agricultural yields: From tonnes to people nourished per hectare. Environmental Research Letters 8(3):034015.

CDC (Centers for Disease Control and Prevention). 2013. Antibiotic Resistance Threats in the United States, 2013. Online. Available at http://www.cdc.gov/drugresistance/threat-report-2013. Accessed October 3, 2014.

CDC. 2014. Diseases Directly Transmitted by Rodents. Online. Available at http://www.cdc.gov/rodents/diseases/direct.html. Accessed August 18, 2014.

Chesapeake Bay Foundation. 2014. Feed Efficiency. Online. Available at http://www.cbf.org/Document.Doc?id=261. Accessed September 3, 2014.

Collier, R. J., and D. E. Bauman. 2014. Update on human health concerns of recombinant bovine somatotropin use in dairy cows. Journal of Animal Science 92(4):1800-1807.

Conte, F. S. 2004. Stress and the welfare of cultured fish. Applied Animal Behavior Science 86(3):205-223.

Cooprider, K. L., F. M. Mitloehner, T. R. Famula, E. Kebreab, Y. Zhao, and A. L. Van Eenennaam. 2011. Feedlot efficiency implications on greenhouse gas emissions and sustainability. Journal of Animal Science 89(8):2643-2656.

Craine, J. M., A. J. Elmore, K. C. Olson, and D. Tolleson. 2010. Climate change and cattle nutritional stress. Global Change Biology 16(10):2901-2911.

Cromwell, G. L. 2009. ASAS centennial paper: Landmark discoveries in swine nutrition in the past century. Journal of Animal Science 87(2):778-792.

Croney, C. C., and R. Anthony. 2010. Engaging science in a climate of values: Tools for animal scientists tasked with addressing ethical problems. Journal of Animal Science 88(13):E75-E81.

Croney C., M. Apley, J. L. Capper, J. A. Mench, and S. Priest. 2012. The ethical food movement: What does it mean for the role of science and scientists in current debates about animal agriculture? Journal of Animal Science 90(5):1570-1582.

Crutzen, P. J., I. Aselmann, and W. Seiler. 1986. Methane production by domestic animals, wild ruminants, other herbivorous fauna, and humans. Tellus 38B(3-4):271-284.

Curran, M. A., M. Mann, and G. Norris. 2005. The international workshop on electricity data for life cycle inventories. Journal of Cleaner Production 13:853-862.

Davis, C. G., D. Harvey, S. Zahniser, F. Gale, and W. Liefert. 2013. Assessing the Growth of U.S. Broiler and Poultry Meat Exports. LPDM-231-01. Washington, DC: USDA Economic Research Service.

Dawkins, M. S. 2006. A user's guide to animal welfare science. Trends in Ecology and Evolution 21(2):77-82.

Dawkins, M. S., C. A. Donnelly, and T. A. Jones. 2004. Chicken welfare is influenced more by housing conditions than by stocking density. Nature 427:342-344.

De Schryver, P., R. Crab, T. Defoirdt, N. Boone, and W. Verstraete. 2008. The basics of bio-flocs technology: The added value for aquaculture. Aquaculture 277(3-4):125-137.

De Silva, S. S., and D. Soto. 2009. Climate change and aquaculture: Potential impacts, adaptation and mitigation. Pp. 151-212 in Climate Change Implications for Fisheries and Aquaculture: Overview of Current Scientific Knowledge, K. Cochrane, C. De Young, D. Soto and T. Bahri, eds. FAO Fisheries and Aquaculture Technical Paper No. 530. Rome: FAO.

de Vries, M., and I. J. M. de Boer. 2010. Comparing environmental impacts for livestock products: A review of life cycle assessments. Livestock Science 128(1-3):1-11.

Dejmek, P. 2013. Fluid Milk Shelf Life—From Hours to Months. Online. Available at http://www.docstoc.com/docs/155035479/At-first-sight_-the-development-of-fluid-milk-shelf-life-in-the-last-sixty. Accessed August 14, 2014.

Diaz, R. J., and R. Rosenberg. 2008. Spreading dead zones and consequences for marine ecosystems. Science 321(5891):926-929.

Donham, K. J., S. Wing, D. Osterberg, J. L. Flora, C. Hodne, K. M. Thu, and P. S. Thorne. 2007. Community health and socioeconomic issues surrounding concentrated animal feeding operations. Environmental Health Perspectives 115(2):317-320.

Drackley, J. K., S. S. Donkin, and C. K. Reynolds. 2006. Major advances in fundamental dairy cattle nutrition. Journal of Dairy Science 89(4):1324-1336.

Drake, B. G., M. A. Gonzàlez-Meler, and S. P. Long. 1997. More efficient plants: A consequence of rising atmospheric CO2. Annual Review of Plant Physiology and Plant Molecular Biology 48:609-639.

Dumas, A., J. Dijkstra, and J. France. 2008. Mathematical modelling in animal nutrition: Centenary review. Journal of Agricultural Science 146(Special Edition 02):123-142.

Eastridge, M. L. 2006. Major advances in applied dairy cattle nutrition. Journal of Dairy Science 89(4):1311-1323.

EC (European Commission). 2014. Eurobarometer Special Surveys. Online. Available at http://ec.europa.eu/public_opinion/archives/eb_special_419_400_en.htm. Accessed August 18, 2014.

EFSA (European Food Safety Authority). 2014. Feed. Online. Available at http://www.efsa.europa.eu/en/topics/topic/feed.htm. Accessed July 25, 2014.

Eggen, A. 2012. The development and application of genomic selection as a new breeding paradigm. Animal Frontiers 2(1):10-15.

Ellis, J. L., J. Dijkstra, E. Kebreab, A. Bannink, N. E. Odongo, B. W. McBride, and J. France. 2008. Aspects of rumen microbiology central to mechanistic modelling of methane production in cattle. Journal of Agricultural Science 146(Special Edition 02):213-233.

EPA (U.S. Environmental Protection Agency). 2014. Inventory of U.S. Greenhouse Gas Emissions and Sinks: 1990-2012. EPA 430-R-14-003. Online. Available at http://www.epa.gov/climatechange/ghgemissions/usinventoryreport.html. Accessed July 21, 2014.

Eshel, G., A. Shepon, T. Makov, and R. Mio. 2014. Land, irrigation water, greenhouse gas, and reactive nitrogen burdens of meat, eggs, and dairy production in the United States. Proceedings of the National Academy of Sciences of the United States of America 111(33):11996-12001.

eXtension. 2014a. Animal Manure Management. Available at https://www.extension.org/animal_manure_management. Accessed August 8, 2014.

eXtension. 2014b. Learn. America's Research-Based Learning Network. Online. Available at https://learn.extension.org. Accessed August 8, 2014.

FAO (Food and Agriculture Organization of the United Nations). 2013. Food Wastage Footprint—Impacts on Natural Resources. Rome: FAO. Online. Available at http://www.fao.org/docrep/018/i3347e/i3347e.pdf. Accessed August 14, 2014.

FAO. 2014a. Global Initiative on Food Loss and Waste Reduction. Online. Available at http://www.fao.org/docrep/015/i2776e/i2776e00.pdf. Accessed August 14, 2014.

FAO. 2014b. The State of the World Fisheries and Aquaculture: Opportunities and Challenges. Rome: FAO. Online. Available at http://www.fao.org/3/a-i3720e.pdf. Accessed September 25, 2014.

Farmland Information Center. 2014. Statistics. Online. Available at http://www.farmlandinfo.org/statistics. Accessed August 18, 2014.

FASS (Federation of Animal Science Societies). 2012. FAIR (Farm Animal Integrated Research) 2012. Online. Available at http://www.fass.org/docs/FAIR2012_Summary.pdf. Accessed August 5, 2014.

FDA (U.S. Food and Drug Administration). 1994. Sec. 683.100: Action Levels for Aflatoxins in Animal Feeds, CPG 7126.33. Online. Available at www.fda.gov/ora/compliance_ref/cpg/cpgvet/cpg683-100.html. Accessed August 4, 2014.

FDA. 2014a. Bovine Spongiform Encephalopathy. Online. Available at http://www.fda.gov/AnimalVeterinary/GuidanceComplianceEnforcement/ComplianceEnforcement/BovineSpongiformEncephalopathy/default.htm. Accessed July 25, 2014.

FDA. 2014b. Safe Feed. Online. Available at http://www.fda.gov/AnimalVeterinary/SafetyHealth/AnimalFeedSafetySystemAFSS/default.htm. Accessed July 25, 2014.

Federation of European Aquaculture Producers. 2013. Concerns: The Amendments proposed for a Regulation on the European Maritime and Fisheries Fund (EMFF). Online. Available at www.feap.info/shortcut.asp?FILE=1201. Accessed October 10, 2014.

Fiore, A. M., D. J. Jacob, B. D. Field, D. G. Streets, S. D. Fernandes, and C. Jang. 2002. Linking ozone pollution and climate change: The case for controlling methane. Geophysical Research Letters 29(19):25-1 to 25-4.

Food Insight. 2014. Food Technology Survey. Online. Available at http://www.foodinsight.org/2014-foodtechsurvey. Accessed October 3, 2014.

Fraser, D. 2001. The "new perception" of animal agriculture: Legless cows, featherless chickens, and a need for genuine analysis. Journal of Animal Science 79(3):634-641.

Fraser, D. 2008a. Toward a global perspective on farm animal welfare. Applied Animal Behaviour Science 113(4):330-339.

Fraser, D. 2008b. Understanding Animal Welfare: The Science in Its Cultural Context. Hoboken, NJ: Wiley-Blackwell.

Fraser, D., J. Mench, S. Millman. 2001. Farm animals and their welfare in 2000. Pp. 87-99 in State of the animals 2001, D. J. Salem and A. N. Rowan, eds. Washington, DC: Humane Society Press.

Fraser, D., I. J. Duncan, S. A. Edwards, T. Grandin, N. G. Gregory, V. Guyonnet, P. H. Hemsworth, S. M. Huertas, J. M. Huzzey, D. J. Mellor, J. A. Mench, M. Spinka, and H. R. Whay. 2013. General principles for the welfare of animals in production systems: The underlying science and its application. Veterinary Journal 198(1):19-27.

Frenzen, P. D., A. Majchrowicz, J. C. Buzby, and B. Imhoff. 2000. Consumer Acceptance of Irradiated Meat and Poultry Products. Issues in Food Safety Economics. Agriculture Information Bulletin 757. Washington, DC: USDA Economic Research Service.

Fuglie, K. O., P. Heisey, J. King, C. E. Pray, K. Day-Rubenstein, D. Schimmelpfennig, S. L. Wang, and R. Karmarkar-Deshmukh. 2011. Research Investments and Market Structure in the Food Processing, Agriculture Input and Biofuel Industries Worldwide. Report 130. Washington, DC: USDA Economic Research Service.

Fulton, J. E. 2012. Genomic selection for poultry breeding. Animal Frontiers 2(1):30-36.

Funston, R. N., and A. F. Summers. 2013. Epigenetics: Setting up lifetime production of beef cows by managing nutrition. Annual Review of Animal Biosciences 1(1):339-363.

Gaggìa, F., P. Mattarelli, and B. Biavati. 2010. Probiotics and prebiotics in animal feeding for safe food production. International Journal of Food Microbiology 141:S15-S28.

Gatlin, D. M., and P. Li. 2008. Use of diet additives to improve nutritional value of alternative protein sources. Pp. 501-522 in Alternative Protein Sources in Aquaculture Diet, C. Lim, C. D. Webster and C. S. Lee, eds. New York: Haworth.

George, S., D. L. Wright, and J. J. Marois. 2013. Impact of grazing on soil properties and cotton yield in an integrated crop-livestock system. Soil and Tillage Research 132:47-55.

Gjedrem, T., N. Robinson, and M. Rye. 2012. The importance of selective breeding in aquaculture to meet future demands for animal protein: A review. Aquaculture 350-353:117-129.

Glencross, B. D., W. E. Hawkins, and J. G. Curnow. 2003. Restoration of the fatty acid composition of red seabream (Pagrus auratus) using a fish oil finishing diet after grow-out on plant oil based diets. Aquaculture Nutrition 9(6):409-418.

Godfray, J., H. C., J. R. Beddington, I. R. Crute, L. Haddad, D. Lawerence, J. F. Muir, J. Pretty, S. Robinson, S. M. Thomas, and C. Toulmin. 2010. Food security: The challenge of feeding 9 billion people. Science 327(5967):812-818.

Goff, H. D., and M. W. Griffiths. 2006. Major advances in fresh milk and milk products: Fluid milk products and frozen desserts. Journal of Dairy Science 89(4):1163-1173.

Gonzalez-Recio, O. 2011. Epigenetics: A new challenge in the post-genomic era of livestock. Frontiers in Genetics 2:106.

Grandin, T. 2010. Auditing animal welfare at slaughter plants. Meat Science 86(1):56-65.

Gustavsson, J., C. Cederberg, U. Sonesson, R. van Otterdijk, and A. Meybeck. 2011. Global Food Losses and Food Waste—Extent, Causes and Prevention. Rome: FAO. Online. Available at http://www.fao.org/docrep/014/mb060e/mb060e00.pdf. Accessed September 25, 2014.

Gous, R. M. 2010. Nutritional limitations on growth and development in poultry. Livestock Science 130(1):25-32.

Grandin, T., and C. Johnson, 2009. Animals Make Us Human: Creating the Best Life for Animals. New York: Houghton-Mifflin Harcourt.

Green, R. D. 2009. ASAS centennial paper: Future needs in animal breeding and genetics. Journal of Animal Science 87(2):793-800.

Handisyde, N. T., L. G. Ross, M. C. Badjeck, and E. H. Allison. 2006. The Effects of Climate Change on World Aquaculture: A Global Perspective. Stirling, UK: Institute of Aquaculture. Online. Avaialble at http://www.ecasa.org.uk/Documents/Handisydeetal.pdf. Accessed September 25, 2014.

Hargreaves, J. A. 2013. Biofloc Production Systems for Aquaculture. Southern Regional Aquaculture Center Publication 4503. Online. Available at http://fisheries.tamu.edu/files/2013/09/SRAC-Publication-No.-4503-Biofloc-Production-Systems-for-Aquaculture.pdf. Accessed August 18, 2014.

Harter, T., J. R. Lund, J. Darby, G. E. Fogg, R. Howitt, K. K. Jessoe, G. S. Pettygrove, J. F. Quinn, J. H. Viers, D. B. Boyle, H. E. Canada, N. DeLaMora, K. N. Dzurella, A. Fryjoff-Hung, A. D. Hollander, K. L. Honeycutt, M. W. Jenkins, V. B. Jensen, A. M. King, G. Kourakos, D. Liptzin, E. M. Lopez, M. M. Mayzelle, A. McNally, J. Medellin- Azuara, and T. S. Rosenstock. 2012. Addressing Nitrate in California's Drinking Water with a Focus on Tulare Lake Basin and Salinas Valley Groundwater. Report for the State Water Resources Control Board to the Legislature. Davis, CA: Center for Watershed Sciences, University of California, Davis. Online. Available at http://groundwaternitrate.ucdavis.edu. Accessed August 17, 2014.

Hatfield, J., G. Takle, R. Grotjahn, P. Holden, R. C. Izaurralde, T. Mader, E. Marshall, and D. Liverman. 2014. Agriculture. Pp. 150-174 in Climate Change Impacts in the United States: The Third National Climate Assessment, J. M. Melillo, T. C. Richmond, and G. W. Yohe, eds. Washington, DC: U.S. Global Change Research Program.

Havenstein, G. B. 2006. Performance changes in poultry and livestock following 50 years of genetic selection. Lohmann Information 41:30-37.

Havenstein, G. B., P. R. Ferket, and M. A. Quershi. 2003. Growth, livability, and feed conversion of 1957 versus 2001 broilers when fed representative 1957 and 2001 broiler diets. Poultry Science 82(10):1500-1508.

Hayward, D. G., D. Nortrup, A. Gardner, and M. Clower, Jr. 1999. Elevated TCDD in chicken eggs and farm-raised catfish fed a diet with ball clay from a southern United States mine. Environmental Research 81(3):248-256.

Heffernan, C., M. Salman, and L. York. 2012. Livestock infectious disease and climate change: A review of selected literature. CAB Reviews 7(011):1-26.

Heisey, P., S. L. Wang, and K. Fuglie. 2011. Public Agricultural Research Spending and Future U.S. Agricultural Productivity Growth: Scenarios for 2010-2050. Economic Brief 17. Washington, DC: USDA Economic Research Service.

Henning, D. R., R. J. Baer, A. N. Hassan, and R. Dave. 2006. Major advances in concentrated and dry milk products, cheese, and milk fat-based spreads. Journal of Dairy Science 89(4):1179-1188.

Hernandez Gifford, J. A., and C. A. Gifford. 2013. Role of reproductive biotechnologies in enhancing food security and sustainability. Animal Frontiers 3(3):14-19.

Herrero, M., D. Grace, J. Njuki, N. Johnson, D. Enahoro, S. Silvestri, and M. C. Rufino. 2013. The roles of livestock in developing countries. Animal 7(S-1):3-18.

Higgs, J. D. 2000. The changing nature of red meat: 20 years of improving nutritional quality. Trends in Food Science and Technology 11(3):85-95.

Hocking, P.M. 2014. Unexpected consequences of genetic selection in broilers and turkeys: Problems and solutions. British Poultry Science 55(1):1-12.

Hocquette, J. F., R. I. Richardson, S. Prache, F. Medale, G. Duffy, and N. D. Scollan. 2005. The future trends for research on quality and safety of animal products. Italian Journal of Animal Science 4(Suppl. 3):49-72.

Hocquette, J. F., S. Lehnert, W. Barendse, I. Cassar-Malek, and B. Picard. 2007. Recent advances in cattle functional genomics and their application to beef quality. Animal 1(1):159-173.

Hoekstra, A. Y. 2012. The hidden water resource use behind meat and dairy. Animal Frontiers 2(2):3-8.

Holling, C. S. 1978. Adaptive Environmental Assessment and Management. Chichester, UK: John Wiley and Sons.

Holm-Nielsen, J. B., T. Al Seadi, and P. Oleskowicz-Popiel. 2009. The future of anaerobic digestion and biogas utilization. Bioresource Technology 100:5478-5484.

Hooper, L. V., and J. L. Gordon. 2001. Commensal host-bacterial relationships in the gut. Science 292(5519):1115-1118.

Hospido, A., and U. Sonesson. 2005. The environmental impact of mastitis: A case study of dairy herds. Science of the Total Environment 343(1-3):71-82.

Hotchkiss, J. H. 2006. Food quality and shelf life. Presentation to the Institute of Food Technologists (IFT) 5th Research Summit, May 7-9, Baltimore, MD. Online. Available at http://www.ift.org/~/media/Knowledge%20Center/Science%20Reports/Research%20Summits/Packaging/Packaging_QualityShelfLife_Hotchkiss.pdf. Accessed August 18, 2014.

Huntingford, F. A., C. Adams, V. A. Braithwaite, S. Kadri, T. G. Pottinger, P. Sandøe, and J. F. Turnbull. 2006. Current issues in fish welfare. Journal of Fish Biology 68(2):332-372.

Howard, C. J., A. Kumar, I. Malkina, F. Mitloehner, P. G. Green, R. G. Flocchini, and M. J. Kleeman. 2010a. Reactive organic gas emissions from livestock feed contribute significantly to ozone production in central California. Environmental Science & Technology 44(7):2309-2314.

Howard, C. J., A. Kumar, F. Mitloehner, K. Stackhouse, P. G. Green, R. G. Folcchini, and M. J. Kleeman. 2010b. Direct measurements of the ozone formation potential from livestock and poultry waste emissions. Environmental Science & Technology 44(7):2292-2298.

Howitt, R. E., J. Medellin-Azuara, D. MacEwan, J. R. Lund, and D. A. Sumner. 2014. Economic Analysis of the 2014 Drought for California Agriculture. Center for Watershed Sciences, University of California, Davis. Online. Available at http://watershed.ucdavis.edu. Accessed August 17, 2014.

Hristov, A. N., J. Oh, J. L. Firkins, J. Dijstra, E. Kebreab, G. Waghorn, H. P. S. Makkar, A. T. Adesogan, W. Yang, C. Lee, P. J. Gerber, B. Henderson, and J. M. Tricarico. 2013. Special topics—Mitigation of methane and nitrous oxide emissions from animal operations: I. A review of enteric methane mitigation options. Journal of Animal Science 91(11):5045-5069.

Hu, Z., J. W. Lee, K. Chandran, S. Kim, and S. K. Khanal. 2012. Nitrous oxide (N2O) emission from aquaculture: A review. Environmental Science and Technology 46(12):6470-6480.

Humane Society of the United States. The Welfare of Animals in the Aquaculture Industry. Online. Available at http://www.humanesociety.org/assets/pdfs/farm/hsus-the-welfare-of-animals-in-the-aquaculture-industry-1.pdf. Accessed October 10, 2014.

Hume, D. A., C. B. A. Whitelaw, and A. L. Archibald. 2011. The future of animal production: Improving productivity and sustainability. Journal of Agricultural Science 149(S-1):9-16.

Institute for Food Technology. 2010. Extrusion Technology in Feed and Food Processing, 2nd Workshop—Feed-to-Food. FP7REGPOT-3. University of Novi Sad. Online. Available at http://fins.uns.ac.rs/uploads/zbornici/Thematic-proceedings-2010.pdf. Accessed August 18, 2014.

Ireland, J. J., R. M. Roberts, G. H. Palmer, D. E. Bauman, and F. W. Bazer. 2008. A commentary on domestic animals as dual-purpose models that benefit agricultural and biomedical research. Journal of Animal Science 86(10):2797-2805.

Jeyanathan, J., C. Martin, and D. P. Morgavi. 2014. The use of direct-fed microbials for mitigation of ruminant methane emissions: A review. Animal 8(2):250-261.

Johnson, A. K. 2009. ASAS centennial paper: Farm animal welfare science in the United States. Journal of Animal Science 87(6):2175-2179.

Johnson, B. J., F. R. B. Ribeiro, and J. L. Beckett. 2013. Application of growth technologies in enhancing food security and sustainability. Animal Frontiers 3(3):8-13.

Johnson, D. E. 2007. Contributions of animal nutrition research to nutritional principles: Energetics. Journal of Nutrition 137(3):698-701.

Johnson, K. A., and D. E. Johnson. 1995. Methane emissions from cattle. Journal of Animal Science 73(8):2483-2492.

Johnson, M. E., and J. A. Lucey. 2006. Major technological advances and trends in cheese. Journal of Dairy Science 89(4):1174-1178.

Johnson, N. L., and V. W. Ruttan. 1997. The diffusion of livestock breeding technology in the U.S.: Observations on the relationships between technical change and industry structure. Journal of Agribusiness 15(1):19-35.

JS Food Brokers. 2014. CO2 Pads for Food Safety. Online. Available at http://www.jsfoodbrokers.com/co2-pads.html. Accessed August 18, 2014.

Kantor, L. S., K. Lipton, A. Manchester, and V. Oliveira. 1997. Estimating and addressing America's food losses. Food Review 20(1):2-12.

Karami, E., and M. Keshavarz. 2010. Sociology of sustainable agriculture. Pp. 19-40 in Sociology, Organic Farming, Climate Change and Soil Science, E. Lichtfouse, ed. Sustainable Agriculture Reviews 3. Dordrecht, The Netherlands: Springer Science.

Kennedy, D. 2014. Building agricultural research. Science 346(6205):13.

Khanal, A., and J. Gillespie. 2013. Adoption and productivity of breeding technologies: Evidence from US dairy farms. AgBioForum 16(1):53-65.

Kobayashi, M., and R. Brummett, 2014. Disease Management in Aquaculture, Forum for Agricultural Risk Management in Development. Online. Available at https://www.agriskmanagementforum.org/content/disease-management-aquaculture. Accessed October 3, 2014.

Lantier, F. 2014. Animal models of emerging diseases: An essential prerequisite for research and development of control measures. Animal Frontiers 4(1):7-12.

Lay, D. C., Jr., R. M. Fulton, P. Y. Hester, D. M. Karcher, J. B. Kjaer, J. A. Mench, B. A. Mullens, R. C. Newberry, C. J. Nicol, N. P. O'Sullivan, and R. E. Porter. 2011. Hen welfare in different housing systems. Poultry Science 90(1):278-294.

Le Boucher, R., M. Vandeputte, M. Dupont-Nivet, E. Quillet, F. Ruelle, A. Vergnet, S. Kaushik, J. M. Allamellou, F. Médale, and B. Chatain. 2013. Genotype by diet interactions in European sea bass (Dicentrarchus labrax L.): Nutritional challenge with totally plant-based diets. Journal of Animal Science. 91(1):44-56.

Ley, R. E., C. A. Lozupone, M. Hamady, R. Knight, and J. I. Gordon. 2008. Worlds within worlds: Evolution of the vertebrate gut microbiota. Nature Review Microbiology 6(10):776-788.

Lippmann, M. 1991. Health effects of tropospheric ozone. Environmental Science & Technology 25(12):1954-1962.

Looft, T., T. Johnson, H. K. Allen, D. O. Bayles, D. P. Alt, R. D. Stedtfeld, W. J. Sul, T. M. Stedtfeld, B. Chai, J. R. Cole, S. A. Hashsham, J. M. Tiedje, and T. B. Stanton. 2012. In-feed antibiotic effects on the swine intestinal microbiome. Proceedings of the National Academy of Sciences of the United States of America 109(5):1891-1696.

Losordo, T. M., M. P. Masser, and J. Rakocy. 1998. Recirculating Aquaculture Tank Production Systems: An Overview of Critical Considerations. Southern Regional Aquaculture Center Publication 451. Online. Available at http://www.lssu.edu/faculty/gsteinhart/GBS-LSSU/BIOL372-Fish%20Culture_files/Recirculating_culture.pdf. Accessed September 26, 2014.

Lundqvist, J., C. de Fraiture, and D. Molden. 2008. Saving Water: From Field to Fork: Curbing Losses and Wastage in the Food Chain. SIWI (Stockholm International Water Institute) Policy Brief. Online. Available at http://www.siwi.org/documents/Resources/Policy_Briefs/PB_From_Filed_to_Fork_2008.pdf. Accessed September 26, 2014.

Lupo, C. D., D. E. Clay, J. L. Benning, and J. J. Stone. 2013. Life-cycle assessment of the beef cattle production system for the northern Great Plains, USA. Journal of Environmental Quality 42(5):1386-1394.

Lyall, J., R. M. Irvine, A. Sherman, T. J. McKinley, A. Nunez, A. Purdie, L. Outtrim, I. H. Brown, G. Rolleston-Smith, H. Sang, and L. Tiley. 2011. Suppression of avian influenza transmission in genetically modified chickens. Science 331:223-226.

MacNeil, M. D., and L. T. Vermeire. 2012. Effect of weather patterns on preweaning growth of beef calves in the northern Great Plains. Journal of Agricultural Science 3(7):929-935.

Mainville, D., M. Farrell, G. Groover, and K. Mundy. 2009. Organic Feed-Grain Markets: Considerations for Potential Virginia Producers. Virginia Cooperative Extension Publication 448-520. Online. Available at http://pubs.ext.vt.edu/448/448-520/448-520_pdf.pdf. Accessed September 11, 2014.

Malone, R. 2013. Recirculating Aquaculture Tank Production Systems: A Review of Current Design Practice. Southern Regional Aquaculture Center Publication 453. Online. Available at http://fisheries.tamu.edu/files/2013/09/SRAC-Publication-No.-453-Recirculating-Aquaculture-Tank-Production-Systems-A-Review-of-Current-Design-Practice.pdf. Accessed September 26, 2014.

Martins, C. I. M., E. H. Eding, M. C. J. Verdegem, L. T. N. Heinsbroek, O. Schneider, J. P. Blancheton, E. Roque d'Orbcastel, and J. A. J. Verreth. 2010. New developments in recirculating aquaculture systems in Europe: A perspective on environmental sustainability. Aquacultural Engineering 43(3):83-93.

Mason, G., and M. Mendl. 1993. Why is there no simple way of measuring animal welfare? Animal Welfare 2(4):301-319.

Mason, G., and J. Rushen, eds. 2008. Stereotypic Animal Behaviour: Fundamentals and Applications to Welfare. Wallingford: CAB International.

Mathews, K. H., Jr., K. G. Jones, M. J. McConnell, and R. J. Johnson. 2013. Trade-adjusted measures of productivity increases in U.S. hog production. Agricultural Systems 114:32-37.

McBride, W., and N. Key. 2013. U.S. Hog Production from 1992 to 2009: Technology, Restructuring, and Productivity Growth. Economic Research Report 158. USDA Economic Research Service. Online. Available at http://www.ers.usda.gov/media/1207987/err158.pdf Accessed September 2, 2013.

McCarthy, A. 2013. Broiler chickens found at grocery retailers grow three times faster than normal chickens, plagued with lameness and disease. Online. PreventDisease.com. Available at http://preventdisease.com/news/13/121913_Broiler-Chickens-Grow-Three-Times-Faster-Than-Normal-Chickens-Plagued-With-Lameness-Disease.shtml. Accessed September 2, 2013.

McNamara, J. P., and S. L. Shields. 2013. Reproduction during lactation of dairy cattle: Integrating nutritional aspects of reproductive control in a systems research approach. Animal Frontiers 3(4):76-83.

Meale, S. J., K. A. Beauchemin, A. N. Hristov, A. V. Chaves, and T. A. McAllister. 2014. Board-invited review: Opportunities and challenges in using exogenous enzymes to improve ruminant production. Journal of Animal Science 92(2):427-442.

Mellor, D. J., E. Patterson-Kane, and K. J. Stafford. 2009. Integrated perspectives: Sleep, developmental stage and animal welfare. Pp. 161-185 in The Sciences of Animal Welfare. Oxford: Wiley-Blackwell.

Mench, J. A. 2008. Farm animal welfare in the U.S.A.: Farming practices, research, education, regulation, and assurance programs. Applied Animal Behaviour Science 113(4):298-312.

Millennium Ecosystem Assessment. 2005. Ecosystems and Human Well-Being: Synthesis. Washington, DC: Island Press.

Mitchell, A. D. 2007. Impact of research with cattle, pigs, and sheep on nutritional concepts: Body composition and growth. Journal of Nutrition 137(3):711-714.

Mitloehner, F. 2014. How high feed efficiency reduces the environmental impact of the dairy. WCDS Advances in Dairy Technology 26:5-14.

Montague, T. G., J. M. Cruz, J. A. Gagnon, G. M. Church, and E. Valen. 2014. CHOPCHOP: A CRISPR/Cas9 and TALEN Web tool for genome editing. Nucleic Acids Research 42:W401-407.

Montes, F., R. Meinen, C. Dell, A. Rotz, A. N. Hristov, J. Oh, G. Waghorn, P. J. Gerber, B. Henderson, H. P. S. Makkar, and J. Dijkstra. 2013. Special topics—Mitigation of methane and nitrous oxide emissions from animal operations: II. A review of manure management mitigation options. Journal of Animal Science 91(11):5070-5094.

Moore, K., and W. W. Thatcher. 2006. Major advances associated with reproduction in dairy cattle. Journal of Dairy Science 89(4):1254-1266.

Moraes, L. E., A. B. Strathe, J. G. Fadel, D. P. Casper, and E. Kebreab. 2014. Prediction of enteric methane emissions from cattle. Global Change Biology 20(7):2140-2148.

Morris, J. G., Jr. 2011. How safe is our food? Emerging Infectious Diseases 17(1):126-128.

Morsy, M. K., H. H. Khalaf, A. M. Sharoba, H. H. El-Tanahi, and C. N. Cutter. 2014. Incorporation of essential oils and nanoparticles in Pullulan films to control foodborne pathogens on meat and poultry products. Journal of Food Science 79(4):M675-M684.

Nafikov, R. A., and D. C. Beitz. 2007. Carbohydrate and lipid metabolism in farm animals. Journal of Nutrition 137(3):702-705.

Nardone, A., and F. Valfrè. 1999. Effects of changing production methods on quality of meat, milk and eggs. Livestock Production Science 59(2-3):165-182.

Nardone, A., B. Ronchi, N. Lacetera, M. S. Ranieri, and U. Bernabucci. 2010. Effects of climate changes on animal production and sustainability of livestock systems. Livestock Science 130(1-3):57-69.

Naylor, R., H. Steinfeld, W. Falcon, J. Galloway, V. Smil, E. Bradford, J. Alder, and H. Mooney. 2005. Agriculture: Losing the links between livestock and land. Science 310(5754):1621-1622.

Naylor, R. L., R. W. Hardy, D. P. Bureau, A. Chiu, M. Elliott, A. Farrell, I. Forster, D. M. Gatlin, R. J. Goldburg, K, Hua, and P. D. Nichols. 2009. Feeding aquaculture in the era of finite resources. Proceedings of the National Academy of Science of the United States of America 106(36):15103-15110.

Neeteson-van Nieuwenhoven, A. M., P. Knap, and S. Avendaño. 2013. The role of sustainable commercial pig and poultry breeding for food security. Animal Frontiers 3(1):52-57.

Neves, D. P., T. M. Banhazi, and I. A. Nääs. 2014. Feeding behaviour of broiler chickens: A review on the biomechanical characteristics. Revista Brasileira de Ciência Avícola 16(2):1-16.

Niemann, H., and W. A. Kues. 2007. Transgenic farm animals: An update. Reproduction, Fertility and Development 19(6):762-770.

Niemann, H., B. Kuhla, and G. Flachowsky. 2011. Perspectives for feed-efficient animal production. Journal of Animal Science 89(12):4344-4363.

NOAA (National Oceanic and Atmospheric Administration). 2014. Basic Questions About Aquaculture. Online. Available at http://www.nmfs.noaa.gov/aquaculture/faqs/faq_aq_101.html. Accessed September 2, 2014.

NOAA and USDA (National Oceanic and Atmospheric Administration and U.S. Department of Agriculture). 2011. The Future of Aquafeeds (Alternative Feeds Initiative). NOAA Technical Memorandum NMFS F/SPO-124. Online. Available at http://www.nmfs.noaa.gov/aquaculture/docs/feeds/the_future_of_aquafeeds_final.pdf. Accessed September 2, 2014.

NRC (National Research Council). 2010. Toward Sustainable Agricultural Systems in the 21st Century. Washington, DC: The National Academies Press.

NRC. 2011. Nutrient Requirements of Fish and Shrimp. Washington, DC: National Academies Press.

NRC. 2012. Nutrients requirements of swine. Washington, DC: The National Academies Press.

NRC. 2014. Advancing Land Change Modeling: Opportunities and Research Requirements. Washington, DC: The National Academies Press.

NRDC (Natural Resources Defense Council). 2014. Pollution from Giant Livestock Farms Threatens Public Health. Online. Available at http://www.nrdc.org/water/pollution/nspills.asp. Accessed September 4, 2014.

NSTC (National Science and Technology Council). 2014. National Strategic Plan for Federal Aquaculture Research (2014-2019). Online. Available at http://www.ars.usda.gov/sp2UserFiles/Place/00000000/NPS/AquacultureResearchStrategicPlan.pdf. Accessed August 18, 2014.

Olden, K., Y. S. Lin, D. Gruber, and B. Sonawane. 2014. Epigenome: Biosensor of cumulative exposure to chemical and nonchemical stressors related to environmental justice. American Journal of Public Health 104(10):1816-1821.

Olesen, I., A. F. Groen, and B. Gjerde. 2000. Definition of animal breeding goals for sustainable production systems. Journal of Animal Science 78(3):570-582.

Olin, P. G. 2011. National Aquaculture Sector Overview United States of America. Fact Sheets. FAO Fisheries and Aquaculture Department. Rome: FAO. Online. Available at http://www.fao.org/fishery/countrysector/naso_usa/en. Accessed September 29, 2014.

Oliver, S. P., D. A. Patel, T. R. Callaway, and M. E. Torrence. 2009. ASAS centennial paper: Developments and future outlook for preharvest food safety. Journal of Animal Science 87(1):419-437.

Oltjen, J., and J. Beckett. 1996. Role of ruminant livestock in sustainable agricultural systems. Journal of Animal Science 74(6):1406-1409.

Olynk, N. J., C. A. Wolf, and G. T. Tonsor. 2012. Production technology option value: The case of rbST in Michigan. Agricultural Economics 43:1-9.

Park, Y. H., S. Y. Hwang, M. K. Hong, and K. H Kwon. 2012. Use of antimicrobial agents in aquaculture. Revue Scientifique et Technique 31(1):189-197.

Pelletier, N. 2008. Environmental performance in the U.S. broiler poultry sector: Life cycle energy use and greenhouse gas, ozone depleting, acidifying and eutrophying emissions. Agricultural Systems 98(2):67-73.

Pelletier, N., P. Lammers, D. Stender, and R. Pirog. 2010a. Life cycle assessment of high- and low-profitability commodity and deep-bedded niche swine production systems in the upper midwestern United States. Agricultural Systems 103(9):599-608.

Pelletier, N., R. Pirog, and R. Rasmussen. 2010b. Comparative life cycle environmental impacts of three beef production strategies in the upper midwestern United States. Agricultural Systems 103(6):380-389.

Pelletier, N., M. Ibarburu, and H. Xin. 2014. Comparison of the environmental footprint of the egg industry in the United States in 1960 and 2010. Poultry Science 93(2):241-255.

Peters, G. M., H. V. Rowley, S. Wiedeman, R. Tucker, M. D. Short, and M. Schulz. 2010. Red meat production in Australia: Life cycle assessment and comparison with overseas studies. Environmental Science & Technology 44(4):1327-1332.

Petersen, A., and D. Green. 2006. Seafood Traceability: A Practical Guide for the U.S. Industry. NC Sea Grant Pub. UNC-SG-06-04. Online. Available at http://seafood.oregonstate.edu/.pdf%20Links/Seafood%20Traceability%20. Accessed December 16, 2014.

Pinder, R. W., P. J. Adams, and S. N. Pandis. 2007. Ammonia emission controls as a cost-effective strategy for reducing atmospheric particulate matter in the eastern United States. Environmental Science & Technology 41(2):380-386.

Place, S. E., and F. M. Mitloehner. 2014. The nexus of environmental quality and livestock welfare. Annual Review of Animal Biosciences 2:555-569.

Polley, H. W., D. D. Briske, J. A. Morgan, K. Wolter, D. W. Bailey, and J. R. Brown. 2013. Climate change and North American rangelands: Trends, projections, and implication. Rangeland Ecology and Management 66(5):493-511.

Price, E. O. 2008. Principles and Applications of Domestic Animal Behavior. New York: CAB International.

Priest, S. H. 2000. U.S. public opinion divided over biotechnology? Nature Biotechnology 18:939-942.

Rappert, S., and R. Müller. 2005. Odor compounds in waste gas emissions from agricultural operations and food industries. Waste Management 25(9):887-907.

Rauw, W. M., E. Kanis, E. N. Noordhuizen-Stassen, and F. J. Grommers. 1998. Undesirable side effects of selection for high production efficiency in farm animals: A review. Livestock Production Science 56(1):15-33.

Reis, L. S., P. E. Pardo, A. S. Camargas, and E. Oba. 2010. Mineral element and heavy metal poisoning in animals. Journal of Medicine and Medical Sciences 1(12):560-579.

Richards, G. P., C. McLeod, and F. S. Le Guyader. 2010. Processing strategies to inactivate enteric viruses in shellfish. Food and Environmental Virology 2(3):183-193.

Rist, L., B. M. Campbell, and P. Frost. 2013. Adaptive management: Where are we now? Environmental Conservation 40(1):5-18.

Roberts, R. M., G. W. Smith, F. W. Baze, J. Cibelli, G. E. Seidel, Jr., D. E. Bauman, L. P. Reynolds, and J. J. Ireland. 2009. Farm animal research in crisis. Science 324(5926):468-469.

Robertson, G. P., T. W. Bruulsema, R. J. Gehl, D. Kanter, D. L. Mauzerall, C. A. Rotz, and C. O. Williams. 2013. Nitrogen-climate interactions in U.S. agriculture. Biogeochemistry 114:41-70.

Robinson, N. 2014. Online. Shelf life extension would reduce food waste. Food Manufacture. Available at http://www.foodmanufacture.co.uk/Manufacturing/Food-waste-could-be-reduced-by-extra-days-shelf-life. Accessed October 30, 2014.

Rodríguez-Martínez, H., and F. Peña Vega. 2013. Semen technologies in domestic animal species. Animal Frontiers 3(4):26-33.

Rosegrant, M. W., J. Koo, N. Cenacchi, C. Ringler, R. Robertson, M. Fisher, C. Cox, K. Garrett, N. D. Perez, and P. Sabbagh. 2014. Food Security in a World of Natural Resource Scarcity: The Role of Agricultural Technologies. Washington, DC: International Food Policy Research Institute.

Rotz, C. A., and T. L. Veith. 2013. Integration of air and water quality issues. Pp. 137-156 in Sustainable Animal Agriculture, E. Kebreab, ed. Wallingford, UK: CAB International.

Rouquette, F. M., Jr., L. A. Redmon, G. E. Aiken, G. M. Hill, L. E. Sollenberger, and J. Andrae. 2009. ASAS centennial paper: Future needs of research and extension in forage utilization. Journal of Animal Science 87(1):438-446.

Rowley, A., and C. Pope. 2012. Vaccines and crustacean aquaculture. Aquaculture 334-337:1-11.

SAIC (Science Applications International Corporation). 2006. Life Cycle Assessment: Principles and Practice. EPA/600/R-06/060. Washington, DC: U.S. Environmental Protection Agency. Online. Available at http://www.epa.gov/nrmrl/std/lca/pdfs/chapter1_frontmatter_lca101.pdf. Accessed October 31, 2014.

Sapkota, A. R., L. Y. Lefferts, S. McKensie, and P. Walker. 2007. What do we feed to food-production animals? A review of animal feed ingredients and their potential impacts on human health. Environmental Health Perspectives 115(5):663-670.

SARE (Sustainable Agriculture Research and Education). 2012. Sustainable Agriculture Research and Education: Grants and Education to Advance Innovations in Sustainable Agriculture. Online. Available at www.sare.org. Accessed September 1, 2014.

Schefers, J. M., and K. A. Weigel. 2012. Genomic selection in dairy cattle: Integration of DNA testing into breeding programs. Animal Frontiers 2(1):4-9.

Schroder, U. 2008. Challenges in the traceability of seafood. Journal of Consumer Protection and Food Safety 3(1):45-48.

Seufert, V., N. Ramankutty, and J. A. Foley. 2012. Comparing the yields of organic and conventional agriculture. Nature 485(7397):229-232.

Shike, D. 2013. Beef cattle feed efficiency. Pp. 3-4 in Proceedings: Driftless Region Beef Conference, January 31-February 1, Dubuque, Iowa. Online. Available at http://www.iowabeefcenter.org/proceedings/DriftlessConference2013.pdf. Accessed September 3, 2014.

Shindell, D., J. C. I. Kuylenstierna, E. Vignati, R. van Dingenen, M. Amann, Z. Klimont, S. C. Anenberg, N. Muller, G. Janssens-Maenhout, F. Raes, J. Schwartz, G. Faluvegi, L. Pozzoli, K. Kupiainen, L. Höglund-Isaksson, L. Emberson, D. Streets, V. Ramanathan, K. Hicks, N. T. K. Oanh, G. Milly, M. Williams, V. Demkine, and D. Fowler. 2012. Simultaneously mitigating near-term climate change and improving human health and food security. Science 335(6065):183-189.

Siegrist, M., T. C. Earle, and H. Gutscher. 2010. Trust in Risk Management: Uncertainty and Skepticism in the Public Mind. Sterling, VA: Earthscan.

Simpson, S. J., and D. Raubenheimer. 1995. The geometric analysis of feeding and nutrition: A user's guide. Journal of Insect Physiology 41(7):545-553.

Soberon, F., E. Raffrenato, R. W. Everett, and M. E. Van Amburgh. 2012. Preweaning milk replacer intake and effects on long-term productivity of dairy calves. Journal of Dairy Science 95(2):783-793.

Soberon, M. A., Q. M. Ketterings, C. N. Rasmussen, and K. J. Czymmek. 2013. Whole farm nutrient balance calculator for New York dairy farms. Natural Sciences Education 42:57-67.

Sofos, J. N. 2009. ASAS centennial paper: Developments and future outlook for postslaughter food safety. Journal of Animal Science 87(7):2448-2457.

Sommerset, I. B., B. Krossoy, E. Biering, and P. Frost. 2005. Vaccines for fish in aquaculture. Expert Review of Vaccines 4(1):89-101.

Spencer, T. E. 2013. Early pregnancy: Concepts, challenges, and potential solutions. Animal Frontiers 3(4):48-55.

Spinner, J. 2014. Meat Market: Active Packaging Boosts Shelf Life and Profits. Online. Available at http://www.foodproductiondaily.com/Packaging/Food-packaging-technology-boosts-meat-shelf-life-and-profitability. Accessed August 18, 2014.

St. Hilaire, S., K. Cranfill, M. A. Mcguire, E. E. Mosley, J. K. Tomberlin, L. Newton, W. Sealey, C. Sheppard, and S. Irving. 2007. Fish offal recycling by the black soldier fly produces a foodstuff high in omega-3 fatty acids. Journal of the World Aquaculture Society 38(2):309-313.

St-Pierre, N. R., B. Cobanov, and G. Schnitkey. 2003. Economic losses from heat stress by U.S. livestock industries. Journal of Dairy Science 86(Suppl.):E52-E77.

St-Pierre, N. R., G. A. Milliken, D. E. Bauman, R. J. Collier, J. S. Hogan, J. K. Shearer, K. L. Smith, and W. W. Thatcher. 2014. Meta-analysis on the effects of sometribove zinc suspension on the production and health of lactating dairy cows. Journal of the American Veterinary Medical Association 245(5):550-564.

Stackhouse-Lawson, K. R., C. A. Rotz, J. W. Oltjen, and F. M. Mitloehner. 2012. Carbon footprint and ammonia emissions of California beef production systems. Journal of Animal Science 90(12):4641-4655.

Stackhouse-Lawson, K. R., M. S. Calvo, S. E. Place, T. L. Armitage, Y. Pan, Y. Zhao, and F. M. Mitloehner. 2013. Growth promoting technologies reduce greenhouse gas, alcohol, and ammonia emissions from feedlot cattle. Journal of Animal Science 91(11):5438-5447.

Stafford, K. J., and D. J. Mellor. 2011. Addressing the pain associated with disbudding and dehorning in cattle. Applied Animal Behavior Science 135(3):226-231.

Steinfeld, H., P. Gerber, T. Wassenaar, V. Castel, M. Rosales, and C. de Haan. 2006. Livestock's Long Shadow: Environmental Issues and Options. Rome: FAO. Online. Available at http://www.fao.org/docrep/010/a0701e/a0701e00.htm. Accessed August 15, 2014.

Stevenson, M. H. 1994. Nutritional and other implications of irradiating meat. Proceedings of Nutrition Society 53(2):317-325.

Steward, D. R., P. J. Bruss, X. Yang, S. A. Staggenborg, S. M. Welch, and M. D. Apley. 2013. Tapping unsustainable groundwater stores for agricultural production in the High Plains Aquifer of Kansas, projections to 2110. Proceedings of the National Academy of Sciences of the United States of America 110(37):E3477-E3486.

Stiglbauer, K. E., K. M. Cicconi-Hogan, R. Richert, Y. H. Schukken, P. L. Ruegg, and M. Gamroth. 2013. Assessment of herd management on organic and conventional dairy farms in the United States. J. Dairy Sci. 96:1290-1300.

Sulc, R. M., and A. J. Franzluebbers. 2014. Exploring integrated crop-livestock systems in different ecoregions of the United States. European Journal of Agronomy 57:21-30.

Sundström, J., A. Albihn, S. Boqvist, K. Ljungvall, H. Marstorp, C. Martiin, K. Nyberg, I. Vågsholm, J. Yuen, and U. Magnusson. 2014. Future threats to agricultural food production posed by environmental degradation, climate change, and animal and plant diseases—a risk analysis in three economic and climate settings. Food Security 6(2):201-215.

Swanson, J. C., Y. Lee, P. B. Thompson, R. Bawden, and J. A. Mench. 2011. Integration: Valuing stakeholder input in setting priorities for socially sustainable egg production. Poultry Science 90(9):2110-2121.

Tacon, A. G. J., M. R. Hasan, and M. Metian. 2011. Demand and Supply of Feed Ingredients for Farmed Fish and Crustaceans: Trends and Prospects. FAO Fisheries Technical Paper 564. Rome: FAO. Available at http://www.fao.org/docrep/015/ba0002e/ba0002e00.htm. Accessed September 29, 2014.

Taheripour, F., C. Hurt, and W. E. Tyner. 2013. Livestock industry in transition: Economic, demographic, and biofuel drivers. Animal Frontiers 3(2):38-46.

Tan, W., D. F. Carlson, C. A. Lancto, J. R. Garbed, D. A. Webster, P. B. Hackett, and S. C. Fahrenkrug. 2013. Efficient nonmeiotic allele introgression in livestock using custom endonucleases. Proceedings of the National Academy of Sciences of the United States of America 110(41):16526-16531.

Tauxe, R. V. 2001. Food safety and irradiation: Protecting the public from foodborne infections. Emerging Infectious Disease 7(3 Suppl.):516-521.

Texas A&M University. 2012. Study Focuses on Feeding Beef Cattle Algae Co-products. Available at http://animalscience.tamu.edu/2012/08/29/study-focuses-on-feeding-beef-cattle-algae-co-products. Accessed August 18, 2014.

Thoma, G., D. Nutter, R. Ulrich, C. Maxwell, J. Frank, and C. East. 2011. National Life Cycle Carbon Footprint Study for Production of US Swine. Online. Available at http://www.pork.org/filelibrary/NPB%20Scan%20Final%20-%20May%202011.pdf. Accessed August 5, 2014.

Thoma, G., J. Popp, D. Nutter, D. Shonnard, R. Ulrich, M. Matlock, D. S. Kim, Z. Neiderman, N. Kemper, C. East, and F. Adom. 2013. Greenhouse gas emissions from milk production and consumption in the United States: A cradle-to-grave life cycle assessment circa 2008. International Dairy Journal 31(Suppl. 1):S3-S14.

Tomich, T. P., S. Brodt, H. Ferris, R. Galt, W. R. Horwath, E. Kebreab, J. H. J. Leveau, D. Liptzin, M. Lubell, P. Merel, R. Michelmore, T. Rosenstock, K. Scow, J. Six, N. Williams, and L. Yang. 2011. Agroecology: A review from a global-change perspective. Annual Review of Environment and Resources 36:193-222.

Tong, P. S. 2009. Unraveling the mysteries of shelf life. Dairy Foods 110(5):68.

Tucker, C. B., J. Mench., and M. A. G. von Keyserlingk. 2013. Animal welfare: An integral component of sustainability. Pp 42-52 in Sustainable Animal Agriculture, E. Kebreab, ed. Boston: CAB International.

Tucker, C., D. Brune, and E. Torrans. 2014. Partitioned pond aquaculture systems. World Aquaculture 45(2):9-17.

USDA (U.S. Department of Agriculture). 1999. USDA Issues Final Rule on Meat and Poultry Irradiation. Online. Available at http://www.fsis.usda.gov/Oa/background/irrad_final.htm. Accessed October 31, 2014.

USDA. 2014. Census of Aquaculture 2013, Vol. 3, Special Studies, Part 2.

USDA Census of Agriculture. 2012a. Census Ag Atlas Maps: Livestock and Animals. Online. Available at http://www.agcensus.usda.gov/Publications/2012/Online_Resources/Ag_Atlas_Maps/Livestock_and_Animals/Livestock,_Poultry_and_Other_Animals/12-M139-RGBDot1-largetext.pdf. Accessed August 7, 2014.

USDA Census of Agriculture. 2012b. Census Full Report. Online. Available at http://www.agcensus.usda.gov/Publications/2012/Full_Report/Volume_1,_Chapter_1_US/st99_1_014_016.pdf. Accessed August 7, 2014.

USDA ERS (U.S. Department of Agriculture Economic Research Service). 2013. Animal Products. Available at http://www.ers.usda.gov/topics/animal-products. Accessed August 18, 2014.

USDA ERS. 2014. Food Availability (Per Capita) Data System. Online. Available at http://www.ers.usda.gov/data-products/food-availability-(per-capita)-data-system.aspx. Accessed September 3, 2014.

USDA ESMIS (U.S. Department of Agriculture Economics, Statistics, and Market Information System). 2012. 1940 Census. Online. Available at http://usda.mannlib.cornell.edu/usda/AgCensusImages/1940/03/07/1300/Table-11.pdf. Accessed August 7, 2014.

USDA FSIS (U.S. Department of Agriculture Food Safety and Inspection Service). 2013. Irradiation and Food Safety Answers to Frequently Asked Questions. Online. Available at http://www.fsis.usda.gov/wps/portal/fsis/topics/food-safety-education/get-answers/food-safety-fact-sheets/production-and-inspection/irradiation-and-food-safety/irradiation-food-safety-faq. Accessed December 15, 2014.

USDA NIFA (National Institute of Food and Agriculture). 2009. Animals. Online. Available at http://www.csrees.usda.gov/nea/animals/animals_all.html. Accessed August 6, 2014.

USDA NIFA. 2013. Animal Agriculture and Climate Change. Online. Available at http://animalagclimatechange.org. Accessed August 8, 2014.

USDA REE (U.S. Department of Agriculture Research, Education, and Economics). 2012. Research, Education, and Economics Action Plan. Online. Available at http://www.usda.gov/documents/usda-ree-science-action-plan.pdf. Accessed August 5, 2014.

U.S. Dairy Export Council. 2014. Export Trade Data. Online. Available at https://www.usdec.org/Why/content.cfm?ItemNumber=82452. Accessed August 15, 2014.

USGS (U.S. Geological Survey). 2014. High Plains Aquifer Water-Level Monitoring Study Area-Weighted Water-Level Change, Predevelopment to 1980, 2000 through 2011. Available at http://ne.water.usgs.gov/ogw/hpwlms/tablewlpre.html. Accessed July 9, 2014.

USMEF (U.S. Meat Export Federation). 2012a. Total U.S. Lamb Exports 2003-2012 (Including Variety Meat). Online. Available at http://www.usmef.org/downloads/Lamb-2003-to-2012.pdf. Accessed August 15, 2014.

USMEF. 2012b. Total U.S. Pork Exports 2003-2012 (Including Variety Meat). Online. Available at http://www.usmef.org/downloads/Pork-2003-to-2012.pdf. Accessed August 15, 2014.

van der Steen, H., G. Prall, and G. Plastow. 2005. Application of genomics to the pork industry. Journal of Animal Science 83(13 Suppl.):E1-E8.

Van Eenennaam, A. L. 2006. What is the future of animal biotechnology? California Agriculture 60(3):132-139.

van Huis, A., J. Van Itterbeeck, H. Klunder, E. Mertens, A. Halloran, G. Muir, and P. Vantomme. 2013. Edible Insects: Future Prospects for Food and Feed Security. FAO Forestry Paper 171. Rome: FAO. Online. Available at http://www.fao.org/docrep/018/i3253e/i3253e.pdf. Accessed September 29, 2014.

Vielma, J., and M. Kankainen. 2013. Offshore Fish Farming Technology in Baltic Sea Production Conditions: Reports of Aquabest Project. Online. Available at http://www.aquabestproject.eu/media/12219/aquabest_10_2013_report.pdf. Accessed August 6, 2014.

Walker, M., M. Diez-Leon, and G. Mason. 2014. Animal Welfare Science: Recent publication trend and future research priorities. International Journal of Comparative Psychology 27(1):80-100.

Wallheimer, B. 2012. Rapidly cooling eggs can double shelf life, decrease risk of illness. Purdue University News Service. Available at http://www.purdue.edu/newsroom/research/2012/120611KeenerCooled.html. Accessed August 18, 2014.

Walsh, J., D. Wuebbles, K. Hayhoe, J. Kossin, K. Kunkel, G. Stephens, P. Thorne, R. Vose, M. Wehner, J. Willis, D. Anderson, S. Doney, R. Feely, P. Hennon, V. Kharin, T. Knutson, F. Landerer, T. Lenton, J. Kennedy, and R. Somerville. 2014. Our changing climate. Pp. 19-67 in Climate Change Impacts in the United States: The Third National Climate Assessment, J. M. Melillo, T. C. Richmond, and G. W. Yohe, eds. Washington, DC: U.S. Global Change Research Program.

Walters, C. J. 1986. Adaptive Management of Renewable Resources. New York: Macmillan.

Walthall, C. L., J. Hatfield, P. Backlund, L. Lengnick, E. Marshall, M. Walsh, S. Adkins, M. Aillery, E. A. Ainsworth, C. Ammann, C. J. Anderson, I. Bartomeus, L. H. Baumgard, F. Booker, B. Bradley, D. M. Blumenthal, J. Bunce, K. Burkey, S. M. Dabney, J. A. Delgado, J. Dukes, A. Funk, K. Garrett, M. Glenn, D. A. Grantz, D. Goodrich, S. Hu, R. C. Izaurralde, R. A. C. Jones, S. H. Kim, A. D. B. Leaky, K. Lewers, T. L. Mader, A. McClung, J. Morgan, D. J. Muth, M. Nearing, D. M. Oosterhuis, D. Ort, C. Parmesan, W. T. Pettigrew, W. Polley, R. Rader, C. Rice, M. Rivington, E. Rosskopf, W. A. Salas, L. E. Sollenberger, R. Srygley, C. Stöckle, E. S. Takle, D. Timlin, J. W. White, R. Winfree, L. Wright-Morton, and L. H. Ziska. 2012. Climate Change and Agriculture in the United States: Effects and Adaptation. USDA Technical Bulletin 1935. Washington, DC: USDA.

Wheeler, T., and C. Reynolds. 2013. Predicting the risks from climate change to forage and crop production for animal feed. Animal Frontiers 3(1):36-41.

White House. 2014. Climate Action Plan: Strategy to Reduce Methane Emissions. Online. Available at http://www.whitehouse.gov/sites/default/files/strategy_to_reduce_methane_emissions_2014-03-28_final.pdf. Accessed August 8, 2014.

WHO and FAO (World Health Organization and Food and Agriculture Organization of the United Nations). 2012. Prevention and Reduction of Food and Feed Contamination. Online. Available at ftp://ftp.fao.org/codex/publications/Booklets/Contaminants/CCCF_2012_EN.pdf. Accessed September 25, 2014.

Widowski, T. M. 2010. The physical environment and its effect on welfare. Pp. 137-164 in The Welfare of Domestic Fowl and Other Captive Birds, I. J. H. Duncan and P. Hawkins, eds:. Dordrecht, The Netherlands: Springer.

Wilkinson, J. M. 2011. Re-defining efficiency of feed use by livestock. Animal 5(7):1014-1022.

WRAP. 2008. The food we waste. Online. Available at: http://www.ifr.ac.uk/waste/Reports/WRAP%20The%20Food%20We%20Waste.pdf. Accessed October 30, 2014.

Wright, J. 2014. Case Study: The Fish Have Landed. SeafoodSource.com. Online. Available at http://www.seafoodsource.com/news/aquaculture/26504-case-study-the-fish-have-landed. Accessed October 3, 2014.

Wu, F., and G. Munkvold. 2008. Mycotoxins in ethanol co-products: Modeling economic impacts on the livestock industry and management strategies. Journal of Agricultural and Food Chemistry 56(11):3900-3911.

Wu, F., D. Bhatnagar, T. Bui-Klimke, I. Carbone, R. Hellmich, G. Munkvold, P. Paul, G. Payne, and E. Takle. 2011. Climate change impacts on mycotoxin risks in U.S. maize. World Mycotoxin Journal 4(1):79-93.

Zain, M. E. 2010. Impact of mycotoxins on humans and animals. Journal of Saudi Chemical Society 15(2):129-144.

4

Global Considerations for Animal Agriculture Research

INTRODUCTION

With a projected world population of nearly 10 billion people by 2050, an unprecedented increase in demand for animal protein including meat, eggs, milk, and other animal products is inevitable. The global challenge will reside in the provision of an affordable, safe, and sustainable food supply. This chapter focuses on global trends in food animal production as it affects the development of technology in food security and sustainability, its linked socioeconomic aspects, and research needs, primarily in developing countries. Animal agricultural research needs on a global scale parallel those outlined in Chapter 3 for the United States, but also need to address the unique challenges specific to middle- and low-income nations worldwide. Those research considerations for sustainability related to the United States outlined in Chapter 3 are also relevant to other nations worldwide; however, often in developing countries, there is a tradeoff between animal production and resultant livelihoods and the environmental and societal impacts of production. Development and dissemination of technologies for improved animal production and wellness and feed/food safety are critical. Because animal products and animal feed are critical components of world food trade, understanding trends in supply and demand for animal products globally is also crucial.

Additionally, corollary needs must be fulfilled with respect to understanding the socioeconomic contexts in which food animals are produced worldwide, such as infrastructural gaps, food insecurity in some world regions, and barriers to technology transfer. One important consideration in discussion of agricultural needs for developing countries is political fragility of many of these nations, particularly those in sub-Saharan Africa. In fact, in a recent study of the most fragile states[1], assessing such factors as demographic pressures, uneven economic development, and poverty and economic decline, six of the most politically unstable places are located in sub-Saharan Africa (Foreign Policy, 2014). This has important consequences for the development and investment in sustainable agricultural activities in these countries. Thornton (2010) provides a summary of the past and projected trends in meat and milk consumption in developing and developed countries from 1980 to 2050. Total meat consumption in developing countries tripled from 47 million to137 million tons between 1980 and 2002. It is projected that from 1980 to 2050, total meat consumption in developed countries will increase from 86 million to 126 million tons while total meat consumption in developing countries will increase from 47 million to 326 million tons. The committee acknowledges that there are regional differences in per capita animal product consumption patterns that are important to consider (Table 4-1); however, it was not able to discuss this in great detail. For background on food animal production in sub-Saharan Africa and South Asia, the committee refers readers to the report by the National Research Council Emerging Technologies to Benefit Farmers in Sub-Saharan Africa and South Asia (NRC, 2008).

4-1 Animal Science Research Priorities

Existing animal science research priorities have been generated primarily in the United States (Chapter 3; FASS, 2012) and the European Union (EU; Figure 4-1; Scollan et al., 2011). The Food and Agriculture Organization of the United Nations (FAO) and the International Livestock Research Institute (ILRI) have ongoing global projects directed at improving food security and reducing poverty in developing countries through the more sustainable use of food animals. In general, animal science research can be grouped into the following broad categories: animal health, food safety, food and feed security, climate change, animal well-being, and water quantity and quality. An EU Animal Task Force identified the following four areas as priorities for research: (1) resource efficiency, (2) responsible livestock farming

---

[1] A fragile region or state has weak capacity to carry out basic governance functions and lacks the ability to develop mutually constructive relations with society (OECD, 2013).

systems, (3) healthy livestock and people, and (4) knowledge exchange toward innovation (Animal Task Force, 2013). The Animal Task Force defined livestock as animals farmed in both agriculture and aquaculture. The goals established by the EU Animal Task Force under each of the main topics and subtopics are provided in Appendix I. These research priorities and subtopics can be applied globally and are also applicable to developing countries (Figure 4-1).

TABLE 4-1 Global and Regional Food Consumption Patterns and Trends

Region	Meat (kg/year: 1997-1999)	Milk (kg/year: 1997-1999)
World	36.4	78.1
Developing countries	25.5	44.6
Near East and North Africa	21.2	72.3
Sub-Saharan Africa[a]	9.4	29.1
Latin America and the Caribbean	53.8	110.2
East Asia	37.7	10.0
South Asia	5.3	67.5
Industrialized countries	88.2	212.2
Transition countries	46.2	159.1

[a] Excludes South Africa
SOURCE: WHO (2014).

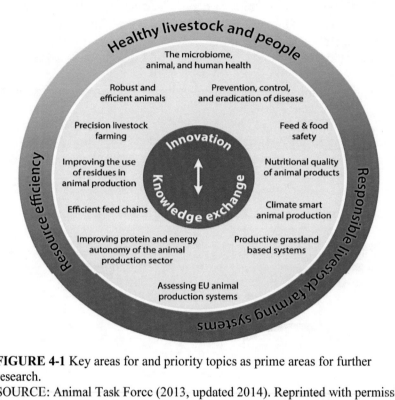

FIGURE 4-1 Key areas for and priority topics as prime areas for further research.
SOURCE: Animal Task Force (2013, updated 2014). Reprinted with permission from the Animal Task Force.

In broad terms, the animal science research goals of the United States and Europe are similar, with a focus on food safety, global food security, sustainability, and animal well-being. These are global goals and are applicable to developing countries as well. A holistic approach where systems analysis, among others, is used to assess the benefits and/or tradeoffs in meeting these goals is advocated. Additionally, a group of scientific experts identified 100 of their most important questions for global agriculture, of which 9 pertained to aquaculture, food animals, and poultry (Pretty et al., 2010), a subset of which includes concerns identified by U.S. and EU animal task forces:

- "What is the appropriate mix of intensification and extensification required to deliver increased production, greenhouse gas reduction and increased ecosystem services?
- What evidence exists to indicate that climate change will change pest and disease incidence? How can intensive food animal systems be

designed to minimize the spread of infectious diseases among animals and the risk of the emergence of new diseases infecting humans?
- How can middle- and small-scale animal production be made suitable for developing countries in terms of environmental impact, economic return and human food supply, and what should be the key government policies to ensure that a balance between the two is implemented?
- How can aquaculture and open water farming be developed so that impacts on wild fish stocks and coastal habitats are minimized?
- What are the priority efficiency targets for food animal production systems (e.g., the appropriate mix of activities in different systems and optimal number and types of animals) that would enable these systems to meet the demand for food animal products in an environmentally sound, economically sustainable and socially responsible way?
- What are the effective and efficient policies and other interventions to reduce the demand for animal products in societies with high consumption levels and how will they affect global trade in food animal products and the competitiveness of smallholder food animal production systems in poor countries?
- In addition to food animal production, how can inland and coastal fish farming contribute to a more sustainable mode of animal protein production in developing countries?
- What are the best means to encourage the economic growth of regional food animal markets, while limiting the effects of global climate change, and what can industrialized countries do to improve the carbon footprint of its food animal sector?
- What are the environmental impacts of different kinds of food animal-rearing and aquaculture systems?"

From a funding perspective, there are several important points to consider for the future of animal sciences research worldwide. In general, the public and private sectors tend to support different kinds of research. A review of the roles of public and private sectors in animal research is found in Fuglie et al. (2008). The public sector is the source of most basic research (i.e., original investigation that advances scientific knowledge but with no immediate application), while the private sector focuses on market-oriented applied and development research (Fuglie et al., 1996). Over the past 20 years, governments throughout the world have continued to reduce spending for research infrastructure development and training (Green, 2009), despite the fact that large-animal research is expensive.

Recommendation 4-1

The committee notes that per capita consumption of animal protein will be increasing more quickly in developing countries than in developed countries through 2050. Animal science research priorities have been proposed by stakeholders in high-income countries, with primarily U.S. Agency for International Development (USAID), World Bank, FAO, the Consultative Group on International Agricultural Research (CGIAR) and nongovernmental organizations (NGOs) individually providing direction for developing countries. A program such as the Comprehensive Africa Agriculture Development Programme (CAADP) demonstrates progress toward building better planning in agricultural development in developing countries, through the composite inclusion of social, environmental, and economic pillars of sustainability.

In addition, for at least the last two decades, governments worldwide have been reducing their funding for infrastructure development and training for animal sciences research. Countries and international funding agencies should be encouraged to adapt an integrated agricultural research system to be part of a comprehensive and holistic approach to agricultural production. A system such as CAADP can be adapted for this purpose.

To sustainably meet increasing demands for animal protein in developing countries, stakeholders at the national level should be involved in establishing animal science research priorities.

Priorities for Research Support

Countries in different regions are affected by different circumstances (e.g., political, demographic, and climatic) that correspondingly impact their animal science research priorities. One priority for research support in this area includes:
- Governments—both domestic and international—as well as international funding organizations, should increase respective commitments to funding animal sciences research and infrastructure that focus on productivity while following an integrated approach (e.g., the CAADP) to meet the increasing needs of global societies.

4-2 Food Security Considerations

Food animal production directly affects food security through food provision and risk reduction, and indirectly as a means of agricultural production, providing employment, income, capital stock, draft power, fertilizer (manure), and energy through burning of manure (Hoddinott and Skoufias, 2003; Erb et al., 2012). A summary of the positive and negative aspects of food animal production on food security can be found in Erb et al. (2012). Additionally, food animal production and marketing can help stabilize the global food supply by acting as a buffer to economic shocks and natural disasters; however, the food supply can be destabilized by animal diseases and bioterrorism (McLeod, 2011). McLeod (2011) describes three food security situations: (1) livestock-dependent societies, (2) small-scale mixed farmers, and (3) city populations. Food animal–dependent societies such as pastoralists (120 million people) and ranchers will face finite global land area that is becoming decreasingly available for grazing, because of population growth and settlement into current land, nature conservation, change in climatic conditions, and declining soil fertility. Small-scale farmers face limits to intensification and difficulties in competing with large-scale producers. They may have to diversify, consider becoming contract farmers, and/or find a high-value niche or specialty markets. Half of the world's population lives in urban areas and falls under the "city population" segment. Feeding this segment is accomplished primarily through intensified animal production systems. This segment is concerned about zoonoses, food safety, environmental pollution and regulations (Birch and Wachter, 2011).

In different parts of the world, food security concerns differ, as do levels of animal production technologies utilized. Amongst Asian nations, China has the largest number of poultry, pigs, goats, and sheep with poultry and pigs accounting for about 50 percent and 80 percent, respectively, of the total food animal production and 66 percent of the total meat production in Asia (Cao and Li, 2013). In 2010, food animals accounted for 30 percent of the total agricultural output in sales in China (Cao and Li, 2013). Rapid growth and technological innovations have resulted in opportunities to improve food security in Asia. The increasing demand for food animal products has become a driver to shift from small-scale farms to large-scale corporations (Cao and Li, 2013). Larger-scale units have proven to be more resilient to market volatility, weather effects, and consumer demand, and more ready to adopt new technologies (Cao and Li, 2013). Krätli et al. (2013) make a case for the necessity of pastoralism in food security in sub-Saharan Africa. They conclude that unless investments are shifted from replacing pastoralism to developing pastoralism on its own terms, food security is jeopardized beyond the limits of the drylands. Pastoralism can turn environmental instability into an asset for food production.

Poultry meat and eggs are and will be very important in meeting food security needs in both developed and especially developing countries (Ianotti et al., 2014). Within the next 10 years, poultry meat is projected to represent 50 percent of the increase in global meat consumption, and global demand for poultry meat and eggs is projected to increase by 63 percent by 2030 (Global Harvest Initiative, 2014). Poultry production for export has been increasing in some of the developing economies (e.g., Thailand and Argentina), while in Africa demand continues to be greater than production capacity, with imports forecasted to increase by 4.8 percent in 2014 (FAO, 2014a). Along with an increase in production, the industry will need to provide food and feed safety (Fink-Gremmels, 2014). Globalization of the feed and food markets will demand a harmonization of feed and food safety standards as well of assurance of food safety throughout the food chain.

In addition to traditional food animal raising, other sources of animal protein are currently utilized worldwide and may play a key role in ensuring protein security in the future. Using raised insect protein as a means of meeting global protein demand has been a topic of recent discussion (Box 4-1).

BOX 4-1
Increasing Insect Protein to Combat Food Insecurity

The FAO received a great deal of publicity after releasing its 2013 report, Edible Insects: Future Prospects for Food and Feed Security. The report describes in detail the environmental and economic benefits of the wider use of insect protein for nutritional purposes. This report establishes the historical role of insects as a protein source for humans, and examines societies that regularly consume insect protein in modern times. Additionally, the FAO provides details on related topics such as the logistics of insect farming (van Huis et al., 2013). It is suggested that insect protein should be further integrated into human diets as well as used for purposes such as animal feed to alleviate food insecurity (van Huis et al.,2013). Despite social aversion to the consumption

> of insects in many Western countries, there are an increasing number of studies examining the benefits and logistics of animal protein consumption. It has been noted that insects contain proteins and fats, as well as nonnegligible quantities of minerals and vitamins (Yi et al., 2013). A chart published by The Economist shortly after the release of the FAO report indicated the environmental and nutritional efficiency benefits of insect consumption, showing that crickets require a lower quantity of feed to produce 1 kg of edible weight than beef, pork, or poultry, and are 80 percent edible, which surpasses other meats (Economist.com, 2013).
>
> The value of insects as a food source differs greatly by culture. In a Ugandan market, it was recorded that purchasing 1 kg of grasshoppers would cost more money than buying 1 kg of beef (van Huis et al., 2013); while in countries such as the United States, the thought of eating insects is a disturbing prospect for many people. As a result, ideas for transforming insect-derived proteins into powders, or making patties constructed by insect protein are options being discussed for Western markets. It is becoming increasingly clear that insects are an efficient and environmentally sound source of protein, leading organizations such as the United Nations to advocate for their adoption.

Aquaculture plays a key role in global animal agriculture and constituted 47 percent of global food fish production in 2010 compared with 9 percent in 1980 (FAO, 2012b). While research is needed to rebuild depleted capture fisheries stocks for many species, aquaculture production will have to continue to grow to meet global demands. Yet, the average annual rate of increase for 2012-2021 is projected to be 2.4 percent compared to 5.8 percent for the previous decade (FAO, 2014d), resulting from freshwater constraints, lack of optimal locations for production, and the increasing costs of fish meal and fish oil for feeds. In 2011, freshwater aquaculture production constituted 70 percent of the total global production (FAO, 2014d). The land-based freshwater constraints may be best addressed by a focus on an increase in marine-based aquaculture production of animal protein (Duarte et al., 2009).

The number of marine and freshwater species that have been successfully cultured is increasing at a much higher rate than that of terrestrial species. The comparatively higher number of species grown (i.e., diversification) results in different arrays of species-dependent resources being allocated, thereby yielding products with a broad range of prices. This diversification aptly succeeds in meeting the different economic levels of the different consumers (Duarte et al., 2009). The diversification also introduces important flexibility in the choice of resource allocation and management practices to meet goals of sustainability through ecosystem management. Troell et al. (2014) discussed the contribution of aquaculture to the resilience to the global food system with impact based on selection of species composition, types of feed inputs, system designs, and corresponding operation. The committee recommends further research into the diversification of species to evaluate what species and/or combination of species feed types and management practices are best for various ecosystems and cultures. In many countries, there is also a lack of understanding and programmatic education relative to the establishment of traceability in the fish supply chain. This factor is very important in achieving consumer confidence. Therefore, educational programs on the importance of traceability in the seafood supply chain needs to be developed and communicated.

Research Priorities

Increased demand for animal protein in developing countries has been a driver to move from extensive to intensive systems. This increased demand is also evident in the aquaculture sector, which constituted 47 percent of global food fish production in 2012 compared to 9 percent in 1980, and is facing land-based freshwater constraints that may be best addressed by a focus on a substantial increase in marine-based production of animal protein. Rapid growth of technological innovations has resulted in opportunities to improve global food security, if the innovations can be introduced and accepted. Research priorities in this area include:
- Research should focus on the identification of socioeconomic, infrastructural, and animal science issues relevant to technology adoption to achieve food security in different nations and world regions.
- Marine based production systems (aquaculture) should take on a more significant focus concerning research funding to meet food security needs in different countries.

4-3 International Trade Considerations

Animal systems occupy 45 percent of the global land area, generate output valued at $1.4 trillion, and account for between 60 and 70 percent of the total global agricultural economy (Thornton et al., 2011; FASS, 2012). These systems employ more than 1.3 billion people globally and directly support the livelihoods of 830 million food-insecure people around the world. In the process, the systems contribute 17 percent of the global kilocalorie consumption and 33 percent of the global protein consumption. The distribution of the global consumption of animal products, however, is skewed to developed countries where consumption is five times higher than in less developed countries. The demand for animal food products in low-income areas of the world is rising rapidly, such as in China, Southeast Asia, and Africa as a result of growth in both population and per capita incomes (Thornton et al., 2011). In addition, the United Nations predicts that 2.5 billion people are expected to be added to urban populations by 2050, with 90 percent of this increase being concentrated in Asia and Africa (United Nations, 2014). As a result, the potential for agricultural production to meet the demands of a growing urbanized population in the developing world is a topic of serious concern.

Nevertheless, that growth is expected to continue to outpace the growth of domestic production in those countries resulting in growing volumes of animal product imports from developed countries. Consumers in those countries will gain from an increasing supply of high-quality, relatively low-priced animal products from imports. Meanwhile, those same imports will represent almost insurmountable competition for many smallholders in developing countries and will negatively impact self-sufficiency, food-security, and employment in those countries. Also, over the longer term, use of precious foreign currency to import animal-source foods, or the feeds with which to produce them, will be in competition with other potential uses of such resources—for minerals, technology, and educational or medical supplies.

Countries with extensive land resources such as the United States, Australia, Brazil, Argentina, and India have a comparative advantage in land-using animal production activities. It is the opinion of the committee that in order to meet the growing world demand for animal products, land resources to produce live animals, along with capital for investment in the necessary technology, systems, and infrastructure will be required to efficiently produce and export those products. Thus, capital-abundant developed countries with extensive land resources have a comparative advantage in the production of animal products over developing countries that may have extensive land resources but lack the necessary capital. At the same time, the demand for nonruminant meat (e.g., poultry and pork) is expected to continue growing faster than the demand for ruminant meat (e.g., beef and lamb). Because nonruminant production requires less land, more intensive production, higher capital resources, and a higher level of technology than the production of grazing ruminants, developed countries will easily dominate world production and export of food animal products in the coming years.

In seafood production, however, developing countries have an apparent comparative advantage over most developed countries. Globally, seafood is the most traded food item, and developing countries play a significant role in exports—61 percent by volume and 54 percent by value in 2012 (FAO, 2014d). U.S. imports of seafood, the largest contributor to the overall U.S. trade deficit, currently exceed $10 billion. The current annual U.S. per capita consumption is approximately 6.4 kg. Of all the seafood consumed in the United States, 91 percent (6.5 million metric tonnes) is imported and approximately one-half is farmed products. The United States produces only about 0.6 percent of the total farmed fish produced globally. The major U.S. seafood imports are shrimp, salmon, and tilapia. Most of the shrimp and tilapia originate from Asia while most of the salmon originates from Chile, Norway, and Canada. For shrimp and tilapia, two of the top seafood items consumed, the United States is highly dependent on production originating from countries in Asia, with China, India, Vietnam, Indonesia, and Bangladesh accounting for 80 percent of the total aquaculture production globally (FAO, 2014d). In fact, a total of 15 countries produce 93 percent of all farmed fish (FAO, 2014d). This unique dependency on imported seafood from a small array of countries in Asia places the United States in a precarious situation in meeting its seafood-based protein demand and increasing per capita consumption. In addition, the major concern of the U.S. seafood consumer is the safety of the seafood imports, particularly those originating from Asian countries.

The growth of trade in animal products is one component in the process of globalization that is revolutionizing world production and trade (Otte et al., 2005). Globalization is pressuring traditional animal systems to modernize through investments in new technology, the adoption of more efficient management systems, and the development of new alliances all along the supply chain. Globalization is forcing global animal systems to search for more efficient ways of producing and delivering products to consumers or risk losing their comparative advantage. Small producers, particularly those in developing countries, often consider such changes to be challenges because their relevance to national supply chains can dissipate over time as large and multinational firms take control of markets if, as is often the case, they lack the capital and knowledge to upgrade their engagement in markets. Despite the competition for domestic animal systems that imports may create, globalization can support producers in importing countries by creating off-farm employment opportunities, particularly for those willing to migrate, regionally, nationally, and even internationally (Otte et al., 2005). Off-farm earnings and remittances can provide rural households with access to the capital necessary to invest as needed to reap at least some benefits of globalizing food animal and meat markets (Otte et al., 2005). At the same time, such infusions of cash can reduce their vulnerability to economic shocks, reduce their relative risk aversion, and promote adoption of new practices and investments related to their food animal enterprises.

The disadvantage of developing countries in the production and export of food animal products is even more pronounced given the global changes in sanitary and phytosanitary (SPS) agreements now in place under the World Trade Organization (WTO). In an effort to protect human, plant, and animal health from transmission of diseases and pests and to establish a mechanism to resolve trade disputes regarding the application of quality and safety standards to animal, plant, and food imports, the Uruguay Round of multilateral trade negotiations established the SPS Agreement in 1995. A major motivation behind that agreement was concern that some countries implement SPS measures primarily as "disguised trade barriers" for more transparent tariffs and other measures to restrict imports proscribed by the WTO agreement (Kogan, 2003; Johnson, 2014). For example, the United States continues to be concerned about the "disguised trade barrier" motivation of the European Union's ban on imports of beef from countries that use growth hormones in cattle production (Johnson, 2014).

To avoid concerns of national sovereignty, the 1995 SPS Agreement allows countries to set the level of SPS protection they deem appropriate, provided that the measures can be justified scientifically and do not unnecessarily impede trade. While the SPS standards are aimed at risk reduction for importing countries, they also impose barriers against exports from developing countries because of the high costs of compliance and a general lack of the resources necessary to exploit the opportunities offered by the Agreement, including the necessary scientific and technical infrastructure (Henson and Loader, 2001). The consequence is a further shifting of the advantage in the production and export of food animals and food animal products to developed countries. According to the authors, developing countries may be increasingly relegated to trade with other developing countries that may accept their SPS standards, which may provide lower levels of disease control and food safety.

Along with increasing integration of world food animal markets, globalization is bringing with it increased scrutiny of the environmental impact of animal production and the treatment of animals around the world. The comparative advantage in food animal production that extensive pastureland or large areas open to conversion pasture (e.g., deforestation) give many countries will likely generate continued extensive animal production in those areas to accommodate growth. The sharp increase in food animal feed prices in recent years is expected to continue adding to the problem by reducing the profitability of intensive production. For those areas, the challenge for the future is to create the necessary incentives and sustainable production system options that will maintain profitability of animal production while lowering its environmental footprint. Without such incentives and profitable production system alternatives, large-scale movements toward intensification and diversification of animal production, particularly in developing countries, is not likely to happen (FAO, 2012a).

The global spread of animal welfare concerns raises similar issues for food animal production, again particularly for developing countries (Box 4-2). The pressures to impose standards on methods of intensive food animal production in developed countries to protect animal welfare are increasingly focused on developing countries. The costs of compliance and the lack of the necessary science and technical assistance to adapt and implement enhanced animal welfare measures in developing countries can serve to put producers, and particularly smaller, less well-capitalized producers, at a global disadvantage in food animal production and in the possibility of more rapid movement toward production intensification. However, they can also provide producers in developing countries with increased global marketing opportunities. For example, Namibia has increased its export market for beef by producing according to EU animal welfare standards and preferences (Bowles et al., 2005).

> **BOX 4-2**
> **Animal Welfare in Developing Countries**
>
> Animal welfare considerations are becoming increasingly important for the global trade in animal products. Increasingly, developing country producers are asked to meet animal welfare standards in order to supply to multinational retailers and distributors, or to meet EU animal welfare standards for export. This creates both market opportunities and challenges in developing countries (Bowles et al., 2005). Many standards (e.g., the standards produced by the World Animal Health Organisation, OIE) are focused on improving animal housing, husbandry, transport, and slaughter across a range of production systems, from intensive to extensive. Following these kinds of standards can benefit producers by increasing production efficiency, for example, via improving animal health and product quality (FAO, 2009). In the least-developed countries in the world, for example, in Southeast Asia and sub-Saharan Africa, animal welfare concerns differ in character and magnitude from those in the more developed countries. They include providing for basic health and care needs of animals to reduce malnourishment and high rates of mortality due to disease, improving handling to reduce injury, facilitating the adoption of technologies that can improve the humaneness of culling and slaughter, and developing/using genetic stocks that are better suited to local/regional disease and environmental challenges. The FAO (2009) recently issued a report identifying the key capacity-building needs for implementing good animal welfare practices in developing countries. These include locally relevant education about the significance of animal welfare for successful animal production, training in skills and behavior related to improving animal welfare, and development of strategic partnerships to foster training and implementation strategies.

Research Priorities

Smallholder livestock, poultry, and aquaculture producers in developing countries are often stymied by lack of access to supply chains and commercial markets. Unfortunately, there is limited information on the full extent of barriers to exports of animal products from developing countries (e.g., WTO SPS regulations)). Globally, fish is the most traded food item, and developing countries play a significant role in exports, 61 percent by volume and 54 percent by value in 2012. Of all the seafood consumed in the United States, for example, 91 percent (6.5 million tonnes) is imported and approximately one-half is farmed products. The United States produces only about 0.6 percent of the total farmed fish produced globally, and increasing per capita consumption may not be met because of reduced exports combined with lack of sufficient access to markets in developing countries. Research priorities in this area include:

- Research is needed to identify and overcome barriers to enable animal producers in developing countries, including smallholders, to more effectively participate in supply chains and commercial markets These investigations should include assessment of the economic and sustainability impacts of barriers to exports from developing countries, and the implications of imports for animal systems into developing countries to guide investments in animal science research, extension, and education.
- Research should be initiated to determine the most effective approaches for the developing countries to comply with SPS and other international regulations, including investments in the necessary scientific and technical infrastructure to allow compliance.
- Policy should be developed to provide sufficient funding for research and related extension programs to stimulate a more intensive and sustainable aquaculture, including mariculture, enterprise globally. International agencies in cooperation with high-income countries should be encouraged to support policy and research approaches to disseminate technology and best practices.

4-4 Smallholder Animal Agriculture

Globally, more than half of all animal commodities and most animal-source foods are produced by the estimated 70 percent of the world's rural poor whose livelihoods depend on animal production (MacMillan, 2014). Thus, future global food security and our ability to meet the future demand for animal protein are dependent to a large extent on smallholder animal production systems. A growing number of rural households in developing countries actually participate in commercial market activities at some level, even though household food production remains their primary goal (Otte et al., 2012; Kelly et al., 2014). For these smallholders and pastoralists, animal production represents an opportunity to earn revenue to supplement subsistence needs and pay for production inputs. The more access these producers have to markets, usually proximity to urban areas in the region, the more opportunities there are for them to take advantage of the ongoing growth of the demand for animal products. In these areas, smallholders may benefit either directly through contract production or by supplementing the supplies of urban food wholesalers and retailers.

In more remote areas where the conditions and infrastructure are not yet suitable for large-scale commercialization of animal production, smallholders and vulnerable populations may benefit from the spillover effects of urban growth, but are more likely to service the needs of local economies for meat, eggs, and milk. In these areas, investments in infrastructure, the extension of training, and the delivery of new technology such as improved genetic material, more efficient production management systems, animal health services, and other modern inputs are likely to generate large social returns (Sanchez et al., 2007; McDermott et al., 2010; Anthony, 2014). Accordingly, some smallholders and pastoralists are enabled to participate to some degree in the benefits of the rapidly growing markets for animal products in their countries (FAO, 2008b).

However, many more smallholders are being marginalized and excluded from the benefits of growing world demand for animal protein. Small-scale livestock producers, the basis for much livestock production worldwide, are often negatively affected by the growth of commercial production of livestock. Attention to this process and the extension services or subsidies that would assist these producers during this transition of production is warranted. A host of factors combine to constrain the integration of smallholders into increasingly globalized live animal and animal product markets, including the diverse roles that animal production plays in developing countries, the risk-averse behavior of low-income producers, and the importance of small stock and aquaculture production technology (Otte et al., 2005; Steinfeld et al., 2006; McDermott et al., 2010).

Introduction and transmission of infectious diseases is tremendously varied by the husbandry type and production methods. Perry and Grace (2009) consider in detail the impacts that food animal diseases, and control of these, can have on rural poverty in poor developing countries. Low-intensity subsistence food animal farming that operates throughout the developing country's rural households, particularly of poultry, sheep, and goats, is critical to sustaining local food supplies. Animals in this type of farming are often kept under marginal conditions with little access to modern disease control, housing, or feed supplementation.

These animals may suffer from uncontrolled diseases and may be in close contact with other species and humans, and potentially in contact with a variety of nondomestic animals. Therefore, the impact of diseases, both endemic and epidemic, on the livelihoods of these stockers can be severe, particularly if there is high mortality or the imposition of animal movement restrictions. Intensive poultry production systems where meat birds are raised in large flocks at high stocking densities have the most efficient feed-to-gain ratios of any animal production system and thus can provide cheap, high-quality animal protein for people in developing countries. Such intensive production systems, however, can only be implemented successfully if the many infectious diseases that would otherwise inflict severe losses can be prevented or controlled (Tomley and Shirley, 2009).

4-4.1 Diverse Economic Roles of Animal Production Livestock, Poultry, and Aquaculture in Smallholder Operations

Animal production plays a critical and well-documented role in contributing to the economic welfare of poor families in rural areas of developing countries (Upton, 2004; Randolph et al., 2007; Otte et al., 2012). Animal production provides meat, milk, eggs, and other products that enhance both the quantity and nutritional quality of the food consumed by poor rural families. Excess animal products can be sold to allow the purchase of additional food and other basic necessities as well as production inputs. Farm animals provide draught power for food production and manure for use as fertilizer or fuel. They also serve as a form of saving, adding to wealth and status during good years and acting as a buffer against lean years. In addition, animal production contributes to the overall economic health and viability of rural communities. The sale of animal products in local markets supports the growth of agricultural support services and related rural enterprises.

In most developing countries, the groups most dependent on animal production for at least a part of their livelihoods include smallholder farmers, subsistence farmers and landless farmers. Thus, the expected future growth in global animal product demand might well be expected to improve the lives of many of the rural poor around the world. Production increases in aquaculture, presumably achieved through systems that are more intensive than existing ones, should contribute to noteworthy increases in employment both on and off farms and contribute to rural development and alleviation of poverty (Allison, 2011). Off-farm employment would include occupations such as feed preparation, production, and processing. The FAO (2014a) estimates that aquaculture currently supports 19 million farm jobs, 96 percent of which are in Asia. Also, with the capture fishery at a maximum sustainable yield and particular fisheries already overfished, culture fisheries have the potential to become an attractive alternative for those who have lost their jobs in the capture fishery. Women should dually benefit from expansion by increases in employment that will lead to a higher income (Williams et al., 2012). The increase in economic status should allow a greater opportunity to purchase fish, a nutritionally high quality of food for themselves and their children.

The most affordable protein food, particularly for those people living in developing countries, comes from inland freshwater fish farming (FAO, 2012b). This protein source is expected to take on the highest importance in the assurance of long-term food security and good nutrition for growing populations in developing countries. The focus should be on aquatic organisms (e.g., fish and crustaceans) at the lower levels of the food chain (e.g., omnivores and herbivores) to ensure a low-cost product. Aquaculture has become an important facet of global food security and is a major source of protein for 17 percent of the world's population and almost 25 percent of the populations in developing countries (Allison, 2011). Overall, fish consumption provides the current 7.2 billion global inhabitants at least 15 percent of their animal protein intake (FAO, 2012b).

The extent to which the expected growth in world animal product production and demand will contribute to the alleviation of poverty and the strengthening of smallholder and family farming in developing countries depends on multiple factors. Many smallholders who depend on animal production as a mainstay of their livelihood are not engaged in commercial markets and instead are focused on survival. They rely on family labor, including children, in essential animal production activities such as herding (FAO, 2013). In addition, many small-scale producers notably operate on very small profit margins. Therefore, obtaining sufficient capital to make improvements in response to the need to meet environmental standards for sustainability is problematic.

A particular barrier to the advancement of animal production in many low-income areas of the world is the exclusion of women from publicly funded training and education opportunities. Women comprise roughly two-thirds of the 400 million poor food animal keepers in the world who rely mainly on animal production for their income (FAO, 2011b; Köhler-Rollefson, 2012). Given the low level of investment in the human capital of women and their key role in many animal production activities in many developing countries, the return to investing in research and information delivery systems focused on women in terms of increases in animal production and the provision of animal products to meet the growing demand for protein is likely quite high (Christoplos, 2010; Meinzen-Dick et al., 2011; Jafry and Sulaiman, 2013; Ragasa et al., 2013). To achieve such a goal, extension agents, in concert with researchers and policy makers, must be trained in gender issues, particularly in the need for gender equality and equal pay for equal work. The vocational training of women is a critical component for meeting the needed increase of animal agricultural production in developing countries. These opportunities would translate to a higher economic status of women in these countries.

For all these reasons, the rapid adoption of new animal production technologies as appropriate, the development of more efficient production systems, the growth of market demand, and related changes that will continue to transform the animal industries in many countries must be developed with an eye toward improving the lives of the many subsistence smallholder animal producers in developing countries. In certain circumstances, those producers may face critical barriers to participating in the potential benefits of a growing animal industry, including rigid social institutions; fragmented markets; and the lack of access to technology, credit, resources, markets, information, and training. However, there is evidence that dairy technologies introduced in coastal Kenya as well as in other eastern African regions can benefit smallholder farmers (Nicholson et al., 1999; CNRIT, 2000).

For these and other reasons, such as climate change, the rapid adoption of new animal production technologies, the development of more efficient production systems, the growth of markets, and related changes that will continue to transform the animal industries in many countries will likely generate winners and losers among small-scale producers (Delgado et al., 2001; Aksoy and Beghin, 2005; Thomas and Twyman, 2005; McMichael et al., 2007). Marginal small-scale food animal producers commonly face critical barriers to participating in the potential benefits of a growing animal industry, including inadequate access to technology, credit, resources, markets, information, and training, and in some cases, maintain values and norms resistant to exogenous forces of change (Ferguson, 1985; Hydén, 2014).

4-4.2 Risk-Averse Behavior of Producers in Developing Countries

A critically important factor affecting the productive performance of animal systems in developing countries is the risky and uncertain nature of subsistence farming and the necessarily risk-averse profile of subsistence producers. In most developing countries across Africa and Asia and elsewhere in the world, agricultural production units are small, and crop and food animal production are dependent on the uncertainties of highly variable rainfall. Average production and yields are typically low. In poor years, producers and their families face the very real prospect of starvation. Consequently, rather than profit maximization being the motivating force in their production- and technology-use decisions as in developed countries, producers in developing countries maximize the family's chances for survival. Given the frequently large risks and uncertainties they face, smallholders in animal production systems are often reluctant to shift from using traditional breeds and production technologies that produce more stable and reliable outcomes to new breeds and production technologies that promise higher yields but entail greater risk of production failure. When survival is at stake, producers will tend to avoid bad years rather than maximize output in good years (Miracle, 1968; Todaro and Smith, 2014). That is, they will choose low levels of technology that combine low mean yields with low variance to alternative technologies that promise higher yields but also pose the risk of greater production variance.

Many programs intended to enhance production and productivity in developing countries end in failure, precisely because they failed to recognize the risk-averse nature of subsistence farmers (Todaro and Smith, 2014). Smallholders in developing countries are not ignorant or irrational when they fail to adopt new technologies that such programs offer them. Rather they are making rational decisions to maximize family survival within the constraints of the institutional, financial, physical, cultural, social, and policy environment within which they live. Consequently, research to identify and mitigate the constraints to new technology adoption faced by small-scale producers in developing countries must go hand in hand with research to develop new production technologies if the production research investments are to realize a reasonable return. Personal food animals can play an important financial role in the developing world (Box 4-3).

BOX 4-3
Livestock as Personal Financial Capital in Low-Income Nations

Livestock held by individuals and families in low-income nations can function in a wide variety of important purposes. In a traditional view, the value of livestock for its holder is through the animal products it can produce, the physical power it can supply, its market value for sale, and/or its use for breeding. It has been documented that investing in livestock is of financial importance for the rural poor, because it allows for the creation of cash flow through the sales of animal products, such as milk or wool, which can cover necessary household expenses, and adds financial and food security in areas where limited or no formal financial services are established (IFAD, 2001). Livestock ownership is often perceived as a sign of wealth and social status in low-income nations, serving as an indicator for one's influence and respect (Randolph et al., 2007). In this role, livestock can serve as social capital for the owner, increasing the owner's agency and social standing. This influence plays a role in efforts that encourage livestock ownership by women, in addition to livestock often being an easier asset for women to acquire and possess in many societies than land or other forms of wealth (Herrero et al., 2013). It is worth noting, however, that despite the views of social status surrounding livestock ownership discussed, in some situations this practice can indicate poverty, since the owners may be holding livestock because of a lack of access to high-quality, arable land.

In the strictly financial sense, livestock take on incredible importance for rural people in developing nations. Livestock serve as assets that are unaffected by inflation, which plagues the currency in many of these countries, as a way for storing monetary value outside of a banking system, and a means for consumption smoothing because the animal can be sold for its value at any time (Randolph et al., 2007). In nations without a developed financial infrastructure, these options can prove to be immensely valuable and present a clear benefit to holding livestock. There have been charitable efforts to increase the prevalence of personal livestock ownership in low-income nations. Heifer International, one of the most prominent of these charities, facilitates the giving of livestock to recipients in the developing world to promote financial empowerment and greater self-sufficiency. Livestock ownership can serve as an effective form of financial capital in developing countries with limited financial infrastructure.

4-4.3 Importance of Small Stock and Aquaculture Production Technology

Investments in new technology development related to small stock such as poultry, pigs, and goats may be more important than such investments in cattle in developing areas of the world. Otte et al. (2012) argue that smallholders are more likely to raise small stock than cattle for various reasons, including the lower capital investment required and their higher efficiency in meat production; however, the production of poultry and pigs is particularly amenable to large-scale, vertically integrated operations. Not surprisingly, much of the growth in both poultry and hogs in developing countries over the last decade has been as the result of efficiencies gained from increased scale of production and vertical integration, the benefits of which have not spread much beyond a relatively few number of enterprises in those countries.

Small-scale operators are also responsible for most aquaculture production in both Asia and Africa, primarily from inland pond culture. More efficient and productive systems need to be established in those countries to meet the needs of protein and other essential nutrients. In an evolution toward the goal of efficient intensification, the value of indigenous management strategies must be respected and maintained when deemed highly effective. Improvement of these systems will contribute to the reduction in poverty and food (nutrition) security. There is a lack of any noteworthy effort to create development projects devoted to and designed for small-scale aquaculture operators. Such efforts can only come to fruition through the encouragement and support of governments (Box 4-4). A possible design for study is a satellite production model whereby a corporate entity provides guidance and information to farmers and purchases outputs from those farmers.

Small-scale (e.g., village or family) poultry production plays an important role in providing for the protein needs of poor rural households in many developing countries. In many of these developing countries, rural poultry accounts for about 80 percent of poultry stocks (Akinola and Essien, 2011; Ngeno, 2014). Village poultry can also be a source of income for poor families, particularly for women (FAO, 2014a), since village chickens are usually managed by women and children. Although the meat and egg output of village chickens is lower than that of intensively raised chickens, inputs are also low. Research that will improve outputs in these systems, or that fosters transitions to more efficient semi-intensive or small-scale intensive systems (FAO, 2014a), is important for improving food security in these countries. Sheldon

(2000) outlines critical research needs for improving small-scale extensive production: (1) effective disease control, for example, by developing efficient, thermostable, low-cost vaccines; (2) low-cost feed supplements based on local ingredients not needed for human nutrition; (3) improving socioeconomic factors that will promote successful development of small-scale poultry businesses; and (4) choice of the best poultry genotypes for use in specific environments, which will require more sustained efforts toward establishing breeding programs that conserve the genetics of locally adapted indigenous breeds (Ngeno, 2014). Research attention also needs to be directed toward adequate flock management under local conditions (Ianotti et al., 2014). Ianotti et al. (2104) estimate that vaccination for Newcastle disease, which is the primary cause of mortality in village flocks, along with modest husbandry improvements, could lead to sevenfold increases in egg production at the household level. The FAO (2014a) emphasizes that technological innovations developed to foster improvements in family poultry production will be most successful if accompanied by hands-on training and capacity building via formation of producer groups.

BOX 4-4
Livestock Policies

Policies around the world that affect livestock sectors can be classified into three groups: (1) price policies, (2) institutional policies, and (3) technical change policies (Upton, 2004). Price policies include trade policies such as import tariffs and export subsidies, exchange-rate policies and domestic price policies such as price supports and consumption subsidies. Although price policies benefit certain groups within countries, such as livestock producers or low-income consumers, they generally have a negative impact on net national welfare. Consequently, since the late 1940s, trading countries have engaged in multiple rounds of negotiations to eliminate price policies and their distortive effects on the world economy. Upton (2004) argues that even though price policies have generally failed to achieve positive gains, there is an argument for limited price policies in certain cases such as subsidies for disaster relief and for promoting the use of new inputs such as vaccines or drugs. Nevertheless, price policies distort the incentive structure in markets and can lead to inefficient outcomes. For example, in the former Soviet republics, the subsidization of livestock production led to a dependence on imported livestock feed and little incentive

> for sustainable grazing or quality forage conservation (Upton, 2004). With the dissolution of the Soviet Union and the loss of subsidies to support livestock production, feed imports ceased, consumer demand for meat declined, and the livestock sector largely collapsed. Consumers turned to imports of meat and milk from Europe and the United States.
>
> Institutional policies are generally considered to be critical for economic development (Upton, 2004). Such policies include those intended to enhance physical and institutional infrastructure. Institutional policies also include the provision of production credit, animal health services, and genetic materials. Policies to promote technological change include support for research and development, the dissemination of information, and the extension of production training to producers. To enhance growth of the livestock sectors in developing countries, Upton (2004) emphasizes the importance not only of research to enhance productivity but also of research on socioeconomic issues related to the institutional constraints faced by producers as well as issues related to livestock processing, marketing, and trade. This will lead to the development of policies of all three types: price, institutional, and technical—to meet the needs of diverse populations worldwide.

Research Priorities

Many, if not most, smallholder animal producers in developing countries worry more on a day-to-day basis about survival than about increasing productivity and profitability of their animal products. This condition may be a critical barrier to sustainable intensification of animal agriculture in developing countries. Small-scale operators are responsible for most aquaculture production in both Asia and Africa, for example, primarily from inland pond culture. However, many aquaculture producers in developing countries are still plagued by poverty and food insecurity. Small-scale poultry production is also important for food security in many developing countries, but is typically low-output due to problems with disease and inadequate husbandry. Technology adoption to improve animal production may be slowed if animal producers do not see an immediate direct benefit in terms of survivability, as well as other factors that are not currently well understood. Additionally, economic constraints may prohibit technological adoption.

Research priorities in this area include:

- Research effort needs to be directed toward the development of technologies that are locally relevant by requiring minimal risk in adoption and augmenting livestock use as a source of wealth and a means of survival during lean times. Donors to developing countries should incorporate community welfare in their considerations of animal science research.
- Research to identify and mitigate the constraints to new technology adoption, such as the lack of access to credit, production resources, markets, information, and training faced by small-scale producers in developing countries, is critical to boost the rates of adoption of new production technologies. The investigation of linkages with supply chains and markets is particularly important for aquaculture.
- Research that examines the economics of sustainable animal production is necessary to determine the optimal strategies for integrating smallholders into global animal product supply chains while mitigating negative associated environmental, social, and other impacts. Higher education and research institutions in the animal science arena should focus on creating a pool of experts capable of adapting science and technology to the local context and promoting local adoption in producing food. In particular, this means changing the current paradigm to include, in addition to teaching and research, the third mission of service to the community and close cooperation with the public and private sectors to contribute to innovation and development. Thus, the land-grant university concept of integrated research, teaching, and outreach functions should be adopted as a model, with adaptation to local circumstances.

4-5 Policy and Certification Systems

Governments impose a wide range of general economic and sector-specific policies to achieve economic and social objectives that critically affect decision making by animal producers. Such interventions include price polices intended to change production and/or consumer behavior, institutional policies intended to affect physical and institutional infrastructure, and technical change policies intended to affect the efficiency of production through research and development and the dissemination of information and training to animal producers (Upton, 2004). In the animal production sectors of most countries, however, policy measures have focused primarily on technical issues related to animal production, animal feeding, and disease control (Otte et al., 2012).

The development and implementation of these policies, however, have failed to consider the economic and institutional constraints facing animal producers in developing countries, such as poor road networks and related infrastructure, limited information about animal diseases, and poor access to health services and production credit (Otte et al., 2012). Animal production sector policies and programs often have been designed by technical staff in food animal departments and NGOs or international organizations who have limited appreciation or understanding of the broader set of policies, markets, and institutional constraints that are relevant for farm-level decision making (Otte et al., 2012). A policy agenda for promoting equitable and efficient growth of the animal production sectors in developing countries that addresses the specific constraints faced particularly by smallholder animal producers proposed by various authors (Dorward et al., 2004a,b; Pica-Ciamarra, 2005; Otte et al., 2012) consists of three major components of animal production:

- *Management policies,* including measures to enhance access to basic production inputs and to help producers cope with risk, natural disasters, and other market shocks;
- *Enhancement policies,* including government facilitation of producer access to animal health services, credit, information, and output markets; and
- *Sustainability policies,* including research, environment-related, and other public measures required to foster the sustainability and competitiveness of animal production over time.

Exactly what animal sector policies should be implemented in a given country is not clear. As Otte et al. (2012) bemoan, there are "no hard and fast rules" for what animal sector policy-making institutions need to do to achieve a given general objective. There are many ways of achieving an objective and many circumstances that determine what might be the optimum policy or policy mix. Often, government authorities opt to implement those policies that are technically feasible, affordable, and politically acceptable (Otte et al., 2012). This second-best approach to policy making at least gets something done, but may not ultimately contribute significantly to development of the animal production sector and may even be welfare reducing because of all the other constraints that producers continue to face (Rodrik, 2007).

The Code of Conduct for Responsible Fisheries, adopted almost 40 years ago, remains the principal directive for achieving global sustainable fisheries and aquaculture (FAO, 2012b). Within the code are guidelines for action based on 4 international plans, 2 strategies, and 28 technical guidelines. All of these guidelines are founded on the ecosystem sensitivity approach. Most countries have established policies and legislation that align with the Code. The idea of responsible fisheries is founded in an array of considerations that include biological, technical, economic, social, environmental, and commercial. For example, in Norway, the Aquaculture Act of 2005 was enacted to regulate the development, expansion, and management of aquaculture for inland waters, marine, and land-based aquaculture. This act promotes "competitiveness and profitability within the context of sustainability and environmental stewardship (FAO, 2010). The governments of developing countries must develop policy with the same objectives in mind.

Over the years, no land-based protein production system has been subjected to such a high level of scrutiny about sustainability as that confronting aquaculture. Being recognized as the fastest growing food production sector in the world, this high degree of oversight is a natural byproduct of its dramatic rise to become a significant contributor to global food security (Tidwell and Allan, 2001). The valid concern about the effect of aquaculture on natural and societal resources has led to third-party certification systems. These systems do not have common standards, but rather focus on a set of sustainability indices that "market" particular NGO interests. In addition, there is high competition among these systems, with success and perceived value being the basis for generating continued financial support (i.e., donations). Nonetheless, these certification systems have served the industry well in marketing to retailers and educating consumers about seafood products from both capture and culture fisheries that heed environmental and socially responsible standards. FAO (2011a) has published technical guidelines for certification of aquaculture production that can be used as a foundation for the development of certification programs by third parties. These guidelines are a natural follow-up to FAO's 1995 Code of Conduct for Responsible Fisheries.

The best certification efforts to realize a positive global impact for sustainable aquaculture production are founded in diverse stakeholder input leading to the adoption of common international standards. In 2006, the World Wildlife Fund introduced a global-based initiative called

Aquaculture Dialogues which consisted of 2,200 multistakeholder participants with the goal of establishing environmental and social standards for nine species of farmed seafood. Aquaculture Dialogues was successful in establishing important standards and also identified the existing problems and their detrimental effects. This substantial effort produced noteworthy results that were provided to the Aquaculture Stewardship Council (ASC), which currently uses these established standards to oversee the certification of management practices of salmon farms. These standards comply with the rigorous guidelines that have been established by the ISEAL (International Social and Environmental Accreditation and Labelling) Alliance.

The standards established by the ASC, in turn, served as a stimulus for the establishment of the Global Salmon Initiative (GSI). The GSI brought together CEOs whose companies represented 70 percent of the global production of salmon and ultimately led to establishment of a salmon farming certification that adhered to the standards established by the ASC. It is hoped that other species-specific industries will mimic these efforts of establishing certification systems whereby retailers and consumers will have the information to make judicious and socially responsible decisions about the purchase of seafood. The importance of appropriate policy measures for achieving growth and development in the animal production sectors of developing countries was emphasized by a recent FAO (2008a) report:

> In the 1990s, an increasing number of development aid
> experts and analysts came to realize that technology
> transfer alone was not going to transform development,
> especially agricultural development, in ways that would
> necessarily be beneficial to the poor. Policy and
> institutional change was identified as a pre-requisite to
> steer agricultural development towards meeting the
> needs of the poor.

By influencing the decisions of producers and consumers, policies and institutions are key drivers of economic growth and development, including in the animal production sector (Otte et al., 2012).

Although the objectives of animal science research are not necessarily to produce policy outcomes, such research can impact and is impacted by policy decisions. For example, research to develop an animal disease vaccine could lead to significant growth in animal production in rural areas where the disease has been endemic. In areas that lack adequate roads and related infrastructure and marketing systems, the increased animal product output will end up in local markets, leading to lower prices and leaving producers potentially worse off than before. In addition, overgrazing and other negative environmental and social consequences may result. Over time, however, the negative pressure on rural incomes, survivability, and sustainability could induce a change in policy to enhance the infrastructure and marketing systems in those areas to allow greater access of rural producers to commercial markets. In turn, the greater access to markets could induce research to develop animal breeds that better meet the needs of commercial markets in urban areas. The bottom line is that for animal science research to have the expected impacts on animal production and on the livelihoods of animal producers, particularly in developing countries, animal science research projects and agendas must recognize and be responsive to the policy environment in which the producers operate. At the same time, research on policies and policy alternatives related to animal production must be incorporated into animal science research projects, particularly those focused on developing-country issues.

4-5.1 Certification and Technology Development and Transfer

Technology has been and will continue to be important in meeting the increasing demands for producing safe, affordable food in an environmentally sustainable manner that is socially accepted. Adopted technological advancements including health, genetics, breeding, nutrition, physiology, management, and food and feed safety in animal agriculture have resulted in improved production, efficiency, and environmental and water footprints for food animals (see Chapter 3). Reproductive technologies (Hernandez Gifford and Gifford, 2013) and technologies to enhance global animal protein production (Lusk, 2013) have contributed to food security and sustainability in developing countries. Technologies used to reduce environmental impact of animal wastes associated with feeding for maximum productivity (Carter and Kim, 2013) and to reduce greenhouse gases (GHGs) through manure management (Montes et al., 2013) are being evaluated.

Many of these advancements that have been adopted in developed countries have had limited success in developing countries. For example, animal breeding has made significant progress in developed countries with nutrition and management developments helping those animals to express their genetic potential. However, these genetically elite animals from developed countries have not performed or even survived when put into developing countries' environment with existing feed, water, disease, and management conditions. Animal breeding and genetic progress should be conducted utilizing in-country indigenous breeds, especially for pastoralists and smallholder farmers. Sustainable intensive operations may successfully utilize breeds and genetic advancements that have global utilization, such as for pigs and poultry.

Typically, developing countries, especially in sub-Saharan Africa, have underinvested in animal science research, infrastructure, and technology. As a result, there has not been much progress in animal health, productivity, and efficiency. For example, mean aquaculture productivity in Africa in 2005, expressed as tons of fish per worker, was 5.5. In contrast, North Africa averaged 8.8 tons per worker compared to 0.5 ton per worker in sub-Saharan Africa (Valderrama et al., 2010). These areas have a high need for aquaculture to meet protein needs but low productivity. The ability to innovate, test, adapt, and adopt technologies and innovations in these countries remains marginal; however, some countries are now realizing the benefits and are investing in the assessment of these technologies. For example, Asia is exploring new technologies to improve yields and product quality, such as the use of feed enzymes, transgenic crops, and breeding of transgenic animals (Cao and Li, 2013). Latin America's beef industry has evolved to incorporate modern technology (Millen and Arrigoni, 2013). Genomic selection for beef cattle breeding in Latin America has made some progress but has had its challenges (Montaldo et al., 2012). Uruguay's sheep production system is a nice case study of how the adoption and intensification of technology results in a more profitable and environmentally friendly sheep production system (Montossi et al., 2013). Rege (2009) provides a discussion of the available biotechnologies with potential application in food animal improvement and identifies those that have been or may be applied in developing countries with special attention to sub-Saharan Africa (Table 4-2). There is a need to build biotechnology capacity in developing countries; however, many of these developing countries especially in sub-Saharan Africa are heavily influenced by the European Union.

TABLE 4-2 Possible Applications of Biotechnology to the Solution of Problems of Food Animal Production in Developing Countries

Problem	Possible Solution1	Scale of Economic Impact	Probable Time to Commercial Use (years)
Animal diseases	New vaccines and new diagnostics	Large	<5
Poorquality forages	Microbial treatment of forages	Moderate	5-10
	Modification of rumen microflora	Moderate/Large	>10
	Genetic improvement of forages and their symbionts	Moderate	5-10
Difficulty of implementing selection programs	Selection of nucleus herds, using AI, ET, embryo sexing	Large	5-10
	Marker-assisted selection	Moderate	5-10
Difficulty of maintaining dairy cattle performance after F1 cross	Use of IVF, ET, and embryo sexing	Large Medium	>10
	Selection among local breeds using AI, MOET	Large	>10
	Development of composite breeds	Large	>10
Cost and environmental challenge to imported cattle	Use of ET to import embryos	Small	<5
Need for increased efficiency in intensive systems	Use of rbST and rpST in dairy and pig production	Large	<5

NOTE: AI, artificial insemination; ET, embryonic transfer; IVF, in vitro fertilization; rbST, recombinant bovine somatotropin; rpST, recombinant porcine somatotropin; MOET, multiple ovulation embryo transfer.
SOURCE: Rege (2009).

In Asia, the development of animal production largely depends on research in animal genetics, nutrition, and feed science. Cao and Li (2013) suggest the following four primary areas be investigated: (1) improved nutrient utilization, (2) enhanced carcass and other food animal product quality via optimized nutrition, (3) reducing excrement waste, and (4) use of biotechnology for improving efficiency of feed and food animal production to increase food security.

Poultry production is also a global industry providing a major source of meat in both developed and developing countries. It faces competition for feed ingredients from other animal industries such as pork and aquaculture and now biofuel. With this increased pressure, the need arises to seek out alternative feed ingredients, increase the digestibility of existing ingredients, and research ways to better utilize the fiber component of feed ingredients (D'Souza et al., 2007). Chadd (2007) discusses the future trends and developments that are needed in poultry nutrition including feed industry, feed manufacturing technology, country focus, feed intake predictability, nutrient relationships, genotype by nutrient interactions, nutrient support of immunocompetence, feed diversity and characterization, pro-nutritional factors, and redefining the systems approach. Besbes et al. (2007) provide a look at the trends for poultry genetic resources on a global basis, including regional distributions of avian breeds, attempted breeding programs for indigenous poultry, development and trends in organized poultry breeding, indigenous and commercial line selection criteria, increased demand for poultry products, increased threat of disease epidemics, environmental issues and climate, increased competition for feed resources, erosion of poultry genetic resources, and developments based on new biotechnologies. The following gaps were highlighted: (1) data on production systems, phenotypes, and molecular markers should be used in an integrated approach to characterization; (2) a comprehensive description of production environments is needed in order to better understand comparative adaptive fitness; (3) field and on-station phenotypic characterization is needed; (4) to facilitate the search for genetic variants, characterization of specific traits of local populations is needed; (5) use of a reference set of microsatellite markers is recommended; and (6) within country, or even within region, all genotyping should be concentrated in a common reference laboratory (Besbes et al., 2007).

Poultry production and the environment are reviewed by Gerber et al. (2008). The paper analyzes the global environmental impacts arising from intensive poultry production and provides technical mitigation options involving farm management, animal waste management, nutrition management, and feed production. One must look beyond the farm level to fully understand the poultry industry's impact on the environment. Researchers worldwide have suggested that the following technologies be researched: (1) use of exogenous enzymes to improve animal productivity (Meale et al., 2014); (2) fundamental research on the biology of birds, microbial genetics, and genetic diversity (Fulton, 2012); (3) mining of genomes and merging of genomic and quantitative approaches (Hocquette et al., 2007; Green, 2009); (4) influences of human technical, societal innovations and environments on each other (Garnett and Godfrey, 2012); (5) conversion of co-proteins into animal protein (Zijlstra and Beltranena, 2013); (6) animal models (Lantier, 2014); (7) major targets for food animal production (Hume et al., 2011); (8) current drivers and future directions for global animal disease dynamics (World Bank, 2012; Perry et al., 2013); (9) interrelatedness of environmental, biological, economic, and social dimensions of zoonotic pathogen emergence (Jones et al., 2013); (10) food animal production including recent trends and future prospects; (11) food animal science and technology as a driver of change (Thornton, 2010); and (12) major gaps in understanding of GHG emissions of seafood products (Parker, 2012).

Animal science research should be focused in world regions with low feed efficiencies and high emission intensities, such as sub-Saharan Africa and parts of South Asia and Latin America with the objective to (1) improve the efficiency of food animal production through improved feeding and management; (2) increase the sustainability of agriculture and enhance food security by researching the shifts between production systems; and (3) conduct research on which food animal should be consumed, how much is consumed, and the production system in which it is raised (Herrero et al., 2013a). Research is needed to evaluate different food animal production systems, different use of resources, tradeoffs, and land-use change (Herrero et al., 2009) and to develop sustainable intensification methods that improve efficiency gains to produce more food without using more land, water, and other inputs (Herrero et al., 2010).

On a global basis, FABRE TP (2006) advocated for a research agenda focused on the genetics and genomics of farmed species, quantitative genetics, data collection and management, operational genetics, breeding program design, numerical biology, genetics of relevant traits, and biology of complex biological systems. Research on reproduction that underpins breeding and the effective dissemination of genetic improvement would benefits all producers. Ideally, it would build genome-to-phenotype predictive models. All world regions need to maintain a continued focus on reproductive biology and the responsible use of biotechnologies.

4-5.2 Technology Improvements

The committee is aware of several examples of animal technologies that have increased production, ranging from poultry production technologies that have dramatically increased production in virtually every region of the world to dairy production that has dramatically increased milk yield per cow. Madan (2005) notes that "the major technologies that are used effectively in livestock production in the developing world include conserving animal genetic resources, augmenting reproduction, embryo transfer and related technologies, diagnosing disease and controlling and improving nutrient availability."

In 2008, the National Research Council published a report that addressed technologies for improving animal health and production in sub-Saharan Africa and South Asia (NRC, 2008). Areas highlighted included reducing preweaning mortality; improving grass and legume forage; using existing and evolving technologies for improving animal germplasm; leapfrogging selective breeding with molecular sampling, including DNA-derived pedigrees; genetically engineering disease resistance in animals; using RNA interference for virus control; using germ cell distribution and spermatogonial stem cell transplantation; and improving health through neonatal passive immunity, animal vaccine development, and animal disease surveillance. This committee supports these previous NRC recommendations.

The Bill & Melinda Gates Foundation chose animal health and genetics to focus on as the greatest opportunities to increase smallholder productivity. Genetics was valued at $83 million, animal health at $20 billion, and the remaining disciplines at $10 billion. Immediate benefits should be seen with the adoption of crossbreeding in cattle and in poultry (Nkrumah, 2014). Genetics and breeding technologies have also been mentioned by others (Hume et al., 2011; Thornton et al., 2011). Hume et al. (2011) predicts that there will be (1) whole-genome selection programs based on linkage disequilibrium for a wide spectrum of traits and genetic selection based on allele sharing rather than pedigree relationships to make breeding value predictions early in the life of the sire; (2) selection will be applied to a wider range of traits, including those that are directed toward environmental outcomes; (3) reproductive technologies will advance to allow acceleration of genetic selection; (4) transgenesis and/or mutagenesis will be applied to introduce new genetic variation or desired phenotypes; and (5) there will be a shift toward more sustainable intensive integrated farming to improve efficiency. To realize the full benefits of the improvements in genetics, advances must also be made in animal health and reproductive health, proper nutrition, management, and animal comfort.

At one time, there was resistance to the use of artificial insemination, which is now commonplace in agriculture and human medicine. As Foote (2002) wrote:

> At the initial stages of attempting to develop AI there were several obstacles. The general public was against research that had anything to do with sex. Associated with this was the fear that AI would lead to abnormalities. Finally, it was difficult to secure funds to support research because influential cattle breeders opposed AI, believing that this would destroy their bull market. The careful field tested research that accompanied AI soon proved to the agricultural community that the technology applied appropriately could identify superior production bulls from lethal genes, would control venereal diseases and did result in healthy calves. Thus, fear was overcome with positive facts. The extension service played an important role in distributing these facts.

Technologies that are deemed safe by regulatory authorities and that have been demonstrated not to have negative effects on animal health and welfare should be reconsidered for global use and adoption. Growth promotants have been safely used in beef cattle production for over 50 years in the United States, and bovine somatotropin has been used in lactating dairy cattle as a productivity enhancer for over two decades in the United States and other countries. These technologies contribute to enhanced food security and sustainability. Increased production, improved feed efficiency and enhanced lean tissue results in a more affordable and desirable animal protein product for the consumer. Adoption of these technologies reduces the carbon and water footprints, reduces GHGs by reducing the number of animals to produce equivalent amounts of product, and reduces the effects of land change to feed the world (Johnson et al., 2013; Neumeier and Mitloehner, 2013). If the use of the growth technologies (i.e., steroidal implants, ionophores and in-feed hormones) were to be withdrawn in the United States, significant negative consequences on the environment and environmental sustainability would occur. Capper and Hayes (2012) stated that without these technologies, land use required to produce 454 million kg of beef would increase by 10 percent. Adopting biotechnological advancements in developing countries is important to achieve food security and environmental sustainability (Rege, 2009). Technologies, tools, and research areas to consider include:

- Technologies to reduce environmental impact of animal wastes associated with feeding and maximum productivity (Carter and Kim, 2013);
- Genetic and omics-based tools (Green, 2009; Niemann et al., 2011; Fulton, 2012; Spencer, 2013);
- Semen technologies and cryopreservation (Rodriquez-Martinez and Vega, 2013), (e.g., rapid adoption of semen sexing in developing countries such as India would accelerate genetic gain toward higher productivity, more efficient herd composition, and less waste (e.g., male calves left to starve);
- Reproductive technologies (e.g., faster genetic progress and improved fertility) (Niemann et al., 2011; Hernandez Gifford and Gifford, 2013);
- Improvement in feed efficiency, especially improved utilization of fiber (Niemann et al., 2011);
- Reduction in maintenance cost per unit of animal protein;
- Better utilization of wastes streams from other industries (e.g., human food processing and biofuels) into animal products;
- Proteomics (Lippolis and Reinhardt, 2008);
- Improved efficiency of water utilization (Beede, 2012; Doreau et al., 2012; Patience, 2012);
- Better utilization and adoption of current technologies such as biotechnology and nanotechnology, and development and adoption of new technologies (Van Eenennaam, 2006);
- Improved forage utilization (Rouquette et al., 2009); and
- Animal health (e.g., vaccines, low-cost accurate diagnostics, antibiotics, formal vaccinology, reverse vaccinology, and vaccine discovery).

Transfer of technology through effective communication is critically important to the adoption of technology. Internationally, extension or advisory personnel can play an important role in improving the productivity of animals through knowledge transfer that can increase productivity, reduce disease, and improve food quality and safety; however, key considerations for food animal advisors, in addition to having up-to-date knowledge and education skills, are the cultural and social norms that can influence access to and trust from animal producers (Meinzen-Dick et al., 2011).

Recommendation 4-5.2

The committee finds that proven technologies and innovations that are improving food security, economics, and environmental sustainability in high-income countries are not being utilized by all developed or developing countries because in some cases they may not be logistically transferrable or in other ways unable to cross political boundaries. A key barrier to technological adoption is the lack of extension to smallholder farmers about how to utilize the novel technologies for sustainable and improved production as well as to articulate smallholder concerns and needs to the research community. Research objectives to meet the challenge of global food security and sustainability should focus on the transfer of existing knowledge and technology (adoption and, importantly, adaptation where needed) to nations and populations in need, a process that may benefit from improved technologies that meet the needs of multiple, local producers. Emphasis should be placed on extension of knowledge to women in developing nations.

> **Research devoted to understanding and overcoming the barriers to technology adoption in developed and developing countries needs to be conducted. Focus should be on the educational and communication role of local extension and advisory personnel toward successful adoption of the technology, with particular emphasis on the training of women.**

Other Research Priorities

Technology development and transfer should be focused on needs of the developing world. In the case of aquaculture, for example, the best certification efforts to realize a positive global impact for sustainable production have been founded in diverse stakeholder input leading to the adoption of common international standards. Other research priorities in this area include:

- Following the example of aquaculture, globalwide certification systems should be developed whereby retailers and consumers will have the information to make judicious and socially responsible decisions about the purchase of more efficient and sustainable animal proteins.
- Existing technologies that are deemed safe and efficacious in the developed world should undergo research evaluations to determine whether alteration is possible to achieve feasible use and efficacy in developing countries.
- Research in genetics and breeding, reproductive technologies, and animal health in conjunction with nutrition, management, and animal welfare required to realize the benefits of the improved genetics and health must be given priority in developing countries.

4-6 Food Losses and Food Waste

Annual global food loss and waste by quantity is estimated to be 30 percent of cereals; 40-50 percent of root crops, fruits, and vegetables; 20 percent of oilseed, meat, and dairy products; and 35 percent of fish (FAO, 2014c). One-third or 1.3 billion tons of the food produced for human consumption is lost or wasted globally (Gustavsson et al., 2011). Food loss and waste are important for multiple reasons. As Buzby and Hyman (2012) summarized: (1) food is needed to feed the growing population and those already in food-insecure areas; (2) food waste represents a significant amount of money and resources; and (3) there are negative externalities (i.e., GHG emissions from cattle production, air pollution and soil erosion, and disposal of uneaten food) throughout the production of the food that affect society and the environment. A recent FAO (2013) report provides a global account of the environmental footprint of food loss and waste along the food chain with a focus on climate, water, land, and biodiversity. Results of the study included the following: (1) global amount of wasted edible food is estimated to be 1.3 gigatonnes (Gtonnes); (2) carbon footprint of food produced and not eaten, without accounting for GHG emissions from land-use change, is estimated to be 3.3 Gtonnes of carbon dioxide equivalents or third to the United States and China as the top emitters; (3) blue-water footprint of food loss is about 250 km3; and (4) uneaten food represents 1.4 billion hectares or about 30 percent of the world's agricultural land area.

Food is lost or wasted throughout the food chain (Figure 4-2). Food losses refer to both qualitative (i.e., reduced nutrient value and undesirable taste, texture, color, and smell) or quantitative (i.e., weight and volume) reductions in the amount of and the value of the food. Food loss represents the edible portion of food available for human consumption that is not consumed. Food waste is a subset of food loss and generally refers to the deliberate discarding of food because of human action or inaction. In developing countries, most food losses occur at the beginning of the food chain because of poor harvest technologies and poor storage and transport facilities, whereas, in developed countries, most of the food loss occurs at the end of the food chain because of wastage at wholesaling and retailing and by consumers (Lundqvist et al., 2008). Since many smallholder farmers live on the edge of food insecurity, any reduction in food loss could quickly have positive impacts on their livelihoods. This section focuses mainly on the food loss and waste of animal products.

Food losses and waste have been evaluated by commodity groups, including meat, fish, and dairy; world areas; and at different phases of

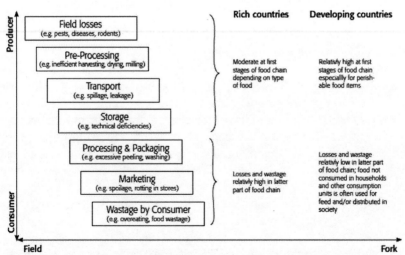

FIGURE 4-2 Areas of food losses and wastage. Illustration by Britt-Louise Andersson, SIWI.
SOURCE: Lundqvist et al. (2008). Reprinted with permission from the Stockholm International Water Institute.

the food chain (Gustavsson et al., 2011). Meat and meat product losses and wastes appeared to be about 20 percent except in sub-Saharan Africa where it was about 30 percent. Developed countries had the most severe food loss and waste at the end of the food chain with large wastage (11 percent) by retailers and consumers, especially in Europe and the United States (Figure 4-3). The relative low level of wastage during animal production (~3 percent) is due to low animal mortality. Losses in all developing regions are more evenly distributed among the food chain segments. The sub-Saharan Africa region had significantly more loss during the animal production phase (15 percent) due to higher animal mortality caused by diseases such as pneumonia, digestive diseases and parasites. All developing countries had more meat loss (10-12 percent) during the processing and distribution phases of the food chain as compared to the developed countries (9 percent).

Fish and seafood losses appeared to be about 30 percent, except for the region of North America and Oceania where food losses were about 50 percent (Figure 4-4). For all three industrialized regions, food losses in production due to discard rates were between 9 and 15 percent of marine catches compared to 6-8 percent in developing countries. High losses (19-24 percent) from the combined processing and distribution stages were explained by the high level of deterioration occurring during fresh-fish and seafood distribution in the developing countries.

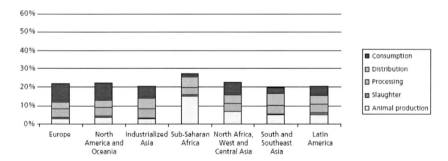

FIGURE 4-3 Initial production lost or wasted for meat products at different stages in the food chain in different regions of the world.
SOURCE: Gustavsson et al. (2011). Reprinted with permission of FAO.

Dairy product food loss and waste were the lowest in Europe and industrialized Asia (11-13 percent), the highest in sub-Saharan Africa (27 percent), and between 21 to 24 percent in the remaining world regions (Figure 4-5). The North America and Oceania region had the highest wastage at consumption of 15 percent compared to 7 percent or less for the other world regions. For all developing regions, milk wastage was relatively high at postharvest handling and storage (6-11 percent) and in distribution (8-10 percent).

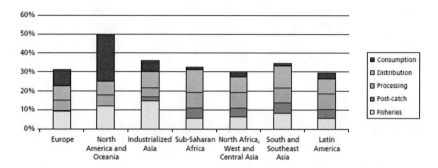

FIGURE 4-4 Initial catch (fish and seafood harvested) discarded, lost, and wasted in different world regions and at different stages in the food chain.
SOURCE: Gustavsson et al. (2011). Reprinted with permission of FAO.

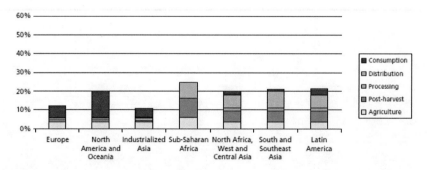

FIGURE 4-5 Initial milk and dairy product lost or wasted for each world region at different stages in the food chain.
SOURCE: Gustavsson et al. (2011). Reprinted with permission of FAO.

Strategies to reduce food losses and waste will differ depending on the animal product and whether the country is developed or developing. Major gains could be expected in developing countries with research focused on (1) animal production in the developing countries and especially in sub-Saharan Africa; (2) postharvest handling and storage, especially with dairy products and fish and seafood; (3) processing and packaging for fish and seafood; and (4) distribution. For developed countries, research needs to focus on the fisheries and the retailer and consumer. All world regions would benefit from research focused on extending the animal product shelf life and product safety.

Hodges et al. (2010) reported on the postharvest losses and wastes in developed and developing countries and strategies to reduce food losses and wastes in each of the two global segments. These strategies included 1) provide incentives to reduce food loss and waste; 2) conduct consumer education campaigns to increase consumer knowledge and awareness; 3) provide more mechanization; 4) adopting/adapting improved technologies; 5) better infrastructure to connect smallholders to markets; 6) more effective value chains that provide financial incentives at the producer level; and 7) public and private sectors sharing investment costs and risks in market-orient interventions. In developing countries, new technologies can be introduced through innovation systems and learning alliances, but direct or indirect benefits need to be clearly visible and measureable (Hodges et al., 2010).

Animal product losses and wastes range from 10 to 50 percent depending on the region of the world and the animal product of focus (e.g., dairy, meat, or seafood product). A reduction in edible animal product waste can have a significant positive impact of meeting the food security needs and improving the environment. This is especially true for developing countries where the demand for food will increase as a result of the surge in projected population growth by 2050.

Research Priorities

Global animal protein product loss and waste ranges between 20 and 30 percent. Across all world regions, caught fish have the highest food loss and waste (30 percent) compared to meat and milk (20 percent). While total animal product loss is similar between developed and developing countries, developed countries produce more waste at the end of the food chain, whereas developing countries produce more waste at the beginning of the food chain. Also, animal product loss varies by type: milk and dairy product loss was consistently greater in developing

countries than in developed countries. One research priority in this area includes:
- A holistic research approach by region and animal product type must be conducted to identify those areas in the food chain where reduction of food loss can be substantial so that the greatest return on investment can be realized.

4-7 Infrastructure Related to Food Security Concerns

4-7.1 Health and Diseases

The World Bank (2012) analyzed and assessed the benefits and cost of controlling zoonotic diseases. Zoonotic diseases account for 70 percent of emerging infectious diseases, and the cost of the six major outbreaks that have occurred between 1997 and 2009 was $80 billion. If the outbreaks had been prevented, the accrued benefits would have averaged $6.7 billion per year. One Health (the collaborative efforts of multiple disciplines working locally, nationally, and globally to attain optimal health for people, animals and the environment) is needed to provide effective surveillance, diagnosis, and control of zoonotic diseases. The annual funding needed for the major zoonotic disease prevention and control system in developing countries to achieve OIE and WHO standards (referred to as One Health Systems) ranges from $1.9 billion to $3.4 billion. These funds would be spent among 60 low-income and 79 middle-income countries. Cost-benefit analysis indicates that an annual investment of $3.4 billion would yield an expected rate of return that ranges between 44 and 71 percent based on half or all mild pandemics being prevented. If only one in five pandemics were prevented, then the rate of return would be 14 percent; however, the funding has been spent with no new funding initiative in place.

Veterinary professionals throughout the world, mainly through their animal health services, are faced with having to fulfill a crucial role in protecting their country's animal health status, providing sound surveillance information on the occurrence of diseases within their territories, and conducting scientifically valid risk analyses to establish justified import requirements. The majority of these tasks and activities require sound research with the aim of identifying approaches and alternatives for disease management.

As a consequence to opening trade and the signing of the General Agreement on Tariffs and Trade SPS Agreement, the world started to take a different shape, especially in the early 1990s. Animal health programs were in the spotlight because the primary issue that would facilitate or impede the trade of animals and their products was their effect on the safety and health of humans, animals, and plants. Comprehensive surveillance, quantitative disease indices, and science-based risk analysis were a few of the new terms that emerged during this critical time. Modifications and adjustments to some existing scientific tools for these demanded activities require further research and transfer of technology to various parts of the world, particularly in the developing countries (Salman, 2009).

During the last two decades, the largest hurdle facing animal health has been the lack of resources available to combat several emerging and reemerging infectious diseases. Because of recent events, particularly those associated with public health, more resources than ever before are currently being directed toward pressing animal heath challenges. The available funds, however, are mainly directed at specific high-profile infectious diseases instead of animal diseases in general. Nevertheless, these resources provide an excellent opportunity to improve the infrastructure of national and global animal health programs. The emergence of diseases that receive the attention of the public and of policy makers requires technically reliable disease investigation and case findings. There also are requirements for a scientifically based approach to trade and assessing risk. Furthermore, international financial institutions have more involvement in shaping government veterinary services and have several requirements to justify plans of action.

Infrastructure is lacking in developing countries, specifically a lack of disease specialists and diagnostic laboratory facilities that would include focus on the etiology of diseases. According to Kelly et al. (2013), there is a major deficit of veterinarians involved in strategic planning at USAID and they were aware of only one veterinarian. The World Bank, with 9,000 employees working in 124 countries of the developing world, has only five veterinarians on staff. FAO has a professional staff of 1,847 that includes 27 veterinarians with only one being a U.S. veterinary college graduate. The Director General of the OIE has noted the inadequacy of veterinary education in most OIE member countries and has emphasized the importance of improving the quality of veterinary education. This is an opportunity for veterinary professionals in developed countries to provide collaborations to meet this unmet need. Translational research activities and outreach programs are required for these developing countries in order to satisfy the international animal health and food safety requirements.

There is consensus that infectious diseases are important and will continue to retain this status. Perry et al. (2013) considered two drivers that exert the greatest influence on food animal disease dynamics: increase in population size and prosperity and demand-driven increase in the consumption of animal products. They identified three trajectories: (1) intensified and worried well of the Western world; (2) intensifying and increasing market-oriented sectors of the developing world hot spots; and (3) smallholder systems dependent on traditional food animal–derived livelihoods (i.e., cold spots). Within each of these trajectories, animal health status and drive summary, animal health risks, animal health service response needs, and key drivers were determined (Table 4-3). The small- and medium-sized emergent intensifiers were identified as the hottest of the hot spots in terms of animal health risks, with high densities of animals occurring in close proximity to people.

In the majority of countries and for good reason, most public funding is given to the concerns of human health with limited focus on the animal health side. This limited funding toward animal health is one of the major factors in the spread of zoonotic diseases globally. A good example of the imbalance of resource allocation is evident in the spread of global avian influenza (AI). Far more resources have been given to the detection and control of spread of AI in human populations with relatively limited resources dedicated to the animal side, even though the disease can be prevented if the focus is on the animal side.

Aquaculture is still developing in Africa, and although it is concentrated in a few countries, it already produces an estimated value of almost $3 billion annually (FAO, 2014a). This level cannot grow and be sustained without research into the problems of disease and effective training to manage disease outbreaks. Mechanisms for biosecurity, including proper surveillance, need to be established. This need was well documented by Reantaso et al. (2002) in an animal health assessment of small-scale aquaculture practices in Southern Lao PDR .

Once new diseases arise, an early response including detection commonly ensues only after a protracted period of time has expired and extensive losses have already occurred. Funding for aquatic animal disease research is lacking (Subasinghe et al., 2001). A substantial amount of USAID funds are currently used to work with producers. However, a more effective approach resides in the development of appropriate infrastructure to establish biosecurity and preventive measure programs (USAID, 2013). Most of the disease problems are caused by poor management practices and the lack of sufficient understanding of the pathogen and the pathogen-host relationship.

TABLE 4-3 Animal Health and Service Response by Trajectory

Trajectory	Animal Health Status and Drivers Summary	Animal Health Risks	Animal Health Service Response Needs	Key Drivers
Intensified and worried well of the Western world	Well-controlled endemic disease	Increased drug resistance	Better surveillance, including for new diseases	Concerns over quality, safety, and animal welfare
	Changing and often stretched private health service to livestock enterprises	Expanded distribution of vector-borne and other pathogens	Appropriate and acceptable disease control measures	Climate change
	Heightened public awareness	Multisector economic impacts of disease incursions or scares	Incentives to develop new animal health products	
	Real/perceived threat from the rest of world			
Intensifying and increasingly market-oriented sectors of the developing world	Increasing intensification and widening of trading partnerships in an environment of endemic disease risk	Endemic disease outbreaks	Greater private-sector response capacity through vertical integration and other models	Livestock revolution (demand-driven intensification)
hot spots	Presence of several major infectious diseases	Inability to prevent and contain disease in the broader county and regional environment	Greater interface with public-sector health authorities	Changing patterns of global trade

GLOBAL CONSIDERATIONS FOR ANIMAL AGRICULTURE RESEARCH 263

Trajectory	Animal Health Status and Drivers Summary	Animal Health Risks	Animal Health Service Response Needs	Key Drivers
	Absence of effective veterinary infrastructure	Unachievable standards imposed by international authorities or trading partners	Greater interface with public-sector health authorities	Urbanization
	Limited voice in national animal health programs	Emergence of new disease		
		Erosion of genetic resources associated with disease resistance		
Smallholder systems dependent on traditional livestock-derived livelihoods (cold spots)	Severity constrained economically	Multiple endemic diseases	Specific services targeted at smallholder and marginal producers	Population growth
	Limited livestock/feed/health resources	Limited or no movement controls	Well-coordinated national systems bringing in NGO, private, and donor-supported services	Climate variability
	Multiple endemic diseases	Provides source of infection to market-oriented trajectory	Particular attention to preparedness and response to shocks	
	Often in harsh environments Inadequate or total absence of animal health services	Highest vulnerability to zoonotic disease		

SOURCE: Perry et al. (2013).

For aquaculture systems, more research needs to be conducted to address the systems-specific loss due to disease. Disease can dramatically affect production. For example, early mortality syndrome in shrimp species, first reported in China in 2009, then spreading to other Asian countries such as Thailand and Vietnam, led to Thai shrimp production falling by 40 percent. In some cases, some farms lost 70 percent of the expected harvest (Waite et al., 2014). Financial losses attributed to disease are substantial. During the 1990s and the first decade of 2000, estimated losses due to some selected diseases in Asia and South America ranged from $15 million to $650 million (Reantaso et al., 2006). Given this magnitude of socioeconomic impact, financial support for research, surveillance, and control programs would provide a substantial return on investment.

There must be a new direction in meeting the challenges of aquatic animal disease because the practices of control and response that are currently being used by producers in developing countries are not based on the biology of the disease organisms but rather on what antibiotics might be conveniently available. Effective technology transfer cannot be realized when critical knowledge is lacking. The identified need is a comprehensive, step-by-step process whereby an understanding of the process is gained through research such that specific and significant control strategies can be developed and implemented. One of the needs to achieve economically viable aquaculture is the technology to address disease problems through research that yields results that lead to the improved health of cultured species.

For diseases in cultured fish, a specific model needs to be developed whereby an understanding of the basic interaction of host, agents, and their environment can be achieved. The zebrafish is an excellent candidate to serve as a model, providing the essential control to establish an understanding of basic disease mechanisms that would lead to the prescription of appropriate therapeutic and ultimately prophylactic solutions whereby disease loss in aquaculture systems is averted. Another example is the crisis facing the Chilean salmon industry resulting from the outbreak of salmonid rickettsial syndrome (Box 4-5). Effective response to disease lies in a proactive approach whereby prevention measures should have a priority over treatment, arising from a basic knowledge of the mechanisms including the interaction of host, disease agent, and environment.

> **BOX 4-5**
> **Chilean Salmon Industry Threatened by Salmonid Rickettsial Syndrome (SRS)**
>
> First reported in Chile in 1989, SRS manifests as an aggressive infection affecting salmon kidneys, spleen, liver, intestine, brain, ovary, and gill integrity (Martinez et al., 2014). The pathogen causing this disease can be transmitted between fish through touch or shared water, resulting in 30-90 percent mortality in Chilean seawater netpen-raised coho salmon (Iowa State University, 2014) and economic losses of up to $10 million per year for the Chilean salmon farming industry (Martinez et al., 2014). Although lower mortality rates have been recorded in other parts of the world, it has been noted that SRS is being found in increasing numbers of farm-raised salmon worldwide (Iowa State University, 2014). New preventive measures such as sea lice baths to curb the spread of SRS have been introduced by regulatory bodies in Chile, and while these measures may help curb mortality, they have also significantly increased the production costs of Chilean salmon with the industry now dependent on prices staying high (Stewart and Tallaksen, 2013). Peter Bridson, the aquaculture research manager for the Monterey Bay Aquarium, was quoted as saying that the challenged posed by SRS is large, and that a vaccine is the only possible solution that does not involve increased antibiotics use (Wright, 2014). Chile used 993,000 pounds of aquaculture antibiotics in 2013, and has attracted concern due to their heavy dependence on the use of these antibiotics (Wright, 2014). The outbreak of SRS poses a significant problem for salmon farming, and an effective vaccine is needed for the Chilean salmon fishing industry.

There is a critical need for the capacity to support and enhance the national animal health programs in developing countries through research on infectious diseases in agricultural animals, especially in using this research as models to gain knowledge about emerging diseases in animals (Lantier, 2014) and for infectious diseases in humans (Roberts et al., 2009; Lanzas et al., 2010). Moreover, in developing countries, there is a limited number of veterinarians who specialize in fish health. A short-term solution lies in establishing fish health inspectors—individuals with doctorates in animal/veterinary science who have earned a fish health certification. Resources are available to provide a mechanism for global fish health certification through the Fish Health Section of the American Fisheries Society.

An example of global prospective of animal health and its importance is the transmissible spongiform encephalopathy (TSE) diseases. TSE, particularly scrapie, has been recognized in animal populations for more than two centuries (Gavier-Widen et al., 2005). There was little emphasis placed on the TSEs until the recognition of bovine spongiform encephalopathy (BSE) in dairy cattle in the United Kingdom in 1986. Several unique characteristics and factors associated with BSE made it a concern for researchers, regulators, policy makers, social scientists, and the general public. Without doubt, BSE has been the most important international veterinary disease in the last 20 years with an impact on national and international economics, trade, and public health (Salman, 2004; Salman et al., 2012).

As with many animal health issues, it is the association of BSE with a neurologic disease of humans, variant Creutzfeldt-Jakob disease (vCJD), that had given it such a high-profile status. The complexity and seriousness of BSE has significantly stimulated those in global animal health to increase the scope of their view and what their role in control of BSE will be. Factors such as the safety of food animal feed supply, the impact on public health, and the effort as well as the expense required to protect the human and pet food supply and the environment from the infectious agent require vast amounts of information from areas not usually dealt with by the veterinary community. The veterinary role in public health maintained for the last three centuries "from stable to table" has expanded to "from conception to consumption." Although scrapie in sheep had not received much emphasis in veterinary research, during the last couple of decades, it has come to be recognized that the prion associated with scrapie is thought to be similar to the agent causing BSE. There is limited knowledge about the persistence of the BSE agent and its impact on contamination of soil, air, water, and plants. We have learned that previous protocols for cleaning, disinfection, and biosecurity measures that were developed to control environmental contamination with viruses and bacteria have to be modified. Conventional methods for cleaning and disinfection require additional steps to inactivate the prion, the etiological agent of BSE. Infected carcasses require different precautionary measures in carcass disposition. Thus, research and technology transfer relevant to these topics are needed. The link of this disease to human health and specifically to the neuropathological manifestations in both humans and animals has built several bridges between human and veterinary medicine. This is reflected in the need for research collaborations by both human and veterinary professionals. The medical community at large now seeks much of their information about this disease through veterinarians and their associates (Salman, 2004).

The strong scientific evidence of the association between the contamination of cattle feed with infected materials and the presence of BSE has highlighted the importance of veterinarians maintaining a role in the decisions and recommendations regarding the nutrition of animals. The role of nutritional advisor for producers has largely been usurped by other fields of specialty. Such scientific and professional activities of veterinarians are gaining more attention from both the veterinary community and users of this type of service, such as food animal producers and feed companies (Salman, 2004).

Grace et al. (2012) identified key hotspots for zoonoses in the world. Diseases were prioritized based on the burden of human disease, impacts on food animal production and productivity, amenability to intervention, and concern about the severity of emergence. They identified 24 zoonoses that hold importance in reference to poor people and focused on 13 of them. Maps of the poor food animal keepers, food animal systems, vulnerability to climate change, and emerging disease events were presented. Key findings of the study included: (1) there is a lack of evidence on zoonoses presence, prevalence, drivers, and impact; (2) literature is one of the best ways of understanding what diseases are present and their impact; (3) recent advances in technology (e.g., biorepositories, genomics, and e-technologies) offer opportunities to improve the understanding of zoonoses epidemiology and control; (4) a relatively small number of countries, such as India, Ethiopia, and Nigeria have a disproportionate share of poor food animal keepers and the corresponding burden of zoonoses; and (5) the relationship between poverty and food animal keeping with emerging zoonotic events has not been obvious, such as the role that bush meat has played in the Ebola outbreak (Box 4-6).

As seafood consumption increases globally, and the proportional contribution of seafood from aquaculture production increases, the possibility of contraction of zoonotic infections increases accordingly. Haenen et al. (2013) reviewed human cases of zoonoses arising from pathogenic organisms affecting fish and shellfish. The committee agrees that areas for further research include: (1) the relationship between controlling zoonoses and benefits to increasing access to emerging markets; (2) the implications of intensification and emerging markets for zoonoses; (3) models for zoonoses control in emerging markets; (4) ecosystem models for management of zoonoses with a wildlife interface; (5) improvement of surveillance for existing and new diseases; (6) impacts of multiple burdens of zoonoses and the ability to better allocate resources; and (7) technologies and innovation for detection, diagnosis, prevention, treatment, and response.

Jones et al. (2013) reviewed the literature on the effect of agricultural intensification and related environmental changes on the risk of zoonoses and found several examples in which agricultural intensification and/or environmental change were associated with an increased risk of zoonotic disease emergence; however, the evidence was not sufficient to judge whether or not the net effect of intensified agricultural production would have enhanced disease emergence or amplification. Future research is needed to address the complexity and interrelationships of environmental, biological, economic, and social dimensions of zoonotic disease emergence and amplification.

BOX 4-6
Fruit Bats as a Possible Source of Ebola Virus Outbreak in Guinea

In March 2014, attention was called to the emergence of a mysterious, highly fatal disease in rural Guinea, which quickly spread to the capital city of Conakry and subsequently to surrounding countries as individuals passed the contagion in this travel hub. Ultimately identified as the Ebola virus, the outbreak likely stemmed from an original case back in December 2013 (Baize et al., 2014). There is currently no direct therapy for Ebola virus, and fatality rates have reached up to 90 percent in the nine recorded outbreaks since the discovery of the disease in 1976 (Gatherer, 2014). Locals in Guinea and throughout much of western Africa often hunt and consume giant fruit bats as bushmeat. Recent studies indicate, however, that this can be an extremely high-risk practice. Specific species of bats have been identified as reservoirs for the virus (Leroy et al., 2005; Gatherer, 2014). This means that the bats carry the disease, but are entirely unaffected and remain without clinical signs. Assessment of the strain in Guinea, which never experienced an outbreak before 2014, indicated that the virus was known to be harbored by bats with the potential to make the jump to humans (Baize et al., 2014; Gatherer, 2014). In an effort to curb the spread, Guinean officials banned the consumption of bats and temporarily shut down the

> weekly markets at which locals purchase bushmeat (BBC, 2014). Despite these measures, the Guinean outbreak spread at an unprecedented rate across three countries, mainly due to the lack of hygiene and appropriate health delivery systems. Consumption of improperly cooked bushmeat likely contributed to one of the most devastating, urban pandemics in recent history (BBC, 2014; Gatherer, 2014). This scenario serves as a harrowing example of the possibilities surrounding cross-species infection; the implications regarding food safety and quality are clear. Additionally, the establishment of effective animal agricultural infrastructure in the region could facilitate dramatic change not only in diet, but also health security for the local population.

Recommendation 4-7.1

Zoonotic diseases account for 70 percent of emerging infectious diseases. The cost of the six major outbreaks that have occurred between 1997 and 2009 was $80 billion. During the last two decades, the greatest challenge facing animal health has been the lack of resources available to combat several emerging and reemerging infectious diseases. The current level of animal production in many developing countries cannot increase and be sustained without research into the incidence and epidemiology of disease and effective training to manage disease outbreaks, including technically reliable disease investigation and case findings. Infrastructure is lacking in developing countries to combat animal and zoonotic diseases, specifically a lack of disease specialists and diagnostic laboratory facilities that would include focus on the etiology of diseases. There is a lack of critical knowledge about zoonoses' presence, prevalence, drivers, and impact. Recent advances in technology offer opportunities for improving the understanding of zoonoses epidemiology and control.

> **Research, education (e.g., training in biosecurity), and appropriate infrastructures should be enhanced in developing countries to alleviate the problems of animal diseases and zoonoses that result in enormous losses to animal health, animal producer livelihoods, national and regional economies, and human health.**

Other Research Priorities

Aquaculture veterinarians are critically lacking in developing countries where aquaculture is an important part of livelihood, and fish diseases result in enormous economic losses. One research priority in this area includes:

- For diseases in cultured fish, research should be directed to the identification of a specific model that can be confidently used as a standard to develop an understanding of the basic interaction of host, agents, and their environment. The lack of veterinarians specifically trained in aquatic diseases should be temporarily alleviated through a certification program in aquatic animal health disease for individuals who are PhDs in animal/veterinary science.

4-7.2 Land-Constraint Considerations

In the United States, where only small amounts of land remain available for large-scale conversion to crop production, future improvement in meat, milk, and egg production must be achieved through enhanced efficiencies and intensification and more effective use of marginal lands for feed and forage production. Such improvements will require continued innovation in genetics and breeding, reproduction, animal health and nutrition, management, and production of feedstuffs. Rangeland and pastureland, which can be poor choices for cropping, will continue to be optimal for use in cow-calf and small-ruminant operations to harvest the grass as meat. Reduced yields of corn, barley, wheat, soybeans, and other feedstuffs arise from using organic production methodologies. Thus, more land would be required in organic farming of these crops to maintain equivalent production compared to use of conventional farming techniques (Seufert et al., 2012). Organic production of meat, milk, and eggs may be viable on a local scale in developed and developing countries, but cannot be sustainable on a national or global scale because of the much lower corn, soybean, and wheat yields (Seufert et al., 2012; Rosegrant et al., 2014) and some animal production yields, such as milk. Stiglbauer et al. (2013) surveyed 192 organic and 100 conventional dairy farms in New York, Oregon, and Wisconsin and reported that cows on organic farms produced 43 percent less milk per day than conventional nongrazing cows and 25 percent less than conventional grazing herds.

In contrast, Badgley et al. (2007) reported higher yields for organically produced grains in developing countries and slightly lower yields in developed countries over conventionally grown crops. They concluded that organic agriculture has the potential to contribute quite substantially to global food supply and that leguminous cover crops could fix enough nitrogen to replace the amount of synthetic fertilizer currently in use. However, the Badgley paper has been criticized for the following: (1) using organic crop yields from systems receiving very large amounts of nitrogen from animal manure compared to lower nitrogen inputs in the conventional system; (2) use of unrepresentative low conventional crop yields in the comparison; (3) failing to consider reduction of yield over time due to rotations with nonfood cover crops; (4) comparison of systems that did not receive the same amount of concern for management practices; (5) inclusion of nonorganic yields in the comparison; (6) multiple counting of high organic yields; and (7) inclusion of unverifiable sources from the grey literature (Seufert et al., 2012). Similarly, Allan Savory and his colleagues have developed and popularized approaches to grassland grazing and have reported local benefits. But there is controversy that needs to be resolved as to whether such approaches are efficacious over the long term or can be scaled up in a way that would have significant global benefits (Savory and Butterfield, 1999; Sherren et al., 2012; Briske et al, 2008, 2014).

Based on the results of Seufert et al. (2012) and Rosegrant et al. (2014) and a constant-to-declining base of U.S. cropland, corn, soybeans, and wheat yields would be significantly less on a regional, national, or global level under an organic program compared to conventional. This deficit is currently small to negligible because less than 1 percent of the U.S. cropland is under organic production. There are local areas where organic methods can produce as much or even more than conventional methods, but the scalability over widely diverse conditions is debatable. Any grain deficit resulting from the use of the organic program would have to be made up through converting more land to crop production or through imports that negatively impact the U.S. trade balance and land use globally. The importing option could have a negative effect on the overall carbon footprint of animal protein production. Land constraints imposed by the continuing growth of crop production for fuel, conversion of cropland to nonagricultural uses, and the impacts of climate change along with government regulations, such as public and environmental policies, will impact the rate of gain needed in improved efficiencies in the production of meat, milk, and eggs. The more constraints that are imposed on the food animal sector, the faster new technological innovations to enhance food animal production will be needed. It is important to have a variety of agricultural systems to provide products for niche markets.

Globally, meat, milk, and egg production can continue to increase using existing technologies as long as there is land available for feedstuffs and the competition for available land relative to biofuel production, environmental limits, or impacts due to changing climate conditions are minimal; however, in developing countries where populations and meat, milk, and egg consumption are predicted to increase, there will be a limited ability to change land use at a rate necessary to meet demand without significant negative environmental effects. Thus, developing countries will either maintain their traditional extensification methods and rely on increased imports of meat, milk, and eggs from developed countries or rely increasingly on imported feedstuffs from developed countries to transition to a more sustainable intensification of their food animal systems to meet food security needs. If developing countries turn to imports of animal protein to meet their food security needs, the United States and other developed countries will become the primary global sources of animal protein and will need to achieve an even greater increase in production efficiency to meet future global animal protein needs. Traditional extensification could result in increased animal protein production through better health, genetics, and feedstuffs. On the other hand, if developing countries transition to a more sustainable intensification of food animal production, infrastructure will need to be improved and an increase in country capacity for animal research to improve production and efficiency while minimizing the environmental impact will be required. The growth of intensive aquaculture systems in developing countries has little possibility unless the availability and level of technology can be effectively provided. Successful intensification can only be achieved through concomitant growth in support facilities and services and the availability of investment funds (Box 4-7).

BOX 4-7
Increasing Sustainable Practices in Thai Shrimp Farming

Shrimp importation has become a substantial component of the U.S. seafood market. The United States imported $3.8 billion worth of shrimp in 2009, with the largest share (35 percent) of imported shrimp originating from Thailand (Miranda, 2010). As a result, the quality of Thai shrimp and the practices of Thailand's shrimping industry are of interest to U.S. consumers. Thailand has a large amount of coastline, which when combined with relatively low wages and government tax incentives, created a favorable environment for the development of a shrimping industry (Szuster, 2006). The adoption of semi-intensive farming

> practices saw a "boom" in Thai shrimp production in the late 1980s and early 1990s (Szuster, 2006), but this industry has since experienced a rise in sustainability concerns. In 1996, the production boom reached a plateau and the shrimp industry began to face obstacles such as viral diseases (Smith, 1999) as well as concerns about the salinization of freshwater and the abandonment of shrimp ponds which made the need for better sustainability practices evident (Dierberg and Kiattisimkul, 1996). Measures such as the substitution of white shrimp instead of the native black tiger shrimp to combat disease concerns (Lebel et al., 2010) and limiting water discharges from shrimp ponds into the surrounding environment (Miranda, 2010) have been effective steps used to decrease the negative environmental effects of shrimp aquaculture. These actions combined with oversight and NGO-containing committees addressing topics such as feed use and biodiversity concerns (World Wildlife Fund, 2014) are moving the Thai shrimp industry in a more sustainable direction, while facilitating intensive farming practices to meet international consumer demand.

Research Priorities

The current rate of transformation of land by agriculture is unsustainable. The suitability of land for animal agriculture will continue to change due to long-term land uses (including uses that compete with animal production) and climate change. In addition, land constraints will affect the rate of gain needed to realize improved efficiencies in the production of meat, milk, and eggs. One research priority in this area includes:
- Improve estimations and projections of land-use constraints regarding global animal protein needs and establish optimal mixes of animal and plant agriculture to meet food security needs.

4-8 Global Environmental Change

Animal agriculture will remain critical for the food and economic security of billions of people around the world over the next 40 years. While the projected increase in both population and incomes will drive demand growth for animal products, animal agriculture will be simultaneously constrained by and contribute to global environmental change. Global environmental change refers to the entirety of changes, both natural and anthropogenic in origin, under way in the earth system.

In the developing world, animal agricultural systems will continue to transition from extensive, pastoral systems to mixed and intensive systems over the next few decades (Herrero et al., 2009). Because intensification typically increases production efficiency, decreases land requirements per calorie produced, and lowers environmental impact intensities (e.g., CO_2-equivalent emissions per kilogram of meat), the transition toward more intensive systems will likely have environmental benefits (Capper, 2011; Rendón-Huerta et al., 2014). There are potential negative impacts, however, on ecosystem services due to intensification, including nutrient loading and pollution due to insufficient land availability to recycle animal waste (Gerber et al., 2013). Additionally, increased animal densities without the development of proper disease surveillance practices and regulations could lead to increased risk of zoonotic disease outbreaks (Herrero and Thornton, 2013).

The need to increase productivity while simultaneously ensuring negative environmental impacts that do not threaten the food security and well-being of future generations has led many to call for sustainable intensification of animal production systems. Sustainable intensification is a new and evolving concept and is about optimizing productivity and a range of environmental and possible other outcomes (Garnett and Godfray, 2012). Barriers to sustainable intensification exist due to constraints of infrastructure, capital, knowledge, and technology transfer (Pretty et al., 2011; Garnett et al., 2013). While the concept of sustainable intensification has been evolving in the past few years, it is now widely acknowledged that the concept encompasses more than just improving productivity and efficiency, but also includes creating the necessary incentives and investments for systems to intensify, and developing regulations and limits for intensifying systems (i.e., animal welfare standards) among other considerations (Herrero and Thornton, 2013). Additionally, when considering the sustainable intensification of animal systems, there are tradeoffs between environmental impacts and the livelihoods of smallholders, which require further research to better inform policy debates (Herrero et al., 2009).

Globally, animal agriculture relies on synthetic fertilizers for crop production including nitrogen and phosphorus. Most nitrogen fertilizer is derived from the Haber-Bosch process, which utilizes molecular nitrogen from the atmosphere. Synthetic nitrogen fertilizer has allowed food productivity to make significant increases over the past several decades; however, the widespread use of synthetic nitrogen fertilizer has also led to an increase in reactive nitrogen in the environment (Gruber and

Galloway, 2008). Increased reactive nitrogen can have negative implications for human and ecosystem health (Townsend et al., 2003). Phosphorus fertilizer is dependent on the mining of phosphate rock, which means synthetic phosphorus fertilizer has resource supply constraints compared to synthetic nitrogen fertilizer (Cordell et al., 2009). In the case of both nutrients, improved feed conversion efficiency and more efficient use and recycling of nutrients in crop production systems, including improved integration of crop and animal production systems, can mitigate negative environmental consequences (Bouwman et al., 2013).

Additionally, as stated in prior sections, global trade is becoming ever more important in animal agriculture and will continue to grow in the next 40 years. A potential unintended consequence of the global trade of animal products and feedstuffs is the movement and concentration of nutrients around the world, consequently altering nutrient cycles and leading to environmental change (e.g., eutrophication; Bouwman and Booij, 1998; Cordell et al., 2009). While the Green Revolution increased the use of new technologies, including fertilizers, by farmers around the world, policy incentives for and knowledge transfer on the judicious use and potential negative consequences of overuse of synthetic fertilizers were lacking in many cases (Pingali, 2012). As a consequence, although the transition to modern agricultural practices increased food production, the transition to those practices also led to environmental damage such as the degradation of water quality, salinization of soils, increased soil erosion, and loss of native habitats (Foley et al., 2005). The environmental tradeoffs of the Green Revolution illustrate the need for a more nuanced, systematic approach to meet future global animal protein demand. A focus on knowledge and technology transfer will be crucial to avoid or mitigate negative environmental consequences of intensifying animal production.

Research Priorities

The infrastructure to address the environmental consequences of animal agriculture is inadequate in many developing countries. Often in developing countries, there is a tradeoff between animal production and resultant livelihoods and environmental and societal impacts of production. Research priorities in this area include:
- Develop and conduct tradeoff analyses of increasing productivity of animal production in regard to environment, environmental change, social issues, and livelihoods.

- Technologies of improved production should be matched by research based on human-environment conditions in the areas where they are to be introduced.

4-8.1 Climate Change and Variability

As discussed from a U.S. perspective in Chapter 3, climate change and variability will present challenges to maintaining or improving the productivity of animal agriculture. Additionally, climate change will affect food security through its impacts on plant agriculture. Bloom et al. (2014) reported that under conditions of elevated atmospheric carbon dioxide, protein concentrations declined an average of 8 percent in wheat, rice, potato, and barley. All of the impacts of climate change and variability will not be negative nor will they be equitably distributed geospatially (De Silva and Soto, 2009; Lobell and Gourdji, 2012). For example, in tropical and subtropical regions, increased temperatures may translate into increase growth in aquaculture systems, while in temperate zones an increase in temperature may exceed the optimal temperature range for currently cultured organisms (De Silva and Soto, 2009). Climate change has significant impacts on feed quantity and quality, animal and rangeland biodiversity, distribution of diseases, management practices, and production systems (Herrero et al., 2009). Adaptation of animal agriculture to climatic change and variability will need to occur. Climate change will have less effect on intensified animal systems than on extensive and intensive grassland systems because of the higher degree of environmental control in many intensive animal housing systems. Droughts will force poor pastoralist and agropastoralist to sell animals and diversity (Gustavsson et al., 2011)

Climate is an important factor in animal diseases and health (Lubroth, 2012); however, our knowledge about the impact of climate change and variability on animal disease and health is deficient (Nardone et al., 2010; Heffernan et al., 2012). Literature reviews have been conducted to address the question of how climate change affects animal disease, health, reproduction, and production systems (Nardone et al., 2010; Heffernan et al., 2012), but no clear conclusions can be drawn on the effects on animal disease and health. Nardone et al. (2010) pointed out that more food animal mortality due to heat stress has been observed with increased temperature, higher incidence of mastitis has been observed during hot weather, and, indirectly, more mycotoxins have been observed in feedstuffs. High environmental temperatures also negatively affect reproductive efficiency and animal performance in both sexes (St-Pierre et al., 2003).

Heffernan et al. (2012) identified the following knowledge gaps that need to be addressed: (1) the role of extra-climate factors on animal health, such as management issues and socioeconomic factors, and how they interplay with climate change impacts;(2) greater cross-fertilization across topics or disciplines and methods within and between the field of animal health and allied subjects; (3) a stronger evidence-base via the increased collection of empirical data in order to inform both scenario planning and future predictions regarding animal health and climate change; and (4) the need for new and improved methods to both elucidate uncertainty and explicate the direct and indirect causal relationships between climate change and animal infections. An integrated approach is needed to make the necessary advancements in the effect of climate change on animal diseases. Nardone et al. (2010) suggested that the following must occur: (1) all animal scientists must collaborate closely with colleagues of other disciplines, first with agronomists, then physicists, meteorologists, engineers, and economists; (2) selection of animals must be oriented toward robustness and adaptability to heat stress; (3) new techniques for cooling systems need to be developed; (4) new indices that are more complete than the thermal heat index need to be developed to evaluate the climatic effects on animal species; and (5) water-conserving technologies need to be developed and applied.

According to Hoffmann (2010), breeding goals may have to be adjusted to account for higher temperatures, lower-quality feed, and greater disease challenge. There may need to be a shift to species and breeds that are better adapted to the climate. Depending on the demand for food, there may need to be a shift to more efficient converters of feed in meat, milk, and eggs such as monogastrics and different breeds of poultry and ruminants. To be able to accomplish this, it is critical that animal genetic diversity be secured and better characterized. This will require a more complete compilation of breed inventories, better characterization of the production environments associated with each breed, more effective conservation measures, genetic improvement targeting adaptive traits in high output and performance traits in locally adapted breeds, increased support for developing countries in their management of animal genetic resources, and wider access to genetic resources and associated knowledge (Hoffmann, 2010). Along with genetic diversity in animal systems, production system diversity will be critical for meeting environmental objectives and for managing risk in the face of climate change and variability, and there is a case to be made for not maximizing production efficiency at all costs everywhere (Herrero and Thornton, 2013). Further information on the effects of climate change on animal agriculture is available in the reviews of Hopkins and del Prado (2007), Tubiello et al. (2007), and Thornton et al. (2009).

Closing productivity gaps could substantially reduce the aggregate environmental impacts of animal agriculture (Steinfeld and Gerber, 2010). Improving productivity does not mean that a transition from one production system to another is necessary, however, because productivity increases can be achieved within many disparate production systems. Productivity in a given system, climate, and region can vary considerably, indicating the potential for improved production practices and technologies to reduce GHG emissions. Gerber et al. (2013) found that if producers within a given system, region, and climate adopted the technologies and practices used by the producers with the 10 percent lowest emission intensity, a 30 percent reduction in GHG emissions could be achieved.

GHG emission intensities tend to be lower for monogastic than ruminant species, and recent trends in shifting consumption toward a higher proportion of monogastric animal species protein relative to ruminant animal protein are projected to continue (Steinfeld and Gerber, 2010). De Vries and de Boer (2010) compared the environmental impacts of the production of beef, pork, chicken, milk, and eggs using life-cycle assessment (LCA). Production of 1 kg of beef used the most land and energy and had the highest global warming potential, followed by the production of pork, chicken, milk, and eggs. The differences in environmental impact among pork, chicken, and beef can be explained by three factors: (1) differences in feed efficiency, (2) differences in enteric methane emission between monogastrics and ruminants, and (3) differences in reproduction rates (de Vries and de Boer, 2010).

The LCA did not include environmental consequences of competition for land between humans and animals or the consequences of land-use changes. LCA is being used as a methodology to evaluate the effects of a food animal system on the environment globally (de Vries and de Boer, 2010). Examples of the use of LCA analysis include beef (Pelletier et al., 2010b; Lupo et al., 2013), dairy (Thoma et al., 2013), swine (Pelletier et al., 2010a; Thoma et al., 2011), broilers (Pelletier et al., 2008; Leinonen et al., 2012a; Prudêncio da Silva et al., 2014), egg production (Leinonen et al., 2012b; Pelletier et al., 2014; Taylor et al., 2014), and aquaculture (Little and Newton, 2010; Cao et al., 2013). Despite LCA's increased use in recent years, differences in methodology can make comparisons across species difficult. The Livestock Environmental Assessment and Performance Partnership (LEAP) is a current effort that is attempting to harmonize LCA methodologies used in animal agricultural assessments. To improve LCA methodology further, there is a need for better models to predict soil nitrogen emissions and carbon storage, as well as better primary farm and environmental emission data from developing nations (Cederberg et al., 2013).

Erb et al. (2012) describe three food animal systems (landless, grassland based, and mixed farming) and their respective sustainability issues. The landless systems produce 72 percent of the global poultry, 55 percent of the pork meat, two-thirds of the global egg supply, and only 5 percent of the global beef. Extensive grassland-based systems provide around 7 percent of the world's global beef, 12 percent of the sheep and goat, and 5 percent of the milk supply. Intensive grassland-based systems produce about 17 percent of the global beef and veal, 17 percent of the sheep and goats, and 7 percent of the global milk supply. Mixed rainfed systems contribute about 53 percent of the global milk supply and 48 percent of the total beef supply. Mixed irrigated systems produce 33 percent of global pork, mutton, and milk production and about 20 percent of the global beef production. Each of these systems has its own sustainability issues. In the landless system, there are concerns that generation of animal wastes and air and water pollution may pose a threat to smallholders regarding market access, animal welfare issues, and zoonoses. In the grassland systems, concerns exist pertaining to degradation of rangeland, effect of droughts, and livestock diseases in the extensive system, and with the intensive grazing systems, concerns about competition for highly productive land with fertile soils, overstocking, and soil degradation are important. Mixed farming systems are the most widely used systems globally. Sustainability concerns in the mixed rainfed systems include zoonoses, concentration of animal waste, and competition for water. Specific sustainability issues in mixed irrigated systems include loss of soil fertility, competition for water, zoonoses, and manure disposal.

Herrero et al. (2013) developed a global, biologically consistent, spatially disaggregated dataset on biomass use, productivity, GHG emissions, and key resource-use efficiencies for the food animal sector, broken down into 28 geographical regions, 8 production systems, 4 animal species (cattle, small ruminants, pigs, and poultry), and three animal products (milk, meat, and eggs). Key findings included the following: (1) cattle account for 77 percent of the GHG emissions, and monogastrics contributed only 10 percent, of which 56 percent of the total emissions was from methane derived from the manure; (2) developing countries contributed 75 percent of the global GHG emissions from ruminants and 56 percent from monogastrics; (3) mixed crop-livestock systems produced 61 percent of the ruminant GHG, grazing systems 12 percent, and urban and other producers the remainder; (4) South Asia, Latin America including the Caribbean, sub-Saharan Africa, Europe, and Russia had the highest total emissions, which was mainly driven by animal numbers and predominant production system; (5) sub-Saharan Africa is the global hotspot for high-intensity GHG emissions as the result of the use of low animal productivity spread across a large area of arid land, low-quality feeds, and feed scarcity; (6) all systems in the developed world have lower emission intensities than those in the developing world; and (7) production of meat and eggs from monogastrics have significantly lower emission intensities than milk and meat from ruminants.

Thornton and Herrero (2010) estimated that the maximum mitigation potential for reducing methane and carbon dioxide emissions from several food animal and pasture management options in the mixed and rangeland-based production systems in the tropics was 7 percent of the global agricultural mitigation potential to 2030. Based on historical adoption rates, however, a 4 percent reduction is more plausible (Thornton and Herrero, 2010). There are numerous reviews written on the role of livestock in food and nutrition security (McLeod, 2011; Smith et al., 2013). Rosegrant et al. (2014) and Ringler et al. (2014) recently published their results from a study looking at the future benefits from alternative agricultural technologies by assessing future scenarios for the potential impact and benefits of these technologies on yield growth and production, food security, demand for food, and agricultural trade. They focused on corn, wheat, and rice. They evaluated the following technologies: (1) no-till, (2) integrated soil fertility, (3) precision agriculture, (4) organic agriculture, (5) nitrogen use efficiency, (6) water harvesting, (7) drip irrigation, (8) sprinkler irrigation, (9) improved varieties such as drought tolerant, (10) improved varieties such as heat tolerant, and (11) crop protection. The technologies that had the greatest positive impact on production and yields were nitrogen use efficiency, heat tolerance, precision agriculture, and no-till. To meet the increasing animal feed demands and challenge of climate change, three things need to happen: (1) increase in crop productivity through increased investment in agricultural research, (2) development and adoption of resource-conserving management, and (3) increased investment in irrigation. Of the eleven technologies, organic agriculture was the only one that consistently showed deceased yields across regions and crops. Production levels of animal products have expanded rapidly in East Asis, Southeast Asia, and Latin America and the Caribbean, but growth in sub-Saharan Africa has been very slow (McLeod, 2011).

Havlík et al. (2014) using the Global Biosphere Management Model (GLOBIOM) simulated livestock system transitions endogenously in response to socioeconomic drivers and climate change mitigation policies. Scenarios were developed to look at livestock, GHG emissions, and food supply relationships to 2030. From 2000 to 2030, global monogastric meat and eggs demand was projected to increase by 63 percent and ruminant meat and milk to increase by 44 and 55 percent, respectively. The increases in demand take into account the dietary shifts in developing countries. The model had 64 percent of all ruminants reared in a mixed system compared to 56 percent in 2000, and 18 percent kept in a grazing system in 2030 compared to 20 percent in 2000. The remaining 18 percent of ruminants would be in other or urban systems. The results support the global transition of food animal systems from an extensive system to a more efficient, intensive system that is less land demanding to provide the best outcome for the reduction of GHG emissions.

Research Priorities

Research that focuses on the impacts of climate change, variability, and extreme weather events on animal production systems is not sufficient for understanding the costs (economic, social, and environmental) to animal production and food security. Current knowledge is lacking for developing adaptive strategies to improve animal agriculture's resiliency to confront the challenges of climate change and variability. Analyses of GHG emissions and other environmental impacts of animal agriculture have commonly paid inadequate attention to economic or social considerations of current production systems, as well as the effects of proposed mitigation strategies and alternative scenarios (including shifts in consumption) on the environment and livelihoods of producers. In this regard, sub-

Saharan Africa warrants special attention owing to its slow growth in animal productivity and high intensities of GHG emissions from food animal production. One research priority in this area includes:

- Sustainable animal production requires systematic assessment of its effects on and impacts of climate change, including mitigation and adaptation strategies. Research priority should be given to those regions that have higher intensity emissions of greenhouse gases, and are especially vulnerable to the effects of climate change and variability.

4-9 Water Security

Freshwater is a vital resource that must be conserved globally by all sectors of society, including agriculture. With increasing population and demand for meat, milk, and eggs, the need for water will increase as well. Currently, agricultural water use accounts for 75 percent of the total global consumption, mainly through crop irrigation (UNEP, 2008). Agriculture has been a key cause for groundwater and surface-water depletion globally (Jury and Vaux, 2005; Gleeson et al., 2012; Scanlon et al., 2012). For example, because of diversion of inlet rivers for the irrigation of farmland, the Aral Sea in Central Asia has declined by 26 meters since the 1960s, exposing the seabed to wind erosion and dust storms that have had negative health consequences for the surrounding populations (Micklin, 2007, 2010).

Where water is already scarce and populations are predicted to grow, providing sufficient water to grow crops and produce meat, milk, and eggs will be even more of a challenge. Availability of freshwater may dictate where food production occurs. Doreau et al. (2012) used LCA to assess water use by food animals. They found that the amount of water use per unit of meat, milk, or eggs ranged widely depending on the food animal systems, the type of water (blue= surface and groundwater; green = water lost from soils by evaporation and transpiration from plants directly from rainfall; gray = theoretical estimate of the amount of water necessary to dilute pollutants) with the production of beef requiring the most, followed by pork, chicken, eggs, and milk. The water footprint will be an important metric now and in the future for measuring the effects of animal production on the environment. Globally, the water footprint of animal production amounts to 2.4 billion m3/year (87 percent green, 6 percent blue, 7 percent gray) with a third of the total due to beef cattle and 19 percent to dairy (Mekonnen and Hoekstra, 2010). The largest fraction (98 percent) of the water footprint of animal products is due to the production of feed for animal consumption, while drinking water for animals, service water, and feed mixing water account for 1.1, 0.8, and 0.03 percent, respectively (Mekonnen and Hoekstra, 2010). As with other nutrients, the movement of animal products and feedstuffs drives the movement of freshwater resources from nations such as the United States, Mexico, Japan, and South Korea (Hoekstra et al., 2012).

Consumptive use of water by aquaculture is difficult to define accurately because of the variety of production systems. Nonetheless, mean use is estimated to be within the same order of magnitude as that of chicken and pork and if expressed as cubic meters per kilogram would be considerably lower (40.4 for global aquaculture) (Waite et al., 2014). Production systems primarily consist of freshwater ponds (56 percent) and the continued increase in inland, freshwater aquaculture may ultimately be confronted with lack of sufficient water and space (land) to meet consumptive demands. Currently, agriculture's share of the use of freshwater globally is approximately 70 percent, and the area of land that remains to increase food production to the anticipated level in 2050 is quite limited (Duarte et al., 2009). Despite increased efficiencies in land and water usage, the future of substantial increases in food and protein from sustainable intensification of aquaculture production may lie in marine habitats. The inevitable demand for freshwater and land to achieve increases in production on land would be ameliorated. Assuming maintenance of current rates of growth (7.5 percent), mariculture production has the potential to achieve levels that will exceed capture fisheries and ultimately reach a level equivalent to or greater than all of animal protein production on land (Duarte et al., 2009).

Research Priorities

Water security is critical for the sustainability of animal agriculture and the global food system; stresses in all stocks of water (e.g., groundwater and reservoirs) have potentially negative effects on animal production. Animal agriculture's greatest impact on water use is through the production of feedstuffs; however, reported estimates of the water footprint of animal agriculture and its subsectors (e.g., beef) exhibit a wide range, often due to different methodology. Despite increased efficiencies in land and water use, the future of substantial increases in food/protein from sustainable intensification of aquaculture production may lie in marine habitats. Animal science research and engagement across value chains (i.e. Sustainable Fisheries Initiative) is an effective

way to drive on-the-ground advancements in sustainable animal systems. Research priorities in this area include:
- Improve understanding of water withdrawal and use for animal production systems, and develop reputable footprint metrics.
- More research needs to be devoted to investigations of the feasibility of sustainable marine based aquaculture systems.

4-10 Global Partnerships and Opportunities for Leveraging Resources and Research in Animal Agriculture

Public–private partnerships (PPPs) can play an important role in leveraging funding for research and fostering technology transfer, particularly in developing countries, although as Ferroni and Castle (2011) note, very few agricultural PPPs exist, and those that do are largely experimental and "form a new field of practice and inquiry for the participants." They also note that "partnerships enable sustainable outcomes that no single party could achieve alone" and can "overcome both the public sector's usually limited ability to take research outputs to market, and the private sector's limited scope for operation where there is no commercially viable market." More broadly, partnerships among stakeholders can create a framework for integrating divergent viewpoints, identifying important research priorities, and fostering transdisciplinary research efforts. The committee has identified several select examples of global partnerships related to animal agriculture. Note that the committee is not commenting on the effectiveness of these particular partnerships, but rather discussing them as part of a broader overview of the types of partnerships that exist related to research and development in animal agriculture, particularly affecting developing countries.

The Consultative Group on International Agricultural Research (CGIAR) is a global partnership of organizations engaged in research with the goal of reducing rural poverty, increasing food security, improving nutrition and health, and sustainably managing natural resources. The ILRI, which is a member of CGIAR, has developed partnerships that address animal research. One of these brought together the public and private sectors to conduct research to develop a vaccine for East Coast fever (ECF), a devastating tickborne bovine disease found throughout parts of Africa. Beginning in 2001, ILRI enlisted the participation of the Institute for Genome Research (United States), the Ludwig Institute of Cancer Research, the University of Victoria

(Canada), Oxford University (UK), the Centre for Tropical Veterinary Medicine (UK), the Weizmann Institute of Science (Israel), and the Kenya Agricultural Research Institute. The private sector was engaged through Merial Ltd., a global company in the animal health field. Although the experimental vaccine developed by the team was only 30 percent effective, the methodologies of vaccine antigen identification and evaluation that were developed are now being explored by several research organizations (NRC, 2009). This project led ILRI to publish a new strategy for its research and development activities that included a focus on partnerships as a central approach (ILRI, 2013).

Two additional multistakeholder partnerships addressing the relationship between livestock and the environment are the Global Agenda for Sustainable Livestock and the Livestock Research Group of the Global Research Agenda for Agricultural Greenhouse Gases, both of which play a role in setting a common agenda for research and policy goals. The FAO also incorporated PPPs into its work on animal agriculture issues. For example, the LEAP Partnership was founded in 2012 and involves stakeholders across the food animal sectors with a shared interest in improving the environmental performance of food animal supply chains. The objective of LEAP is to develop a globally harmonized LCA methodology and guidelines that reflect a consensus among multistakeholder partners. This approach helps foster awareness and global adoption of these methodologies. The approach highlights areas of missing or limited data where further research needs to be conducted to enhance the accuracy of the values generated. This partnership promotes an exchange of information, technical expertise, and research geared toward improving and harmonizing the way in which food animal food chains are assessed and monitored. The main focus of LEAP is on the development of broadly recognized sector-specific guidelines (metrics and methods) for monitoring environmental impact of the food animal sector that will result in a better understanding and management of the key factors influencing the sector's performance. Three stakeholder groups— the private sector, FAO member countries, and NGOs—participate in this work. Private-sector participation includes producer and processor organizations engaged in the food animal supply chain representing various subsectors such as feed, pork, beef and lamb, poultry meat and eggs, and dairy (FAO, 2014c). This partnership resulted in the development of the Global Feed Life Cycle Assessment guidelines, considered the "first feed-specific LCA guidelines that reflect a consensus among partners in the multi-stakeholder process" (FAO, 2014c).

An integrated European Union–based partnership of global significance was the project "Integration of Animal Welfare in the Food Quality Chain: From Public Concern to Improved Welfare and Transparent Quality," commonly referred to as Welfare Quality (Blokhuis et al., 2013). The Welfare Quality project funded research by 44 institutes and universities in 20 countries. The goals were to better understand the concerns and attitudes of consumers, retailers, and producers toward animal welfare, to develop and implement robust strategies for on-farm animal welfare monitoring and information systems, and to define and implement species-specific strategies to improve animal welfare on farms. Transdisciplinary teams of social and natural scientists were assembled to address these goals. Research institutes from Chile, Brazil, Mexico, and Uruguay were involved with a focus not only on understanding consumer attitudes toward animal welfare in Latin America but also on improving welfare during transport and slaughter to help producers in those countries meet EU animal welfare standards for export. A broad array of stakeholders, including animal producers and breeders, retailers, policy makers, and consumer and other nongovernmental groups were involved in project discussions and research activities. Accomplishments from the project included numerous scientific articles and reports, fact sheets and articles for laypersons, and the first comprehensive set of on-farm animal welfare assessment tools for use by producers, auditors, and certifiers (Welfare Quality, 2014).

Another international partnership is the International Committee for Animal Recording (ICAR)[1], an international nonprofit, which was created to promote the development and improvement of the activities of performance recording and the evaluation of farm livestock. ICAR, which includes a global membership, establishes rules and standards and specific guidelines for the purpose of identifying animals and the registration of their parentage, recording their performance and their evaluation, and publishing findings. Another partnership that focuses on global agricultural issues is the BREAD Ideas Challenge, funded by the National Science Foundation and the Bill & Melinda Gates Foundation. This challenge provides small funding opportunities to support

---

[1] http://www.icar.org.

innovative ideas by researchers working on agricultural issues facing smallholders.

Although many of the partnerships above have been productive, there are challenges associated with forging such relationships, including the possibility of increased administrative costs, cumbersome decision making, and the possibility that research focus will suffer (Spielman and von Grebmer, 2004). Another issue is that, where projects are funded either mainly by private sources or with joint public-private funding, the focus of the research will be on more near-term issues with potential product outcomes. In contrast, public research funding can focus on long-term issues and issues that provide a greater good that may not be product oriented, such as the development of net energy systems, animal genome sequencing, creation of databases, complex sustainability issues, microbial collections, and gene banks to maintain animal genetic diversity.

Despite these challenges, successful partnerships in agriculture can be a productive means of maximizing synergies between sectors. Ferroni and Castle (2011) describe several factors critical to the success of agricultural partnerships, including "contracts, planning, inter-partner relationships and the distribution of tasks." The authors describe four key pillars to effective partnerships: initial partnering, priority setting, contractual arrangements, and transparency. Regarding initial partnering, to maximize skills and resources, selection must begin with

> a realistic assessment of an organization's own strengths and weaknesses. It is then essential to invest considerable research in identifying organizations most likely to benefit from and add to this profile. Secondly, developing agreed upon priorities is also key; discussions of these priorities should include not only the desired main goals and milestones within the project, and the order in which they are to be tackled, but also their positions on each organization's own internal priority list. Next, contracts must describe the division of tasks and the distribution and use of any commercial rights emerging in connection with the project. Finally, transparency is key as partners need to understand and respect each other's communication requirements. This includes communication around privacy and institutional competitiveness, as well as for scientific information-

sharing amongst public sector researchers, public awareness-building about new technology and products, and the fulfillment of public reporting obligations (Ferroni and Castle, 2011).

The International Food Policy Research Institute conducted a study of agricultural PPPs for innovation in Latin America (Hartwich et al., 2007), and identified six main conclusions regarding developing effective partnerships:

- "Capacity strengthening in partnership building is specific to the value chains and actors it involves.
- Capacity strengthening for partnership building goes beyond traditional training to include horizontal learning among the partners; it a continuous process that does not suit a one-size-fits-all approach and requires that needs be identified while taking all partners into consideration.
- Determining when to enter into a partnership depends on the partners' analytical skills and the information available on technological and market opportunities; participation in diagnostic exercises strengthens the capacity of partners to enter into present and future partnerships.
- The choice of appropriate capacity strengthening measures depends on the existing level of cohesion among the potential partners; for example, awareness building may not be necessary if talks about potential collaboration are already occurring.
- Strengthening partnership-building capacity should predominantly focus on identifying and exploring common interests among potential partners through a variety of tools that help clarify interests in terms of technology development, production, and sales. Third-party catalyzing agents are necessary to bring partners together, motivate them, provide information, and organize space for negotiations.
- It is important to have at least one visionary leader among the partners, be it in the private sector or in the public research community. The leader supplies the capacity for sectoral analysis in the partnership and can help to clarify and communicate the advantages the partnership offers. The leader is also important in motivating and attracting potential partners. The internal leader may also eventually take over the initiative from the external promoter, but a gradual transfer process is the most successful option."

The benefits of partnerships, particularly for advancing animal science research, are many, including the ability to leverage resources, knowledge, technology, and human capital.

REFERENCES

Akinola, L. A. F., and A. Essien. 2011. Relevance of rural poultry production in developing countries with special reference to Africa. World's Poultry Science Journal 67:697-705.

Aksoy, M. A., and J. C. Beghin, eds. 2005. Global Agricultural Trade and Developing Countries. Washington, DC: The World Bank. Online. Available at http://siteresources.worldbank.org/INTPROSPECTS/Resources/GATfulltext.pdf. Accessed August 21, 2014.

Allison, E. 2011. Aquaculture, Fisheries, Poverty and Food Security. Working Paper 2011-65. Penang, Malaysia: WorldFish Center. Online. Available at http://www.worldfishcenter.org/resource_centre/WF_2971.pdf. Accessed September 30, 2014.

Animal Task Force. 2013. Research and Innovation for a Sustainable Livestock Sector in Europe. Animal Task Force White Paper. Online. Available at http://www.animaltaskforce.eu/Portals/0/ATF/documents%20for%20scare/ATF%20white%20paper%20Research%20priorities%20for%20a%20sustainable%20livestock%20sector%20in%20Europe.pdf. Accessed June 16, 2014.

Anthony, R. 2014. Integrating ethical considerations into animal science research. Presentation at the Second Meeting on Considerations for the Future of Animal Science Research, May 12.

Badgley, C., J. Moghtader, E. Quintero, E. Zakem, M. J. Chappell, K. Aviles-Vazquez, A. Samulon, and I. Perfecto. 2007. Organic agriculture and the global food supply. Renewable Agriculture and Food Systems 22(2):86-108.

Baize, S., D. Pannetier, L. Oestereich, T. Rieger, L. Koivogui, N. Magassouba, B. Soropogui, M. S. Sow, S. Keïta, H. De Clerck, A. Tiffany, G. Dominguez, M. Loua, A. Traoré, M. Kolié, E. R. Malano, E. Heleze, A. Bocquin, S. Mély, H. Raoul, V. Caro, D. Cadar, M. Gabriel, M. Pahlmann, D. Tappe, J. Schmidt-Chanasit, B. Impouma, A. K. Diallo, P. Formenty, M. Van Herp, and S. Günther. 2014. Emergence of Zaire Ebola virus disease in Guinea. New England Journal of Medicine. 371:1418-1425. Online. Available at http://www.nejm.org/doi/pdf/10.1056/NEJMoa1404505. Accessed October 13, 2014.

BBC. 2014. Ebola: Liberia confirms cases, Senegal shuts border. News, March 31. Online. Available at http://www.bbc.co.uk/news/world-africa-26816438. Accessed October 13, 2014.

Beede, D. K. 2012. What will our ruminants drink? Animal Frontiers 2(2):36-43.

Besbes, B., M. Tixier-Boichard, I. Hoffmann, and G. L. Jain. 2007. Future Trends for Poultry Genetic Resources. Online. Available at http://www.fao.org/Ag/againfo/home/events/bangkok2007/docs/part1/1_8.pdf. Accessed October 23, 2014.

Birch, E. L., and S. M. Wachter, eds. 2011. Global Urbanization. Philadelphia: University of Pennsylvania Press.

Blokhuis, H., B. Jones, M. Miele, and I. Veissier. 2013. Assessing and improving farm animal welfare: The way forward. Pp. 215-221 in Improving Farm Animal Welfare. Science and Society Working Together: The Welfare Quality Approach, H. Blokhuis, M. Miele, I. Veissier, and B. Jones, eds. Wageningen, The Netherlands: Wageningen Academic.

Bloom, A. J., M. Burger, B. A. Kimball, and P. J. Pinter, Jr. 2014. Nitrate assimilation is inhibited by elevated CO_2 in field-grown wheat. Nature Climate Change 4(6):477-480.

Bouwman, A. F., and H. Booij. 1998. Global use and trade of feedstuffs and consequences for the nitrogen cycle. Nutrient Cycling in Agroecosystems 52(2-3):261-267.

Bouwman, L., K. K. Goldewijk, K. W. Van Der Hoek, A. H. W. Beusen, D. P. Van Vuuren, J. Willems, M. C. Rufino, and E. Stehfest. 2013. Exploring global changes in nitrogen and phosphorus cycles in agriculture induced by livestock production over the 1900-2050 period. Proceedings of the National. Academy of Sciences of the United States of America 110(52):20882-20887.

Bowles, D., R. Paskin, M. Gutiérrez, and A. Kasterine. 2005. Animal welfare and developing countries: Opportunities for trade in high-welfare products from developing countries. Scientific and Technical Review of the Office International des Epizooties (Paris). 24(2):783-790.

Briske, D. D., J. D. Derner, J. R. Brown, S. D. Fuhlendorf, W. R. Teague, K. M. Harvstad, R. L. Gillen, A. J. Ash, and W. D. Willms. 2008. Rotational grazing on rangelands: Reconciliation of perception and experimental evidence. Rangeland Ecology & Management 61:3-17.

Briske, D. D., A. J. Ash, J. D. Derner, and L. Huntsinger. 2014. Commentary: A critical assessment of the policy endorsement for holistic management. Agricultural Systems 125:50-53.

Buzby, J. C., and J. Hyman. 2012. Total and per capita value of food loss in the United States. Food Policy 37(5):561-570.

Cao, L., J. S. Diana, and G. A. Keoleian. 2013. Role of life cycle assessment in sustainable aquaculture. Reviews in Aquaculture 5(2):61-71.

Cao, Y., and D. Li. 2013. Impact of increased demand for animal protein products in Asian countries: Implications on global food security. Animal Frontiers 3(3):48-55.

Capper, J. L. 2011. The environmental impact of beef production in the United States: 1977 compared with 2007. Journal of Animal Science 89(12):4249-4261.

Capper, J. L., and D. L. Hayes. 2012. The environmental and economic impact of removing growth-enhancing technologies from U.S. beef production. Journal of Animal Science 90(10):3527-3537.

Carter, S. D., and H. Kim. 2013. Technologies to reduce environmental impact of animal wastes associated with feeding for maximum productivity. Animal Frontiers 3(3):42-47.

Cederberg, C., M. Henriksson, and M. Berglund. 2013. An LCA researcher's wish list—data and emission models needed to improve LCA studies of animal production. Animal. 7(Suppl. 2):212-219.

Chadd, S. 2007. Future Trends and Developments in Poultry Nutrition. Rome: FAO. Online. Available at http://www.fao.org/ag/AGAInfo/home/events/bangkok2007/docs/part1/1_7.pdf. Accessed October 23, 2014.

Christoplos, I. 2010. Mobilizing the Potential of Rural and Agricultural Extension. Rome: FAO. Online. Available at http://www.fao.org/docrep/012/i1444e/i1444e.pdf. Accessed September 4, 2014.

CNRIT (Center for Natural Resource Information Technology). 2000. Extrapolation of the impact assessment of smallholder dairy technology in Kenya to adjacent East African countries. Section 3 in Impact Methods to Predict and Assess Contributions of Technology (IMPACT). Texas A&M University. Online. Available at http://cnrit.tamu.edu/IMPACT/section3all.pdf. Accessed September 11, 2014.

Cordell, D., J. O. Drangert, and S. White. 2009. The story of phosphorus: Global food security and food for thought. Global Environmental Change 19(2):292-305.

De Silva, S. S., and D. Soto. 2009. Climate change and aquaculture: Potential impacts, and mitigation. Pp. 151-212 in Climate Change Implications for Fisheries and Aquaculture: Overview of Current Scientific Knowledge, K. Cochrane, C. De Young, D. Soto, and T. Bahri, eds. FAO Fisheries and Aquaculture Technical Paper No. 530. Rome: FAO.

de Vries, M., and I. J. M. de Boer. 2010. Comparing environmental impacts for livestock products: A review of life cycle assessments. Livestock Science 128(1-3):1-11.

Delgado, C. L., M. W. Rosegrant, H. Steinfeld, S. K. Ehui, and C. Courbois. 2001. Livestock to 2020: The next food revolution. Outlook on Agriculture 30(1):27-29.

Dierberg, F., and W. Kiattisimkul. 1996. Issues, impacts, and implications of shrimp aquaculture in Thailand. Environmental Management 20(5):649-666.

Doreau, M., M. S. Corson, and S. G. Wiedemann. 2012. Water use by livestock: A global perspective for a regional issue? Animal Frontiers 2(2):9-16.

Dorward, A., S. Fan, J. Kydd, H. Lofgren, J. Morrison, C. Poulton, N. Rao, L. Smith, H. Tchale, S. Thorat, I. Urey, and P. Wobst. 2004a. Institutions and policies for pro-poor agricultural growth. Development Policy Review 22(6):611-622.

Dorward, A., J. Kydd, J. Morrison, and I. Urey. 2004b. A policy agenda for pro-poor agricultural growth. World Development 32(1):73-89.

D'Souza, D., S. Bourne, A. Sacranie, and A. Kocher. 2007. Global Feed Issues Affecting the Asian Poultry Industry. Online. Available at http://www.fao.org/ag/AGAInfo/home/events/bangkok2007/docs/part1/1_10.pdf. Accessed October 23, 2014.

Duarte, C. M., M. Holmer, Y. Olsen, D. Soto, N. Marbà, J. Guiu, K. Black, and I. Karakassis. 2009. Will the oceans help feed humanity? BioScience 59(11):967-976.

Economist.com. 2013. Grub's up: Why eating more insects might be good for the planet and good for you. May 14. Available at http://www.economist.com/blogs/graphicdetail/2013/05/daily-chart-11.

Erb, K. H., A. Mayer, T. Kastner, K. E. Sallet, and H. Haberl. 2012. The Impact of Industrial Grain Fed Livestock Production on Food Security: An Extended Literature Review. Online. Available at http://www.fao.org/fileadmin/user_upload/animalwelfare/the_impact_of_industrial_grain_fed_livestock_production_on_food_security_2012.pdf. Accessed June 17, 2014.

FABRE TP (FABRE Technology Platform). 2006. Sustainable Farm Animal Breeding and Reproduction: A Vision for 2025. Online. Available at www7.inra.fr/content/download/6638/73070/.../visionFABRETPdef.pdf. Accessed August 12, 2014.

FAO (Food and Agriculture Organization of the United Nations). 2008a. Pro-Poor Livestock Policy and Institutional Change: Case Studies from South Asia, the Andean Region and West Africa. Rome: FAO. Online. Available at http://www.fao.org/ag/againfo/programmes/en/pplpi/docarc/pplpi-casestudy01.pdf. Accessed August 18, 2014.

FAO. 2008b. The Global Livestock Sector—A Growth Engine. Rome: FAO. Online. Available at ftp://ftp.fao.org/docrep/fao/010/ai554e/ai554e00.pdf. Accessed August 12, 2014.

FAO. 2009. How to Feed the World in 2050. Rome: FAO. Online. Available at http://www.fao.org/fileadmin/templates/wsfs/docs/expert_paper/How_to_Feed_the_World_in_2050.pdf. Accessed August 18, 2014.

FAO. 2010. National Aquaculture Legislation Overview (NALO) Fact Sheets. FAO Fisheries and Aquaculture Department. Updated December 14, 2010.

FAO. 2011a. Technical Guidelines on Aquaculture Certification. Rome: FAO. Available at http://www.fao.org/docrep/015/i2296t/i2296t00.htm.

FAO. 2011b. The Role of Women in Agriculture. ESA Working Paper 11-02. Rome: FAO. Online. Available at http://www.fao.org/docrep/013/am307e/am307e00.pdf. Accessed August 12, 2014.

FAO. 2012a. The Outlook for Agriculture and Rural Development in the Americas: A Perspective on Latin America and the Caribbean. Santiago, Chile: Economic Commission for Latin America and the Caribbean (ECLAC) and FAO. Online. Available at http://www.fao.org/docrep/019/as167e/as167e.pdf. Accessed September 25, 2014.

FAO. 2012b. The State of World Fisheries and Aquaculture. Rome: FAO. Online. Available at http://www.fao.org/docrep/016/i2727e/i2727e.pdf. Accessed October 13, 2014.

FAO. 2013. Children's Work in the Livestock Sector: Herding and Beyond. Rome: FAO. Online. Available at http://www.fao.org/docrep/017/i3098e/i3098e.pdf. Accessed November 3, 2014.

FAO. 2014a. Family Poultry Development: Issues, Opportunities, and Constraints. FAO Animal Production and Health Working Paper 12. Online. Available at http://www.fao.org/docrep/019/i3595e/i3595e.pdf. Accessed October 30, 2014.

FAO. 2014b. Global Initiative on Food Loss and Waste Reduction. Online. Available at http://www.fao.org/docrep/015/i2776e/i2776e00.pdf. Accessed August 14, 2014.

FAO. 2014c. Livestock Environmental Assessment and Performance (LEAP) Partnership. Online. Available at http://www.fao.org/partnerships/leap/en. Accessed August 8, 2014.

FAO. 2014d. The State of the World Fisheries and Aquaculture. Rome: FAO Fisheries and Aquaculture Department. Online. Available at http://www.fao.org/3/a-i3720e.pdf. Accessed September 25, 2014.

FASS (Federation of Animal Science Societies). 2012. FAIR (Farm Animal Integrated Research) 2012. Online. Available at http://www.fass.org/docs/FAIR2012_Summary.pdf. Accessed August 5, 2014.

Ferguson, J. 1985. The bovine mystique: Power, property and livestock in rural Lesotho. Man 20(4):647-674.

Ferroni, M., and P. Castle. 2011. Public-private partnerships and sustainable agricultural development. Sustainability 3:1064-1073.

Fink-Gremmels, J. 2014. The future of poultry production: Meeting the challenges of food safety and food security with declining resources. Presentation at European Poultry Conference 2014. Online. Available at http://en.engormix.com/MA-poultry-industry/meat-industry/articles/the-future-poultry-production-t3232/471-p0.htm. Accessed October 23, 2014.

Foley, J. A., R. DeFries, G. P. Asner, C. Barford, G. Bonan, S. R. Carpenter, F. S. Chapin, M. T. Coe, G. C. Daily, H. K. Gibbs, J. H. Helkowski, T. Holloway, E. A. Howard, C. J. Kucharik, C. Monfreda, J. A. Patz, I. C. Prentice, N. Ramankutty, and P. K. Snyder. 2005. Global consequences of land use. Science 309:570-574.

Foote, R. H. 2002. The history of artificial insemination: Selected notes and notables. Journal of Animal Science 80(Suppl. 2):1-10.

Foreign Policy. 2014. Fragile States Index. Online. Available at http://www.foreignpolicy.com/fragile-states-2014. Accessed August 19, 2014.

Fuglie, K., N. Ballenger, K. Day, C. Klotz, M. Ollinger, J. Reilly, U. Vasavada, and J. Yee. 1996. Agricultural Research and Development: Public and Private Investments Under Alternative Markets and Institutions. Agricultural Economic Report 735. Washington, DC: U.S. Department of Agriculture Economic Research Service.

Fuglie, K. O., C. Narrod, and C. Neumeyer. 2008. Public and private investments in animal research. Pp. 117-151 in Public-Private Collaboration in Agricultural Research: New Arrangements and Economic Implications, K. O. Fuglie and D. E. Schimmelpfennig, eds. Ames: Iowa State University.

Fulton, J. E. 2012. Genomic selection for poultry breeding. Animal Frontiers 2(1):30-36.

Garnett, T., and H. C. J. Godfray. 2012. Sustainable Intensification in Agriculture: Navigating a Course Through Competing Food Systems Priorities: A Report on a Workshop. Food Climate Research Network and the Oxford Martin Programme on the Future of Food, University of Oxford. Online. Available at http://www.fcrn.org.uk/sites/default/files/SI_report_final.pdf. Accessed June 18, 2014.

Garnett, T., M. C. Appleby, A. Balmford, I. J. Bateman, T. G. Benton, P. Bloomer, B. Burlingame, M. Dawkins, L. Dolan, D. Fraser, M. Herrero, I. Hoffmann, P. Smith, P. K. Thornton, C. Toulmin, S. J. Vermeulen, and H. C. J. Godfray. 2013. Sustainable Intensification in agriculture: Premises and policies. Science 341(6141):33-34.

Gatherer, D. 2014. The 2014 Ebola virus disease outbreak in West Africa. Journal of General Virology 95(Pt. 8):1619-1624.

Gavier-Widen, D., M. J. Stack, T. Baron, A. Balachandran, and M. Simmons. 2005. Diagnosis of transmissible spongiform encephalopathies in animals: A review. Journal of Veterinary Diagnostic Investigation 17(6):509-527.

Gerber, P., C. Opio, and H. Steinfeld. 2008. Poultry production and the environment—a review. Pp. 379-405 in Proceedings of the International Conference Poultry in the Twenty-First Century: Avian Influenza and Beyond, O. Thieme and D. Pilling, eds. Rome: FAO. Online. Available at http://www.fao.org/3/a-i0323e/i0323e02.pdf. Accessed October 23, 2014.

Gerber, P. J., H. Steinfeld, B. Henderson, A. Mottet, C. Opio, J. Dijkman, A. Falcucci, and G. Tempio. 2013. Tackling Climate Change Through Livestock: A Global Assessment of Emissions and Mitigation Opportunities. Rome: FAO. Online. Available at http://www.fao.org/docrep/018/i3437e/i3437e.pdf. Accessed August 28, 2014.

Gleeson, T., Y. Wada, M. F. P. Bierkens, and L. P. H. van Beek. 2012. Water balance of global aquifers revealed by groundwater footprint. 488(7410):197-200.

Global Harvest Initiative. 2014. The 2014 Global Agricultural Productivity Report. Online. Available at http://www.globalharvestinitiative.org/GAP/2014_GAP_Report.pdf. Accessed October 31, 2014.

Grace, D., F. Mutua, P. Ochungo, R. Kruska, K. Jones, L. Brierley, L. Lapar, M. Said, M. Herrero, P. M. Phuc, N. B. Thao, I. Akuku, and F. Ogutu. 2012. Mapping of Poverty and Likely Zoonoses Hotspots. Zoonoses Project 4. Report to the UK Department for International Development. Nairobi, Kenya: ILRI (International Livestock Research Institute). Online. Available at https://cgspace.cgiar.org/bitstream/handle/10568/21161/ZooMap_July2012_final.pdf. Accessed October 1, 2014.

Green, R. D. 2009. Future needs in animal breeding and genetics. Journal of Dairy Science 87(2):793-800.

Gruber, N., and J. N. Galloway. 2008. An Earth-system perspective of the global nitrogen cycle. Nature 451(7176):293-296.

Gustavsson, J., C. Cederberg, U. Sonesson, R. van Otterdijk, A. Meybeck. 2011. Global Food Losses and Food Waste – Extent, Causes and Prevention. Rome: FAO. Online. Available at http://www.fao.org/docrep/014/mb060e/mb060e00.pdf. Accessed September 25, 2014.

Haenen, O. L. M., J. J. Evans, and F. Berthe. 2013. A review of human cases of zoonoses derived from pathogenic organisms affecting fish and shellfish. Revue Scientifique et Technique 32(2):497-507.

Hartwich, F., J. Tola, A. Engler, C. González, G. Ghezan, J. M. P. Vázquez-Alvarado, J. A. Silva, J. J. Espinoza, and M. V. Gottret. 2007. Building Public–Private Partnerships for Agricultural Innovation: Food Security in Practice. Washington, DC: IFPRI.

Havlík, P., H. Valin, M. Herrero, M. Obersteiner, E. Schmid, M. C. Rufino, A. Mosnier, P. K. Thornton, H. Böttcher, R. T. Conant, S. Frank, S. Fritz, S. Fuss, F. Kraxner, and A. Notenbaert. 2014. Climate change mitigation through livestock system transitions. Proceedings of the National Academy of Sciences of the United States of America 111(10):3709-3714.

Heffernan, C., M. Salman, and L. York. 2012. Livestock infectious disease and climate change: A review of selected literature. CAB Reviews 7(011):1-26.

Henson, S., and R. Loader. 2001. Barriers to agricultural exports from developing countries: The role of sanitary and phytosanitary requirements. World Development 29(1):85-102.

Hernandez Gifford, J. A., and C. A. Gifford. 2013. Role of reproductive biotechnologies in enhancing food security and sustainability. Animal Frontiers 3(3):14-19.

Herrero, M., and P. K. Thornton. 2013. Livestock and global change: Emerging issues for sustainable food systems. Proceedings of the National Academy of Sciences of the United States of America 110(52):20878-20881.

Herrero, M., P. K. Thornton, P. Gerber, and R. S. Reid. 2009. Livestock, livelihoods and the environment: Understanding the trade-offs. Current Opinion in Environmental Sustainability 1(2):111-120.

Herrero, M., P. K. Thornton, A. M. Notenbaert, S. Wood, S. Msangi, H. A. Freeman, D. Bossio, J. Dixon, M. Peters, J. van de Steeg, J. Lynam, P. Parthasarathy Rao, S. MacMillan, B. Gerard, J. McDermott, C. Sere, and M. Rosegrant. 2010. Smart investments in sustainable food production: Revisiting mixed crop-livestock systems. Science 327(5967):822-825.

Herrero, M., P. Havlík, H. Valin, A. Notenbaert, M. C. Rufino, P. K. Thornton, M. Blummel, F. Weiss, D. Grace, and M. Obersteiner. 2013. Biomass use, production, feed efficiencies, and greenhouse gas emissions from global livestock systems. Proceedings of the National Academy of Sciences of the United States of America 110(52):20888-20893.

Hocquette, J. F., S. Lehnert, W. Barendse, I. Cassar-Malek, and B. Picard. 2007. Recent advances in cattle functional genomics and their application to beef quality. Animal 1(1):159-173.

Hoddinott, J., and E. Skoufias. 2003. The Impact of PROGRESA on Food Consumption. International Food Policy Research Institute (IFPRI) Discussion Paper 150. Online. Available at http://www.ifpri.org/sites/default/files/publications/fcndp150.pdf. Accessed August 19, 2014.

Hodges, R. J., J. C. Buxby, and B. Bennett. 2010. Postharvest losses and waste in developed and less developed countries: Opportunities to improve resource use. Journal of Agricultural Science 149(Suppl. 1):37-45.

Hoekstra, A. Y., M. M. Mekonnen, A. K. Chapagain, R. E. Mathews, and B. D. Richter. 2012. Global monthly water scarcity: Blue water footprints versus blue water availability. PLoS One 7(2):e32688.

Hoffmann, I. 2010. Climate change and the characterization, breeding and conservation of animal genetic resources. Animal Genetics 41(Suppl. 1):32-46.

Hopkins, A., and A. del Prado. 2007. Implications of climate change for grassland in Europe: Impacts, adaptations and mitigation options: Review. Grass and Forage Science 62(2):118-126.

Hume, D. A., C. B. A. Whitelaw, and A. L. Archibald. 2011. The future of animal production: Improving productivity and sustainability. Journal of Agricultural Science 149(Suppl. 1):9-16.

Hydén, G. 2014. The economy of affection revisited: African development management in perspective. Occasional Paper 17:53-75.

Ianotti, L., C. K. Lutter, D. A. Bunn, and C. P. Stewart. 2014. Eggs: The uncracked potential for improving maternal and young child nutrition among the world's poor. Nutrition Reviews 72:355-368.

IFAD (International Fund for Agricultural Development). 2001. Rural Poverty Report 2001: The Challenge of Ending Rural Poverty. London: Oxford University Press.

ILRI (International Livestock Research Institute). 2013. Livestock Research for Food Security and Poverty Reduction: ILRI Strategy 2013–2022. Online. Available at https://cgspace.cgiar.org/handle/10568/27796. Accessed August 21, 2014.

Iowa State University. 2014. Fish Diseases: Piscirickettsiosis. Online. Available at http://eeda.cfsph.iastate.edu/11Jan/S_FishDiseases/fishdz0100.php?frames=frameless&layout=new. Accessed October 18, 2014.

Jafry, T., and R. Sulaiman V. 2013. Gender inequality and agricultural extension. Journal of Agricultural Education and Extension 19(5):433-436.

Johnson, B. J., F. R. B. Ribeiro, and J. L. Beckett. 2013. Application of growth technologies in enhancing food security and sustainability. Animal Frontiers 3(3):8-13.

Johnson, R. 2014. Sanitary and Phytosanitary (SPS) and Related Non-tariff Barriers to Agricultural Trade. Report No. R43450. Washington, DC: Congressional Research Service. Available at http://nationalaglawcenter.org/wp-content/uploads//assets/crs/R43450.pdf. Accessed October 1, 2014.

Jones, B. A., D. Grace, R. Kock, S. Alonso, J. Rushton, M. Y. Said, D. McKeever, F. Mutua, J. Young, J. McDermott, and D. U. Pfeiffer. 2013. Zoonosis emergence linked to agricultural intensification and environmental change. Proceedings of the National Academy of Sciences of the United States of America 110(21):8399-8404.

Jury, W. A., and H. Vaux, Jr. 2005. The role of science in solving the world's emerging water problems. Proceedings of the National Academy of Sciences of the United States of America 102(44):15715-15720.

Kelly, A. M., J. D. Ferguson, D. T. Galligan, M. Salman, and B. I. Osburn. 2013. One Health, food security, and veterinary medicine. Journal of the American Veterinary Medical Association 242(6):739-743.

Kelly, A., B. Osburn, and M. Salman. 2014. Veterinary medicine's increasing role in global health. Lancet Global Health 2(7):e379-e380.

Kogan, L. A. 2003. Looking Behind the Curtains: The Growth of Trade Barriers That Ignore Sound Science. Washington, DC: National Foreign Trade Council, Inc.Online. Available at http://www.wto.org/english/forums_e/ngo_e/posp47_nftc_looking_behind_e.pdf#page=1&zoom=auto,-265,646. Accessed October 1, 2014.

Köhler-Rollefson, I. 2012. Invisible Guardians: Women Manage Livestock Diversity. FAO Animal Production and Health Paper 174. Rome: FAO. . Online. Available at http://www.fao.org/docrep/016/i3018e/i3018e00.pdf. Accessed September 25, 2014.

Krätli, S., C. Huelsebusch, S. Brooks, and B. Kaufmann. 2013. Pastoralism: A critical asset for food security under global climate change. Animal Frontiers 3(1):42-50.

Lantier, F. 2014. Animal models of emerging diseases: An essential prerequisite for research and development of control measures. Animal Frontiers 4(1):7-12.

Lanzas, C., P. Ayscue, R. Ivanek, and Y. T. Grohn. 2010. Model or meal? Farm animal populations as models for infectious diseases of humans. Nature Reviews Microbiology 8(2):139-148.

Lebel, L., R. Mungkung, S. H. Gheewala, and P. Lebel. 2010. Innovation cycles, niches and sustainability in the shrimp aquaculture industry in Thailand. Environmental Science and Policy 13(4):291-302.

Leinonen, I., A. G. Williams, J. Wiseman, J. Guy, and I. Kyriazakis. 2012a. Predicting the environmental impacts of chicken systems in the United Kingdom through a life cycle assessment: Broiler production systems. Poultry Science 91(1):8-25.

Leinonen, I., A. G. Williams, J. Wiseman, J. Guy, and I. Kyriazakis. 2012b. Predicting the environmental impacts of chicken systems in the United Kingdom through a life cycle assessment: Egg production systems. Poultry Science 91(1):26-40.

Leroy, E. M., B. Kumulungui, X. Pourrut, P. Rouquet, A. Hassanin, P. Yaba, A. Delicat, J. T. Paweska, J. P. Gonzalez, and R. Swanepoel. 2005. Fruit bats as reservoirs of Ebola virus. Nature 438(7068):575-576.

Lippolis, J. D., and T. A. Reinhardt. 2008. Centennial paper: Proteomics in animal science. Journal of Animal Science 86(9):2430-2441.

Little, D., and R. Newton. 2010. Towards Product Carbon Footprinting for Aquaculture Products. Global Aquaculture Alliance. Online. Available at http://www.gaalliance.org/update/GOAL10/Little.pdf. Accessed August 11, 2014.

Lobell, D. B., and S. M. Gourdji. 2012. The influence of climate change on global crop productivity. Plant Physiology 160(4):1686-1697.

Lubroth, J. 2012. Climate change and animal health. Pp. 63-70 in Building Resilience for Adaptation to Climate Change in the Agriculture Sector: Proceedings of a Joint FAO/OECD Workshop, A. Meybeck, J. Lankoski, S. Redfern, N. Azzu, and V. Gitz, eds. Rome: FAO.

Lundqvist, J., C. de Fraiture, and D. Molden. 2008. Saving Water: From Field to Fork: Curbing Losses and Wastage in the Food Chain. SIWI (Stockholm International Water Institute) Policy Brief. Online. Available at http://www.siwi.org/documents/Resources/Policy_Briefs/PB_From_Filed_to_Fork_2008.pdf. Accessed October 1, 2014.

Lupo, C. D., D .E. Clay, J. L. Benning, and J. J. Stone. 2013. Life-cycle assessment of the beef cattle production system for the Northern Great Plains, U.S.A. Journal of Environmental Quality 42(5):1386-1394.

Lusk, J. L. 2013. Role of technology in the global economic importance and viability of animal protein production. Animal Frontiers 3(3):20-27.

MacMillan, S. 2014. Jimmy Smith on closing big livestock yield gaps in developing countries. ILRI News, May 5. Online. Available at http://news.ilri.org/2014/05/05/jimmy-smith-on-closing-big-livestock-yield-gaps-in-developing-countries. Accessed August 13, 2014.

Madan, M. L. 2005. Animal biotechnology: Applications and economic implications in developing countries. Revue Scientifique et technique 24(1):127-139.

Martinez, V., J. Larenas, P. Smith, A. Jedlicki, M. Kent, and S. Lien. 2014. Genetics and genomics of salmonid rickettsial syndrome. Presentation at 2nd International Conference on Integrative Salmonid Biology, Vancouver, June 10-12. Online. Available at http://www.icisb.org/sites/default/files/abstracts/PRESENTATIONS_TIRSDAG/VictorMartinez.pdf. Accessed November 3, 2014.

McDermott, J. J., S. J. Staal, H. A. Freeman, M. Herrero, and J. A. Van de Steeg. 2010. Sustaining intensification of smallholder livestock systems in the tropics. Livestock Science 130(1):95-109.

McLeod, A., ed. 2011. World Livestock 2011: Livestock in Food Security. Rome: FAO. Online. Available at http://www.fao.org/docrep/014/i2373e/i2373e.pdf. Accessed September 25, 2014.

McMichael, A. J., J. W. Powles, C. D. Butler, and R. Uauy. 2007. Food, livestock production, energy, climate change, and health. Lancet 370(9594):1253-1263.

Meale, S. J., K. A. Beauchemin, A. N. Hristov, A. V. Chaves, and T. A. McAllister. 2014. Board-invited review: Opportunities and challenges in using exogenous enzymes to improve ruminant production. Journal of Animal Science 92(2):427-442.

Meinzen-Dick, R., A. Quisumbing, J. Behrman, P. Biermayr-Jenzano, V. Wilde, M. Noordeloos, C. Ragasa, and N. Beintema. 2011. Engendering Agricultural Research, Development, and Extension. Washington, DC: International Food Policy Research Institute.

Mekonnen, M. M., and A. Y. Hoekstra. 2010. The Green, Blue and Grey Water Footprint of Farm Animals and Animal Products, Value of Water Research Report Series No. 48. Delft, The Netherlands: UNESCO-IHE. Online. Available at http://www.waterfootprint.org/Reports/Report-48-WaterFootprint-AnimalProducts-Vol1.pdf. Accessed October 1, 2014.

Micklin, P. 2007. The Aral Sea disaster. Annual Review of Earth and Planetary Sciences 35:47-72.

Micklin, P. 2010. The past, present, and future Aral Sea. Lakes & Reservoirs: Research & Management 15(3):193-213.

Millen, D. D., and M. D. B. Arrigoni. 2013. Drivers of change in animal protein production systems: Changes from "traditional" to "modern" beef cattle production systems in Brazil. Animal Frontiers 3(3):56-61.

Miracle, M. P. 1968. Subsistence agriculture: Analytical problems and alternative concepts. American Journal of Agricultural Economics 50(2):292-310.

Miranda, I. T. 2010. Farmed Pacific White Shrimp, Litopenaeus vanname, Thailand. Final Report. Seafood Watch. Online. Available at http://www.seachoice.org/wp-content/uploads/2011/10/MBA_SeafoodWatch_ThaiFarmedShrimpReport.pdf. Accessed October 6, 2014.

Montaldo, H. H., E. Casas, J. Bento Sterman Ferraz, V. E. Vega-Murillo, and S. I. Roman-Ponce. 2012. Opportunities and challenges from the use of genomic selection for beef cattle breeding in Latin America. Animal Frontiers 2(1):23-29.

Montes, F., R. Meinen, C. Dell, A. Rotz, A. N. Hristov, J. Oh, G. Waghorn, P. J. Gerber, B. Henderson, H. P. S. Makkar, and J. Dijkstra. 2013. Special Topics—Mitigation of methane and nitrous oxide emissions from animal operations: II. A review of manure management mitigation options. Journal of Animal Science 91(11):5070-5094.

Montossi, F., E. Barbieri, G. Ciappesoni, A. Ganzabal, G. Banchero, S. Luzardo, and R. San Julian. 2013. Intensification, diversification, and specialization to improve the competitiveness of sheep production systems under pastoral conditions: Uruguay's case. Animal Frontiers 3(3):29-35.

Nardone, A., B. Ronchi, N. Lacetera, M. S. Ranieri, and U. Bernabucci. 2010. Effects of climate changes on animal production and sustainability of livestock systems. Livestock Science 130(1-3):57-69.

Ngeno, K. 2014. Indigenous chicken genetic resources in Kenya: Their unique attributes and conservation options for improved use. World's Poultry Science Journal 70:173-184.

Neumeier, C. J., and F. M. Mitloehner. 2013. Cattle biotechnologies reduce environmental impact and help feed a growing planet. Animal Frontiers 3(3):36-41.

Nicholson, C. F., P. K. Thornton, L. Mohammed, R. W. Muinga, D. M. Mwamachi, E. H. Elabasha, S. J. Staal, and W. Thorpe. 1999. Smallholder Dairy Technology in Coastal Kenya. An Adoption and Impact Study. ILRI Impact Assessment Series No. 5. Nairobi (Kenya): International Livestock Research Institute.

Niemann, H., B. Kuhla, and G. Flachowsky. 2011. Perspectives for feed-efficient animal production. Journal of Animal Science 89(12):4344-4363.

Nkrumah, D. 2014. Growing sustainable smallholder livestock productivity for world's poorest people. Presentation at the First Meeting on Considerations for the Future of Animal Science Research, March 10, Washington, DC.

NRC (National Research Council). 2008. Emerging Technologies to Benefit Farmers in Sub-Saharan Africa and South Asia. Washington, DC: National Academies Press.

NRC. 2009. Enhancing the Effectiveness of Sustainability Partnerships: Summary of a Workshop. Washington, DC: The National Academies Press.

OECD (Organisation for Economic Co-operation and Development). 2013. Fragile States 2013: Resource Flows and Trends in a Shifting World. Online. Available at http://www.oecd.org/dac/incaf/FragileStates2013.pdf. Accessed October 31, 2104.

Otte, J., A. Costales, and M. Upton. 2005. Smallholder Livestock Keepers in the Era of Globalization. Pro-Poor Livestock Policy Initiative, Living from Livestock Research Report RR Nr.05-06. University of Reading, Reading, UK. Online. Available at: http://www.fao.org/ag/againfo/programmes/en/pplpi/docarc/rep-0506_globalisationlivestock.pdf. Accessed October 1, 2014.

Otte, J., A. Costales, J. Dijkman, U. Pica-Ciamarra, T. Robinson, V. Ahuja, C. Ly, and D. Roland-Holst. 2012. Livestock Sector Development for Poverty Reduction: An Economic and Policy Perspective—Livestock's Many Virtues. Rome: FAO. Online. Available at http://www.fao.org/docrep/015/i2744e/i2744e00.pdf. Accessed September 25, 2014.

Parker, R. 2012. Review of Life Cycle Assessment Research on Products Derived from Fisheries and Aquaculture: A Report for Seafish as Part of the Collective Action to Address Greenhouse Gas Emissions in Seafood. Final Report. Online. Available at http://www.seafish.org/media/583639/seafish_lca_review_report_final.pdf. Accessed August 11, 2014.

Patience, J. F. 2012. The importance of water in pork production. Animal Frontiers 2(2):28-35.

Pelletier, N. 2008. Environmental performance in the U.S. broiler poultry sector: Life cycle energy use and greenhouse gas, ozone depleting, acidifying and eutrophying emissions. Agricultural Systems 98(2):67-73.

Pelletier, N., P. Lammers, D. Stender, and R. Pirog. 2010a. Life cycle assessment of high- and low-profitability commodity and deep-bedded niche swine production systems in the upper midwestern United States. Agricultural Systems 103(9):599-608.

Pelletier, N., R. Pirog, and R. Rasmussen. 2010b. Comparative life cycle environmental impacts of three beef production strategies in the upper midwestern United States. Agricultural Systems 103(6):380-389.

Pelletier, N., M. Ibarburu, and H. Xin. 2014. Comparison of the environmental footprint of the egg industry in the United States in 1960 and 2010. Poultry Science 93(2):241-255.

Perry, B., and D. Grace. 2009. The impact of livestock disease and their control on growth and development process that are pro-poor. Philosophical Transactions of the royal Society B: Biological Sciences 364(1530):2643-2655.

Perry, B. D., D. Grace, and K. Sones. 2013. Current drivers and future directions of global livestock disease dynamics. Proceedings of the National Academy of Sciences of the United States of America 110(52):20871-20877.

Pica-Ciamarra, U. 2005. Livestock Policies for Poverty Alleviation: Theory and Practical Evidence from Africa, Asia and Latin America. PPLPI Working Paper No.27. Pro Poor Livestock Policy Initiative. Rome: FAO. Online. Available at http://www.fao.org/ag/againfo/programmes/en/pplpi/docarc/execsumm_wp27.pdf. Accessed October 3, 2014.

Pingali, P. L. 2012. Green Revolution: Impacts, limits, and the path ahead. Proceedings of the National Academy of Sciences of the United States of America 109(31):12302-12308.

Pretty, J., W. J. Sutherland, J. Ashby, J. Auburn, D. Baulcombe, M. Bell, J. Bentley, S. Bickersteth, K. Brown, J. Burke, H. Campbell, K. Chen, E. Crowley, I. Crute, D. Dobbelaere, G. Edwards-Jones, F. Funes-Monzote, H. C. J. Godfray, M. Griffon, P. Gypmantisiri, L. Haddad, S. Halavatau, H. Herren, M. Holderness, A. M. Izac, M. Jones, P. Koohafkan, R. Lal, T. Lang, J. McNeely, A. Mueller, N. Nisbett, A. Noble, P. Pingali, Y. Pinto, R. Rabbinge, N. H. Ravindranath, A. Rola, N. Roling, C. Sage, W. Settle, J. M. Sha, L. Shiming, T. Simons, P. Smith, K. Strzepeck, H. Swaine, E. Terry, T. P. Tomich, C. Toulmin, E. Trigo, S. Twomlow, J. K. Vis, J. Wilson, and S. Pilgrim. 2010. The top 100 questions of importance to the future of global agriculture. International Journal of Agricultural Sustainability 8(4):219-236.

Pretty, J., C. Toulmin, and S. Williams. 2011. Sustainable intensification in African agriculture. International Journal of Agricultural Sustainability. 9(1):5-24.

Prudêncio da Silva, V., H. M. G. van der Werf, S. R. Soares, and M. S. Corson. 2014. Environmental impacts of French and Brazilian broiler chicken production scenarios: An LCA approach. Journal of Environmental Management 133:222-231.

Ragasa, C., G. Berhane, F. Tadesse, and A. S. Taffesse. 2013. Gender differences in access to extension services and agricultural productivity. Journal of Agricultural Education and Extension 19(5):437-468.

Randolph, T. F., E. Schelling, D. Grace, C. F. Nicholson, J. L. Leroy, D. C. Cole, M. W. Demment, A. Omore, J. Zinsstag, and M. Ruel. 2007. Invited review: Role of livestock in human nutrition and health for poverty reduction. Journal of Animal Science 85(11):2788-2800.

Reantaso, M. B., B. Sengvilaykham, A. Cameron, and P. Chanratchakool. 2002. A survey of the socio-economic impact of aquatic animal disease on small-scale aquaculture production and reservoir/capture fisheries in southern Lao PDR. Pp. 235-252 in Primary Aquatic Animal Health Care in Rural, Small-scale, Aquaculture Development, J. R. Arthur, M. J. Phillips, R. P. Subasinghe, M. B. Reantaso, and I. H. MacRae, eds. FAO Fisheries Technical Paper No. 406. Rome: Food and Agriculture Organization.

Reantaso, M. B., R. P. Subasinghe, and R. Van Anrooy. 2006. Application of risk analysis in aquaculture. FAO Aquaculture Newsletter 35:20-26.

Rege, J. E. O. 2009. Biotechnology Options for Improving Livestock Production in Developing Countries, with Special Reference to Sub-Saharan Africa. Online. Available at http://www.fao.org/wairdocs/ilri/x5473b/x5473b05.htm. Accessed June 19, 2014.

Rendón-Huerta, J. A., J. M. Pinos-Rodríguez, J. C. García-López, L. G. Yáñez-Estrada, and E. Kebreab. 2014. Trends in greenhouse gas emissions from dairy cattle in Mexico between 1970 and 2010. Animal Production Science 54(3):292-298.

Ringler, C., N. Cenacchi, J. Koo, R. Robertson, M. Fisher, C. Cox, N. Perez, K. Garrett, and M. Rosegrant. 2014. Sustainable agricultural intensification: The promise of innovative farming practices. Pp. 43-52 in 2013 Global Food Policy Report, A. Marble and H. Fritschel, eds. Washington, DC: International Food Policy Research Institute.

Roberts, R. M., G. W. Smith, F. W. Baze, J. Cibelli, G. E. Seidel, Jr., D. E. Bauman, L. P. Reynolds, and J. J. Ireland. 2009. Farm animal research in crisis. Science 324(5926):468-469.

Rodrik, D. 2007. Normalizing Industrial Policy. Prepared for the Commission on Growth and Development. Online. Available at http://www.hks.harvard.edu/fs/drodrik/Research%20papers/Industrial%20Policy%20_Growth%20Commission_.pdf. Accessed November 3, 2014.

Rodriquez-Martinez, H., and F. P. Vega. 2013. Semen technologies in domestic animal species. Animal Frontiers 3(4):26-33.

Rosegrant, M. W., J. Koo, N. Cenacchi, C. Ringler, R. Robertson, M. Fisher, C. Cox, K. Garrett, N. D. Perez, and P. Sabbagh. 2014. Food Security in a World of Natural Resource Scarcity: The Role of Agricultural Technologies. Washington, DC: International Food Policy Research Institute.

Rouquette, F. M., Jr., L. A. Redmon, G. E. Aiken, G. M. Hill, L. E. Sollenberger, and J. Andrae. 2009. ASAS centennial paper: Future needs of research and extension in forage utilization. Journal of Animal Science 87(1):438-446.

Salman, M. D. 2004. BSE: How does it impact veterinary education? Journal of Veterinary Medical Education 31(2):87-88.

Salman, M. D. 2009. The role of veterinary epidemiology in combating infectious animal diseases on a global scale: The impact of training and outreach programs. Preventive Veterinary Medicine 92(4):284-287.

Salman, M. D., V. Silano, D. Heim, and J. Kreysa. 2012. Geographical BSE risk assessment and its impact on disease detection and dissemination. Preventive Veterinary Medicine 105(4):255-264.

Sanchez, P., C. Palm, J. Sachs, G. Denning, R. Flor, R. Harawa, B. Jama, T. Kiflemariam, B. Konecky, R. Kozar, E. Lelerai, A. Malik, V. Modi, P. Mutuo, A. Niang, H. Okoth, F. Place, S. E. Sachs, A. Said, D. Siriri, A. Teklehaimanot, K. Wang, J. Wangila, and C. Zamba. 2007. The African millennium villages. Proceedings of the National Academy of Sciences of the United States of America 104(43):16775-16780.

Savory, A., and J. Butterfield. 1999. Holistic Management. Washington, DC: Island Press.

Scanlon, B. R., C. C. Faunt, L. Longuevergne, R. C. Reedy, W. M. Alley, V .L. McGuire, and P. B. McMahon. 2012. Groundwater depletion and sustainability of irrigation in the U.S. High Plains and Central Valley. Proceedings of the National Academy of Sciences of the United States of America 109(24):9320-9325.

Scollan, N. D., P. L. Greenwood, C. J. Newbold, D. R. Y. Ruiz, K. J. Shingfield, R. J. Wallace, and J. F. Hocquette. 2011. Future research priorities for animal production in a changing world. Animal Production Science 51(1):1-5.

Seufert, V., N. Ramankutty, and J. A. Foley. 2012. Comparing the yields of organic and conventional agriculture. Nature 485(7397):229-234.

Sheldon, B. L. 2000. Research and development in 2000: Directions and priorities for the world's poultry science community. Poultry Science 79:147-158.

Sherren, K., J. Fischer, and I. Fazey. 2012. Managing the grazing landscape: Insights for agricultural adaptation from a mid-drought photo-elicitation study in the Australian sheep-wheat belt. Agricultural Systems 106:72-83.

Smith, J., K. Sones, D. Grace, S. MacMillan, S. Tarawali, and M. Herrero. 2013. Beyond milk, meat, and eggs: Role of livestock in food and nutrition security. Animal Frontiers 3(1):6-13.

Smith, P. T., ed. 1999. Towards Sustainable Shrimp Culture in Thailand and the Region. Proceedings of a Workshop, Hat Yai, Songkhla, Thailand, October 28-November 1, 1996. ACIAR Proceedings No. 90. Online. Available at http://aciar.gov.au/files/node/2196/pr90_pdf_11112.pdf.

Spencer, T. E. 2013. Early pregnancy: Concepts, challenges, and potential solutions. Animal Frontiers 3(4):48-55.

Spielman, D. J., and K. von Grebmer. 2004. Public-Private Partnerships in Agricultural Research. International Food Policy Research Institute (IFPRI). Series 113. Online. Available at http://www.ifpri.org/publication/public-private-partnerships-agricultural-research?print. Accessed August 22, 2014.

Steinfeld, H., and P. Gerber. 2010. Livestock production and the global environment: Consume less or produce better? Proceedings of the National Academy of Sciences of the United States of America 107(43):18237-18238.

Steinfeld, H., T. Wassenaar, and S. Jutzi. 2006. Livestock production systems in developing countries: Status, drivers, trends. Revue Scientifique et Technique 25(2):505-516.

Stewart, J., and E. Tallaksen. 2013. High disease-control costs eat chilean salmon farmers' margins. Undercurrent News: August 12. Online. Available at http://www.undercurrentnews.com/2013/08/12/high-disease-control-costs-eat-chilean-salmon-farmers-margin. Accessed October 6, 2014.

St-Pierre, N. R., B. Cobanov, and G. Schnitkey. 2003. Economic losses from heat stress by U.S. livestock industries. Journal of Dairy Science 86(Suppl.):E52-E77.

Subasinghe, R. P., M. G. Bondad-Reantaso, and S. E. McGladdery. 2001. Aquaculture development, health and wealth. Pp. 167-191 in Aquaculture in the Third Millennium. Technical Proceedings of the Conference on Aquaculture in the Third Millennium, Bangkok, Thailand, February 20-25, 2000, R. P. Subasinghe, P. Bueno, M. J. Phillips, C. Hough, S. E. McGladdery, and J. R. Arthur, eds. Bangkok: Network of Aquaculture Centers in Asia-Pacific and Rome: Food and Agriculture Organization.

Szuster, B. W. 2006. A review of shrimp farming in central Thailand and its environmental Implications. Pp. 155-164 in Shrimp Culture: Economics, Market, and Trade, P. Leung and C. Engle, eds. Ames, IA: Blackwell.

Taylor, R. C., H. Omed, and G. Edwards-Jones. 2014. The greenhouse emissions footprint of free-range eggs. Poultry Science 93(1):231-237.

Thoma, G., D. Nutter, R. Ulrich, C. Maxwell, J. Frank, and C. East. 2011. National Life Cycle Carbon Footprint Study for Production of US Swine. Online. Available at http://www.pork.org/filelibrary/NPB%20Scan%20Final%20-%20May%202011.pdf. Accessed August 5, 2014.

Thoma, G., J. Popp, D. Nutter, D. Shonnard, R. Ulrich, M. Matlock, D.S. Kim, Z. Neiderman, N. Kemper, C. East, and F. Adom. 2013. Greenhouse gas emissions from milk production and consumption in the United States: A cradle-to-grave life cycle assessment circa 2008. International Dairy Journal 31(Suppl. 1):S3-S14.

Thomas, D. S., and C. Twyman. 2005. Equity and justice in climate change adaptation amongst natural-resource-dependent societies. Global Environmental Change 15(2):115-124.

Thornton, P. K. 2010. Livestock production: Recent trends, future prospects. Philosophical Transactions of the Royal Society B: Biological Sciences 365(1554):2853-2867.

Thornton, P. K., and M. Herrero. 2010. Potential for reduced methane and carbon dioxide emissions from livestock and pasture management in the tropics. Proceedings of the National Academy of Sciences of the United States of America 107(46):19667-19672.

Thornton, P. K., J. van de Steeg, A. M. Notenbaert, and M. Herrero. 2009. Impacts of climate change on livestock and livestock systems in developing countries: A review of what we know and what we need to know. Agricultural Systems 101(3):113-127.

Thornton, P., M. Herrero, and P. Ericksen. 2011. Livestock and Climate Change. Livestock Exchange Issue Brief 3. Online. Available at https://cgspace.cgiar.org/bitstream/handle/10568/10601/IssueBrief3.pdf. Accessed August 11, 2014.

Tidwell, J. H., and G. L. Allan. 2001. Fish as food: Aquaculture's contribution. EMBO Reports 2(11):958-963.

Todaro, M. P., and S. C. Smith. 2014. Economic Development, 12th ed. Upper Saddle River, NJ: Pearson.

Tomley, F. M., and M. W. Shirley. 2009. Livestock infectious diseases and zoonoses. Philosophical Transactions of the Royal Society B: Biological Sciences 364(1530):2637-2642.

Troell, M., R. Naylor, M. Metian, M. Beveridge, P. Tyedmers, C. Folke, H. Osterblom, A. de Zeeuw, M. Scheffer, K. Nyborg, S. Barrett, A-S. Crepin, P. Ehrlich, S. Levin, T. Xepapadea, S. Polasky, K, Arrow, A. Gren, N. Kautsky, S. Taylor, and B. Walker. 2014. Does aquaculture add resilience to the global food system? Proceedings of the National Academy of Science of the United States of America 111(37):13257-13263.

Townsend, A. R., R. W. Howarth, F. A. Bazzaz, M. S. Booth, C. C. Cleveland, S. K. Collinge, A. P. Dobson, P. R. Epstein, E. A. Holland, D. R. Keeney, M. A. Mallin, C. A. Rogers, P. Wayne, and A. H. Wolfe. 2003. Human health effects of a changing global nitrogen cycle. Frontiers in Ecology and the Environment 1(5):240-246.

Tubiello, F. N., J. F. Soussana, and S. M. Howden. 2007. Crop and pasture response to climate change. Proceedings of the National Academy of Sciences of the United States of America 104(50):19686-19690.

UNEP (United Nations Environment Programme). 2008. Vital Water Graphics: An Overview of the State of the World's Fresh and Marine Waters, 2nd Ed. Online. Available at http://www.grida.no/publications/vg/water2. Accessed June 17, 2014.

United Nations. 2014. World Urbanization Prospects: The 2014 Revision, Highlights. ST/ESA/SER.A/352. Department of Economic and Social Affairs, Population Division. Available at http://esa.un.org/unpd/wup/Highlights/WUP2014-Highlights.pdf.

Upton, M. 2004. Policy Issues in Livestock Development and Poverty Reduction. Pro-Poor Livestock Policy Initiative Policy Brief. Online. Available at http://www.fao.org/ag/againfo/programmes/en/pplpi/docarc/pb_wp10.pdf. Accessed August 13, 2014.

USAID. 2013. Sustainable Fisheries and Responsible Aquaculture: A Guide for USAID Staff and Partners. Available at http://www.usaid.gov/documents/1865/fisheries-and-aquaculture-guide.

Valderrama, D., N. Hishamuda and X. W. Zhou. 2010. Estimating employment in world aquaculture. FAO Fisheries and Aquaculture Newsletter 45:24-25.

Van Eenennaam, A. L. 2006. What is the future of animal biotechnology? California Agriculture 60(3):132-139.

van Huis, A., J. Van Itterbeeck, H. Klunder, E. Mertens, A. Halloran, G. Muir, and P. Vantomme. 2013. Edible Insects: Future Prospects for Food and Feed Security. FAO Forestry Paper 171. Rome: FAO. Online. Available at http://www.fao.org/docrep/018/i3253e/i3253e.pdf. Accessed September 29, 2014.

Waite, R, M. Beveridge, R. Brummett, S. Castine, N. Chaiyawannakarn, S. Kaushik, R. Mungkung, S. Nawapakpilai, and M. Phillips. 2014. Increasing Productivity of Aquaculture. Working Paper, Installment 6 of Creating a Sustainable Food Future. Washington, DC: World Resources Institute.

Welfare Quality. 2014. Science and Society Improving Animal Welfare. Online. Available at http://www.welfarequality.net. Accessed August 11, 2014.

Williams, M. J., R. Agbayani, R. Bhujel, M. G. Bondad-Reantaso, C. Brugère, P. S. Choo, J. Dhont, A. Galmiche-Tejeda, K. Ghulam, K. Kusakabe, D. Little, M. C. Nandeesha, P. Sorgeloos, N. Weeratunge, S. Williams, and P. Xu. 2012. Sustaining aquaculture by developing human capacity and enhancing opportunities for women. Pp. 785-874 in Farming the Waters for People and Food: Proceedings of the Global Conference on Aquaculture 2010, September 22–25, 2010, Phuket, Thailand, R. P. Subasinghe, J. R. Arthur, D. M. Bartley, S. S. De Silva, M. Halwart, N. Hishamunda, C. V. Mohan, and P. Sorgeloos, eds. Bangkok: Network of Aquaculture Centers in Asia-Pacific and Rome: Food and Agriculture Organization.

WHO (World Health Organization). 2014. Global and Regional Food Consumption Patterns and Trends. Online. Available at http://www.who.int/nutrition/topics/3_foodconsumption/en/index4.html. Accessed October 28, 2014.

World Bank. 2012. People, Pathogens and Our Plant, Vol. 2. The Economics of One Health. Report No. 69145-GLB. Washington, DC: The World Bank. Online. Available at http://www-wds.worldbank.org/external/default/WDSContentServer/WDSP/IB/2012/06/12/000333038_20120612014653/Rendered/PDF/691450ESW0whit0D0ESW120PPPvol120web.pdf. Accessed October 1, 2014.

World Wildlife Fund. 2014. Certifying Sustainable Shrimp Farming Practices in Thailand. Online. Available at http://wwf.panda.org/about_our_earth/search_wwf_news/?222810/Certifying-sustainable-shrimp-farming-practices-in-Thailand. Accessed October 18, 2014.

Wright, J. 2014. Case Study: The Fish Have Landed. SeafoodSource.com.Online. Available at http://www.seafoodsource.com/news/aquaculture/26504-case-study-the-fish-have-landed. Accessed October 3, 2014.

Yi, L., C. M. M. Lakemond, L. M. C. Sagis, V. Eisner-Schadler, A. van Huis, and M. A. J. S. van Boekel. 2013. Extraction and characterization of protein fractions from five insect species, Food Chemistry 141(4):3341-3348.

Zijlstra, R. T., and E. Beltranena. 2013. Swine convert co-products from food and biofuel industries into animal protein for food. Animal Frontiers 3(2):48-53.

5

Capacity Building and Infrastructure for Research in Food Security and Animal Sciences

INTRODUCTION

This chapter discusses capacity building in the animal sciences via research, research transfer, and undergraduate and graduate education, and the infrastructural changes needed to effectively support these activities. The focus is mainly on land-grant universities because they are the primary institutions conducting integrated research, research outreach, and instructional activities in the animal sciences in the United States. Data collected by the committee about the 128 U.S. universities with animal science programs entered into the U.S. Department of Agriculture (USDA) Food and Agricultural Education Information System (FAEIS) database showed that in 2012, 90 percent of the 909 faculty in animal science departments were at land-grant institutions (1862, 1890, or 1994), with the remainder at non-land-grant institutions and none from private institutions. The committee recognizes the value of animal science training and research provided by non-land-grant institutions, as well as by non-U.S. institutions. Neither the committee's mandate nor the areas of expertise of the committee members allowed comparisons of capacity-building activities in U.S. and non-U.S. institutions. It was, however, recognized that several U.S. higher education institutions have established bilateral or multilateral agreements with non-U.S. institutions in the field of animal sciences, and that these agreements can be beneficial and effective in enhancing and improving the quality of research and training in animal sciences. The committee also recognizes the critical role that private industry plays in conducting research and providing research outreach as a part of their technical service teams. There were insufficient data available to the committee in order for this contribution to be evaluated, but the critical role of public–private partnership building for the future of animal science research, outreach, and training is recognized and discussed.

5-1 Research in Animal Sciences

Federally, most animal science research is funded by the USDA via two mechanisms: competitive grant funding and intramural allocations to the Agricultural Research Service (ARS). The Hatch Act of 1887 transformed the Bureau of Agriculture established by President Lincoln in 1862 into the USDA, with its primary emphasis on agricultural research. The Act mandated that USDA sponsor extramural agricultural research to solve the food challenges that the country was, and would be, facing. It provided funding to federal laboratories and state agricultural colleges through a formula based on each state's share of the rural and farm populations, and authorized federal funds for the development of agricultural research at land-grant institutions (FA-RM, 2014). These "formula funds" focused on applied, mission-oriented programs, teaching, and extension (Roberts et al., 2009) and provided the backbone on which research in animal agriculture was based; however, these formula funds have declined significantly over time. Huffman and Evenson (2006) demonstrated a decline in formula funding (in constant dollars) for agricultural research and extension of 57 percent from 1980 to 2003. Overall revenue to state agricultural experiment stations actually increased by about 21 percent during this same period, mainly due to increased funding from non-Cooperative State Research Service (CSRS)-Cooperative State Research, Education, and Extension Service (CSREES) federal government research funds, contracts and grants, as well as commodity group and foundation funding. Hatch funding continued to decline in constant dollars from 2003-2008; it then increased in 2007 and 2008, but thereafter declined again to levels similar to 2003 (Figures 5-1 and 5-2). Huffman and Evenson (2006) found that formula funding has a greater impact on agricultural research productivity than competitive grant funding. Advantages of formula funding are that it can provide steady support for core research and be used to address issues that are of local importance. In addition is has greater "purchasing power" given that it is not subject to university overhead costs.

The National Research Council has repeatedly warned about gross underfunding of the basic sciences of agriculture (NRC, 1972, 1989, 2000). Despite these warnings, Congress has failed to act in a significant way as overall funding has decreased for animal systems (animal science/animal health research; CRIS codes RPA 301-315) from 1998 to 2011 (Tables 5-1a and 5-1b; Box 5-1). Overall funding for animal systems has shown large annual fluctuations, mainly due to variations in funding received from other federal (i.e., non-USDA) funding sources. In general, there was an overall increase in funding in real terms compared to 1998 dollars until 2010 when funding dropped and actually fell below 1998 dollars. Despite this lack of congressional priority and the declining role that USDA funding has played in terms of its contribution to the overall public funding portfolio for agricultural research (NRC, 2014), USDA has funded many notable advances in animal health, food safety, genetic improvements, reproductive efficiencies, nutrient utilization, and animal production systems.

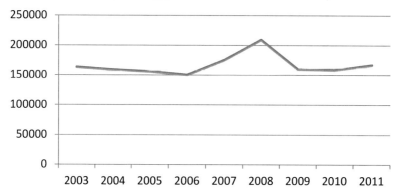

FIGURE 5-1 Hatch funding in real dollars (2003-2011).
SOURCE: Data from USDA NIFA (2014a).

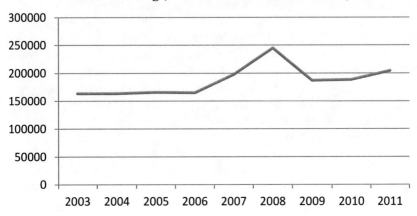

FIGURE 5-2 Hatch funding in nominal dollars (2003-2011).
SOURCE: Data from USDA NIFA (2014a).

TABLE 5-1a Source of Funds for Animal Systems Research in FY 1998-2004 as reported by Current Research Information System for 15 Fields[a], Funds by Fiscal Year (thousands)

Source	1998	1999	2000	2001	2002	2003	2004
USDA[b]	104,760	117,121	122,219	133,855	144,408	152,445	158,970
Other USDA [c]	10,682	11,004	12,970	18,856	22,878	23,713	22,848
CSRESS ADM[d]	77,336	74,978	79,088	83,637	87,178	91,870	104,658
State funds	289,771	302,520	304,970	315,566	296,144	299,943	303,177
Other Nonfederal[e]	160,103	151,063	167,922	180,900	187,261	197,655	206,117
Other federal[f]	110,999	151,320	156,458	181,552	219,499	232,644	251,848
Total	753,651	808,007	843,627	914,366	957,368	998,270	1,047,618
Total (real 1998 dollars)	753,651	790,547	798,555	841,567	867,431	884,337	903,979

TABLE 5-1b Source of Funds for Animal Systems Research in FY 2005-2011 as reported by Current Research Information System for 15 Fields[a], Funds by Fiscal Year (thousands)

Source	2005	2006	2007	2008	2009	2010	2011
USDA[b]	155,255	164,302	170,509	171,826	168,783	188,986	177,277
Other USDA[c]	25,486	26,732	26,462	28,529	30,675	38,803	33,973
CSRESS ADM[d]	101,655	102,087	103,609	111,270	100,582	49,807	56,055
State funds	300,279	324,549	321,557	344,337	322,229	260,619	250,917
Other Nonfederal[e]	189,522	211,619	212,568	221,621	234,579	196,896	201,635
Other federal[f]	232,070	495,340	524,503	538,184	226,764	185,632	221,293
Total	1,004,268	1,324,629	1,359,209	1,415,767	1,083,612	920,743	941,150
Total (real 1998 dollars)	838,175	1,071,004	1,068,529	1,071,838	823,302	688,268	681,995

[a]CRIS reporting categories RPA 301-315 (reproduction, nutrition, genetics, animal genome, animal physiology, environmental stress, animal production and management, improved animal products, animal disease, external parasites and pests, internal parasites, toxicology, and animal welfare).

[b]Regular USDA appropriations used for in-house research by USDA research agencies and centers (excludes CSREES programs) (Form AD-418 field 131).
[c]Other USDA: expenditure of funds received by state agriculture experiment stations (SAESs) and other cooperating institutions from contracts, grants, or cooperative agreement with one of the USDA research agencies other than CSREES. Identification of awarding agency is not collected. (Form AD-419 field number 219).
[d]CSREES ADM: expenditure of formula and grant funds administered by CSREES and distributed to SAESs and other cooperating institutions (OCIs). Programs included are National Research Initiative, Hatch, McIntire-Stennis, Evans-Allen, Animal Health, Special Grants, Competitive Grants, Small Business Innovation Research Grants, and other CSREES grant programs (Form AD-419 field 31).
[e]Other nonfederal: expenditures by USDA agencies, SAESs, and OCIs of funds received from sources outside federal government, such as industry grants and sale of products (self-generated).
[f]Other federal: expenditures by USDA agencies, SAESs, and OCIs of funds received from federal sources outside USDA through contracts, grants, and cooperative agreements directly with other federal agencies. Sponsoring agencies may include National Science Foundation, Department of Energy, Department of Defense, Agency for International Development, National Institutes of Health, Public Health Service, Department of Health and Human Services, National Aeronautics and Space Administration, and Tennessee Valley Authority. (Form AD-418 field number 332/Form AD-419 field number 332 minus field 219).
SOURCE: USDA-CSREES; NRC (2005)Beginning in 2009, the National Institute for Food and Agriculture (NIFA) was established to replace the USDA's CSREES. The NIFA program that allocates competitive funding is the Agricultural and Food Research Initiative (AFRI). NIFA investment in animal science research was essentially stagnant in real dollars between 2003 and 2012 (Table 5-2), with a mean of $114,584,000 per year (CV of 8.2 percent) (Table 5-3; Appendix J). The highest priority was dairy production (17 percent of the funding) followed by aquaculture (16 percent), beef production (16 percent), poultry production (12 percent), and swine production (11 percent).

TABLE 5-2 Dollars Directed Toward Animal Science Research by NIFA

Year	Actual dollars ($1,000)	Real dollars ($1,000), 2003 as Base
2003	99,497	99,497
2004	117,915	114,856
2005	117,492	110,694
2006	116,183	106,040
2007	106,398	94,420
2008	130,556	111,574
2009	116,015	99,501
2010	114,117	96,294
2011	124,363	101,729
2012	103,299	82,785

SOURCE: USDA (2014b).

TABLE 5-3 Average Annual NIFA Investment in Animal Science Research Between 2003 and 2012, Mean from 2003 to 2012

Animal Science Investment ($1000) by Species	Mean	CV, %	% of Total
Dairy, production	$19,700	12.7	17
Aquaculture	$18,653	25.9	16
Beef, production	$17,966	8.9	16
Poultry, production	$14,249	16.2	12
Swine, production	$12,317	24.2	11
Other	$10,264	40.3	9
Goats, food & hair	$4,987	26.1	4
Sheep, production	$4,773	27.8	4
Dairy, food	$2,995	40.2	3
Horses, ponies, and mules	$2,911	21.8	3
Beef, Food (Meat)	$2,054	27.1	2
Poultry food (meat & eggs)	$1,706	30..8	1
Swine, food	$1,430	26,1	1
Sheep food & wool	$581	27.1	1
Total	$114,584	8.2	100

SOURCE: USDA (2014b).

The NIFA investment by discipline or knowledge area was also flat in terms of relative percentage allocations over the 10-year period between 2003 and 2012 (Table 5-4; Appendix K). Animal health was the highest priority (23 percent of the funding) followed by genetic improvement/genome (17 percent); reproductive performance (12 percent); nutrient utilization (12 percent); animal management systems (10 percent); animal physiology (6 percent); food (5 percent); economic, marketing, trade, policy (4 percent); improved animal products (3 percent); animal welfare (3 percent); environmental stress (1 percent); facilities and engineering (1 percent); waste disposal (1 percent); and nonfood products and maintenance (1 percent).

The USDA ARS allocation declined in constant dollars from FY 2010 to FY 2014 (Figures 5-3 and 5-4), and the research priority area percentage allocations essentially remained static (Appendix L); ARS research priorities were different from those of NIFA. Animal agriculture research investment (with a mean of $241,538,400 per year) was the highest for beef (25 percent of the animal and aquaculture investment) followed by poultry (19 percent), dairy (18 percent), swine (14 percent), aquaculture (14 percent), sheep (6 percent), and other animal research (4 percent). Of the total USDA ARS appropriations (with a mean $1,109,389,200 per year), animal agriculture represented 22 percent of the total allocation compared to 38 percent for plant research (Table 5-5; Appendix L).

TABLE 5-4 Average Annual NIFA Investment in Animal Science Knowledge Areas, 2003-2012

Knowledge Areas	Mean from 2003 to 2012		
	Mean ($1,000)	CV, %	% of Total
Animal health	26,805	14.8	23
Genetic improvement/genome	19,524	15.7	17
Reproductive performance	14,319	11.2	12
Nutrient utilization	13,942	12.6	12
Animal management systems	11,599	10.4	10
Animal physiology	7,086	13.8	6
Food	5,305	18.6	5
Economic, marketing, policy	4,011	20.8	4
Animal welfare	3,729	55.7	3
Improved animal products	3,722	47.3	3
Environmental stress	1,660	22.8	1
Facilities, engineering	1,209	54.4	1
Waste disposal	1,019	32.4	1
Nonfood products & maintenance	652	48.0	1
Total	114,584	8.2	100

SOURCE: USDA (2014b).

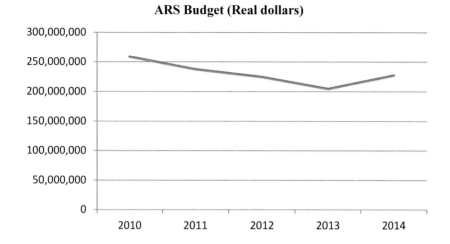

FIGURE 5-3 ARS budget in real dollars (FY 2010-2014). SOURCE: USDA (2014a).

Overall, animal science research funding was flat, with a decrease in purchasing power over the past few decades and with little change in the allocation of funds among species and knowledge areas. Underfunding animal science research has long-lasting consequences, including a decrease in faculty positions and animal, dairy, and poultry science departments; reduction in infrastructure, trained students, and industry and government jobs; and reduced innovations needed to address challenges. Once lost, it will take a greater investment with a longer time lag before productive research can be regained, if possible. In 2008, the USDA competitive grants program, the National Research Initiative, was replaced with a new competitive grants programs called the Agriculture and Food Research Initiative (AFRI). This program has greater potential for receiving funding because allocations are stipulated within the Farm Bill. Increased coordination among federal funding agencies could also stimulate additional funding for animal sciences research, such as the current USDA-NIH Dual Purpose Dual Benefit program (Box 5-2)

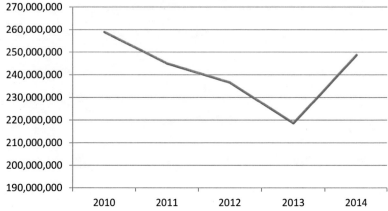

FIGURE 5-4 ARS budget in nominal dollars (FY 2010-2014).
SOURCE: USDA (2014a).

TABLE 5-5 Average Annual USDA ARS Appropriations by Knowledge Area from FY 2010 to FY 2014

Knowledge Area	Mean per Year, $	% of Appropriation	% of Animal & Aquaculture Research
Animal health	$65,977,200	6	27
Food safety	$53,633,800	5	22
Food animal production	$47,809,800	4	20
Aquaculture	$30,457,800	3	13
Entomology	$11,490,200	1	5
Quality and utilization of agricultural products	$9,025,400	1	4
Pasture, forage, and rangeland systems, agricultural system competitiveness	$8,364,400	1	4
Agricultural and industrial byproducts	$5,124,000	1	2
Manure and byproduct utilization	$3,723,200	0	2
Human nutrition	$3,559,600	0	1
Biorefining	$808,200	0	0

Knowledge Area	Mean per Year, $	% of Appropriation	% of Animal & Aquaculture Research
Sustainability	$337,600	0	0
Bioenergy and energy alternatives	$320,400	0	0
Climate change, soils, emissions	$283,600	0	0
Plant genetic resources, genomics, and genetic improvement	$202,400	0	0
Crop protection and quarantine	$129,400	0	0
Global change	$122,600	0	0
Air quality	$104,600	0	0
Water availability and management	$64,200	0	0
Total research in support of animals	$241,538,400	22	100
ARS total appropriation	$1,109,389,200		

SOURCE: USDA (2014a).

The lack of change in percentage allocations among animal science research priorities within USDA, despite the changing pressures facing animal agriculture research, suggests a need for a more structured and consistent research priority planning process that involves representative stakeholders. The NRC (2014) review of AFRI noted that "AFRI does not have clearly articulated plans to guide its priority-setting management processes and interagency collaboration," and recommended that AFRI develop a strategic plan to identify such priorities and create a framework for assessing the program's progress. The committee strongly endorses this recommendation, as well as the report's recommendation that there be an external advisory council that can provide guidance and validate AFRI strategic directions.

> **BOX 5-2**
> **Dual Purpose for Dual Benefit**
>
> The Eunice Kennedy Shriver National Institute of Child Health and Human Development (NICHD) and the National Institute of Food and Agriculture (USDA NIFA) posted a request for applications (RFA) in August 2013 seeking proposals related to research on agriculturally important animal species, with estimated funding of $5 million (NIH, 2014). The goal is to encourage research using agriculturally relevant animals that will benefit both human and animal populations. The proposals for research were due in September 2014, and the RFA background text acknowledged the progress related to human diseases and reproductive medicine that results from agricultural animal science research. This proposal seeks to specifically gain knowledge related to assisted reproduction technologies and stem-cell biology, metabolism, the developmental origin of adult disease, and infectious diseases, which are noted as being "high priority" issues in the fields of both biomedicine and agriculture (NIH, 2014). The RFA emphasizes the benefits to be obtained from these areas for both human and animal health, and how joint funding initiatives such as this can facilitate cross-disciplinary and comparative research that benefits both medicine and agriculture.

Two existing models for collecting information essential to the research prioritization process that could be of value are the Council for Agricultural Science and Technology (CAST) and the European Food Safety Authority (EFSA). CAST is a nonprofit organization in the United States comprising scientific societies and many individual, student, company, nonprofit, and associate society members. CAST's board is composed of representatives of the scientific societies, commercial companies, nonprofit or trade organizations, and a board of directors. The primary work of CAST is the publication of task force reports, commentary papers, special publications, and issue papers written by scientists from many disciplines. Through its publications, CAST is able to address research needs, including the priority of food animal sciences research, by composing teams representing various stakeholders.

EFSA is the keystone of European Union risk assessment regarding food and feed safety including agricultural production as a whole. EFSA is an independent European agency funded by the European Union (EU) budget that operates separately from the European Commission, European Parliament, and EU member states. In close collaboration with national authorities and in open consultation with its stakeholders, EFSA provides independent scientific advice and clear communication on existing and emerging risks. EFSA has been a scientific entity in producing technical scientific reports of various topics relevant to food, feed, and food security. Because of the structure and procedures utilized by EFSA, these reports have provided scientifically reliable, mostly independent scientific opinions about the topics and have been used as a basis for developing research priorities and synthesizing systems approaches for research in animal sciences.

Priorities for Infrastructure

Federal formula funding for agricultural research has declined significantly since the 1980s, thereby exerting a significant impact on research and the infrastructure needed to support that research. Since 1972, the NRC has repeatedly warned about the consequences of underfunding of agricultural sciences. Although overall funding for animal science and animal health research has actually increased during that time, this increase has mainly been derived from increases in non-USDA federal funding and non-federal funding. These sources are subject to large annual fluctuations and do not provide the steady source of support needed for animal science research to achieve the long-term goal of sustainable intensification. USDA competitive funding for research, as well as ARS funding, showed a decreasing trend in real dollars during the last decade. Underfunding animal science research has long-lasting consequences, including a decrease in faculty, postdoctoral, and graduate student positions; loss or consolidation of many animal, dairy, and poultry science departments; and the lagging of improvement/enhancement of other essential infrastructure that is critical to the development of innovations to address challenges. In addition, the number of industry and government jobs have continued to decrease. Recovery from these current funding trends and associated consequences will require a greater investment with a longer time lag before productive research levels can be realized.

The USDA, CSREES/NIFA, and ARS research priority areas have remained essentially unchanged in terms of percentage allocation of funding during the last decade. Sustainability-related topics have chronically received only a small proportion of total allocations. Priorities for infrastructure for this area include:

- There is an imminent need to revitalize animal agriculture research infrastructure (human and physical resources) through a series of strategic planning approaches.
- The percentage allocation of public funding by agencies including USDA ARS, CSREES/NIFA, and ARS should be reprioritized by species, taking into account the long-term projected consumer demand for that animal product and the potential for reducing the environmental impact contributed by animal agriculture, with a focus on basic research.

5-2 Research Outreach in the Animal Sciences

To facilitate agricultural research outreach to stakeholders throughout the United States, the Smith-Lever Act of 1914 created a unique entity, the Cooperative Extension (CE) System. The Act provided funding for CE activities by creating a partnership between the USDA, the state land-grant universities and governments, and local governments (i.e., city and county). USDA (via NIFA, formerly via CSREES) delivers congressionally appropriated funds to support CE activities annually to each state land-grant university. The amount of these funds is based on population-based formulas as well as funding for specific programs, and states are asked to match this funding.

CE programs are largely administered through county and regional extension offices, with the land-grant universities determining the funding allocations. The way in which formula funds are used is influenced by state and local needs as well as the national priorities established by NIFA (USDA NIFA, 2014b). The early emphasis of CE was on increasing agricultural productivity, and the CE has been the nucleus for the application and advancement of technology in agricultural production; however, as the number of farms in the United States declined, priorities shifted. The current focus is much broader, covering six major areas (USDA NIFA, 2014b): natural resources, family and consumer sciences, 4-H youth development, agriculture, leadership development, and community and economic development. Nationally the majority of current CE full-time equivalents (FTEs) are directed toward the first three areas, although there is a great deal of regional variation (Wang, 2014).

Despite this broader mandate, funding for CE has declined in real terms, which has in turn affected the capacity to deliver extension (Wang, 2014). Although the federal appropriation for CE has continued to increase, in real dollars there has been a decrease in total federal funding for extension from a peak of $778.83 million in 1973 to $333.17 million in 2008; formula funding had decreased from a peak of $505.82 million in 1980 to $228.24 million by 2008 (Wang, 2014). State funding for extension has continued to grow, and from 2000 to 2005 made up approximately 80 percent of the total extension budget overall (Wang, 2014). Regardless, this pattern of funding has been insufficient to maintain the strength of extension programs, with the number of extension FTEs nationally declining significantly by 22 percent from 1980 to 2010, with some regions experiencing greater declines than others. A factor that may have exacerbated the decline in overall university extension specialist FTEs is the trend toward 9-month appointments and split (e.g., research/teaching/extension) appointments for faculty in colleges of agriculture at land-grant universities. The Internet has also had an important impact on extension and outreach, possibly reducing some infrastructure requirements. For example, clients can obtain information from distant sources rather than having to rely on the universities located within their state or region; a farmer or rancher in one state can access information via the Internet as easily from another state's land-grant university as from their state's university (Britt et al., 2008).

Quantifying the value of such diverse extension programs is challenging. Wang (2014) has summarized some recent statistics about the economic impacts of agricultural extension. Agricultural extension activities have a high rate of return, with literature estimates ranging from 16 to 110 percent. Extension was estimated to contribute to 7.3 percent of annual agricultural productivity growth from 1949 to 2002 via improving farm production efficiencies. The loss of extension capacity will impact the ability of CE not only to help animal agriculture address the upcoming challenges related to food production and food security nationally and globally, but to bridge the communication gap between producers and stakeholders about animal agriculture.

Priorities for Infrastructure

CE has been particularly hard hit by decreases in federal funding allocations, with the number of CE FTEs declining by 22 percent from 1980 to 2010, and the underfunding impedes progress in animal science research enterprises. This loss of extension capacity will impact the ability of CE to not only address the upcoming challenges of national and global food security, but also to bridge communication gaps to educate both the public and stakeholders about animal agriculture production strategies and technology. One priority for infrastructure in this area includes:

- CE funding should increase to levels that are commensurate with animal science research and technology transfer needs. Its important communication role should be upgraded and improved to meet varied and changing demands of technology transfer.

5-3 Education in the Animal Sciences

Education and training are the main ways to transfer knowledge, skills, and attitudes in research. They are critical to capitalizing on the diversity of animal science research needed to improve national and global food security. Higher education institutions need to prepare current and future generations to meet labor market demands that increasingly require the ability to adapt and adopt emerging technologies in order to actively participate in the global knowledge economy.

Although higher education in animal sciences is generally recognized as a critical component of building human capacity for research, it has recently struggled to attract sufficient resources because of several internal academic factors as well as funding for this specific area. For many years, the largest support for education infrastructure in agriculture at land-grant universities in the United States was federal agencies, with some state funds. The limited support created by this environment diminishes the role of higher education within the public and private sectors, impacts staffing levels at universities, and affects the way higher education investments in animal sciences are designed. The U.S. higher education component for professional veterinary medicine was exceptional in receiving attention and funding until recently. This attention was mainly due to the field's engagement with companion animals, but was not relevant to other animal species, including food animals. New approaches, businesses, and business models for building human capital for higher education have evolved but with limited utilization by universities and other research entities in the field of animal sciences (APLU, 2014a). The future outcome of this need for transformation of human capital investment in research for animal sciences is challenging, but ignoring it will lead to missed opportunities.

Recently the role of the private sector has become increasingly important in terms of animal agriculture science and technology job creation, underscoring the necessity and urgency of meaningful engagement between higher education and the private sector to ensure the relevance of education and training and the absorption of graduates. The food production industry is expanding both horizontally and vertically. Institutes of higher education will need to modify their tactical approaches in the disciplines of animal sciences to address the demand and supply sides of the food system by engaging students in food-related disciplines. The support of higher education not just by public funding but by the private sector is critical to ensure that innovation, research, and its job creations are served. This may require shifting the existing paradigms in education and in conducting research.

The priority for investing in higher education has changed over time. Although higher education in agriculture in general, including animal science disciplines, was in fashion in the 1950s and 1960s, it subsequently fell out of favor. In general the focus of public funding in the 1970s through the 1980s was on essential needs for rural development related to policy reform, with limited emphasis on higher education. In the 1990s and until now, investments in human capital have been more directly related to technology, with limited inputs in agriculture education (APLU, 2014a).

5-3.1 Past, Current, and Future Drivers for Human Capital

Animal science has been a popular subject among college students for most of the 20th century. Early training and education in animal sciences focused mainly on improving husbandry, nutrition, and breeding. The historical demand for efficiency in animal production led to a research focus in animal science departments on nutrition, breeding, physiology, and genetics, which were research activities that were integrated in several ways with the training and education of undergraduate and graduate students. Animal science departments in U.S. universities, and their faculty members, were the drivers for expansion in basic science and the establishment of various departments such as food science, biochemistry, genetics, biostatistics, nutritional science, and veterinary science. The interaction among research, education, and outreach was well demonstrated in the early activities of these departments; however, animal science departments are now in a state of flux due to the changing face of animal agriculture and basic research needs in animal biology, public interest in the food system, changes in the job market, and evolving interests of students (and consequently future faculty) pursuing animal science degrees.

Animal science continues to be a popular undergraduate major. Undergraduate enrollments in animal science disciplines (e.g., general animal science; animal breeding, health, and nutrition; dairy science; food animal management; and poultry science) have steadily increased since 1987, with 5,000 undergraduate animal science degrees conferred in 2012 (Figure 5-5). In 2012, animal science students made up approximately 7 percent of all students enrolled in food and agriculture departments (FAEIS, 2014).

In contrast, there has been a decline in the number of M.S. and Ph.D. degrees in animal science conferred, from approximately 600 M.S. graduates annually in 1987 to 450 in 2012 and 200 Ph.D. graduates to 150 in the same years (Figure 5-6), although the numbers of both have remained essentially flat since the mid-1990s. It is unclear what these figures reflect about the current numbers of graduate students with expertise or emphasis in the animal sciences, because universities may also confer disciplinary (e.g., physiology, immunology, animal behavior, and nutrition) or interdisciplinary (e.g., agricultural sustainability) advanced degrees, with students nevertheless trained by animal science faculty members. The committee was unable to find statistical information about the numbers of animal science–oriented students receiving these kinds of disciplinary or interdisciplinary degrees.

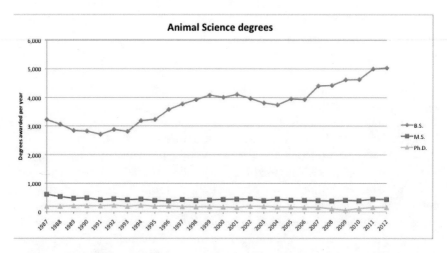

FIGURE 5-5 Number of B.S., M.S., and Ph.D. degrees awarded over a 25-year period.
SOURCE: Knapp, 2014. Summary of Animal Science degree data prepared for the National Research Council.

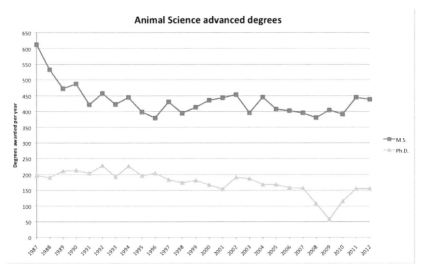

FIGURE 5-6 Trend over time in the number of M.S. and Ph.D. degrees awarded in animal sciences.
SOURCE: Knapp, 2014. Summary of Animal Science degree data prepared for the National Research Council.

Faculty headcount in animal science also remained flat from 2007 until 2010 despite an 8 percent increase in undergraduate enrollments during this period (FAEIS, 2012).

There has been a significant shift over time in the demographics of the student population. By 2010, the vast majority of bachelor's (76 percent) and master's (57 percent) students were female, as were 50 percent of doctoral students (FAEIS, 2012). A recent survey of animal science departments revealed some other elements of change at the undergraduate level (Buchanan, 2008). There has been an increase in the proportion of urban students, with fewer students now intending to return to a family farm and more intending to apply to veterinary school. Students are now interested in a wider diversity of animals, especially companion and exotic animals and horses, leading some animal science departments to establish curriculum subspecializations within these areas. To address student interests and emerging knowledge needs, course offerings have also expanded into less traditional animal science disciplines and areas such as animal behavior, animal ethics, contemporary issues, biotechnology, and molecular biology. These nontraditional undergraduates may also require different types of introductory courses (e.g., basic animal handling courses) than the animal science students of past decades. Student interest in veterinary medicine also suggests that students would benefit from courses on topics such as animal health and the human–animal bond (Britt et al., 2008).

Undergraduate student research also provides a key opportunity to develop the next generation of animal scientists. If students do not continue on into postgraduate education, the experience can still be valuable and help improve science literacy. Universities have a variety of methods for encouraging and rewarding students for participation in research, including paid fellowships, undergraduate research conferences, and research specialization designations for their bachelor's degree. Undergraduate research is also facilitated at the annual meetings of the animal science societies (e.g., American Society of Animal Science, American Dairy Science Association, Poultry Science Association) with competitions for undergraduates for presenting research or quiz bowls to demonstrate their knowledge of the animal sciences.

5-3.2 Capacity Building for Employment in the Agricultural Industries

A major challenge for undergraduate and professional education, including animal science and veterinary medicine institutions, is ensuring that students are sufficiently trained in the skills needed for the job market. A survey of employer perceptions of agricultural sciences graduates of U.S. land-grant universities indicated that students are better prepared in technical skills than communication skills (Alston et al., 2009). Although this survey did not include the profession of veterinary medicine, there are also skill gaps for veterinarians entering the food animal industry. An established technical officer in the animal agriculture allied industry stated that "effective veterinary care is essential in food animal production." While traditional veterinary medicine has focused on individual animal diagnosis, treatment, and health management at the herd or flock level, in modern animal production systems, animals are raised on hundreds or even thousands of individual farms. The health management of these farms is highly interrelated within a production system, between production systems, and across other farms in the region. Today, to be more fully prepared to contribute in modern food animal agriculture, veterinary professionals need training, skills, and understanding to ensure that they provide proper care for individual animals while maintaining effective population health management. In addition to the traditional training that veterinarians receive as part of a 4-year professional degree, enhanced focus in the areas of epidemiology, statistical analysis of data, risk analysis, and effective communication would be beneficial. "Adequate training and skills in these disciplines would better prepare new veterinarians for greater contribution to modern food animal agriculture production" (Terry Coffey, Smithfield Foods, personal communication).

Higher education institutions and technical schools in the United States are coping with the changing societal views of food production and the value of food animal to society, as well as changes in public attitudes toward and experience with animals (Fraser, 2001, 2008). For the majority of people, animals are viewed as companions, and thus institutions and technical schools of agriculture have increased their emphasis on nonlivestock species. Nevertheless, private industry has maintained and advanced research in food production over the entire production cycle from conception to consumption because of society's demand for higher quality, more efficiently produced, and safer food. Although the education and training for current students in these U.S. animal and food science departments are of high quality, these students may lack sufficient knowledge to cope with the necessary requirements and tasks of the private food industry. For example, employment opportunities within the poultry industry have shifted from primary production to the processing plant, largely due to increases in on-farm production efficiencies that have reduced labor needs (Thaxton et al., 2003). In addition, the poultry industry views technical competence in poultry science as a less-important attribute for employment than good business and communication skills (Pardue, 1997). Courses and expertise in business and food processing or food safety generally lie outside of animal science departments, which are more focused on core scientific disciplines and animal production. This mismatch between industry needs and traditional animal science curricula, along with decreasing enrollments, has been identified as a driver for the decline in the number of poultry science departments in the United States from 45 departments in the 1940s to only 6 today (Thaxton et al., 2003). Ensuring that animal science graduates continue to be employed in the agricultural industries is critical for the future animal science research because these graduates are able to recognize and communicate industry research needs and opportunities to academic scientists.

5-3.3 Research Gaps and Limitations of the Current Higher Education Systems

As discussed in other sections of the report, meeting the challenges of sustainability over the next 40 years will require more and different animal science research, extension, and education efforts. Several of the challenges of sustainability will require in-depth, specialized knowledge and advancement of the basic sciences of animal production systems; however, transdisciplinary research that incorporates other disciplines and the social sciences will also be necessary to simultaneously address the environmental, economic, and social facets of sustainability. The findings from a 1996 NRC report, Colleges of Agriculture at the Land Grant Universities: Public Service and Public Policy, are still highly relevant to the animal sciences—curricula need to be developed and expanded to reflect a contemporary view of the agrifood system and promote interdisciplinary, experiential, and systems-based learning. The more recent empirical evidence of the contributions of education in general to economic growth finds that cognitive skills and school quality are important in explaining economic growth. Therefore, policies that improve educational quality and raise educational outcomes are also important for improving income distribution. The higher education component should improve cognitive skills as part of undergraduate curriculum; such an approach is insufficiently emphasized in higher education in animal sciences across the world.

A Web-based Delphi exercise of academic experts conducted by University of California, Davis researchers (Parr et al., 2007) suggested some important elements that should be included in teaching about agricultural sustainability, with respect to plant and animal agriculture. The first is that course content needs to be broad and multidisciplinary to include information about agroecological processes; environmental impacts; food system–environment interfaces; nutrient cycling; relationship between agriculture, environment, and community, including vulnerable communities; and social and economic impacts (Anthony, 2014). In recent years there has been a remarkable increase in the number of undergraduate and graduate programs with "sustainability" in their title, demonstrating high student interest in this area. This suggests that courses that provide this kind of integrated content related to the animal sciences could not only be beneficial for animal science students, but could interest students outside of the major and help to promote an understanding of animal agriculture among non–animal science students and faculty.

The second point made by Parr et al. (2007) is that the curriculum needs to provide opportunities for experiential learning, group projects, and evaluation of case studies. Bawden (1992) describes an undergraduate program developed at Hawksbury College, an agricultural vocational college in Australia, to break through traditional silos and increase application by developing such a systems-based learning approach. During the first year, students were required to participate in the analysis of seven agricultural development projects that progressively increased in complexity in terms of their solutions. Examples were an initial project that required primarily applying simple technical solutions to an agricultural problem, with difficulty increasing as assignments progressed to exploring a paradox in rural development. In the second year, students collaborated with a commercial producer, participating in day-to-day tasks and strategic decisions. In the final year, students worked as part of collaborative projects that typically involved multiple stakeholders. One example of such a project was an animal health evaluation involving the Department of Agriculture, local farmers, and rural land protection boards. Bawden (1992) notes that this teaching activity had knock-on effects in that it increased the engagement and expertise of the faculty in dealing with complex issues, as well as their involvement in extension activities.

Making these kinds of curricular changes requires that animal science faculty be more broadly trained, for example, in ethics as it applies to value judgments and the assessment of risk, and in pedagogical methods that encourage discussion and analysis of complex problems (Schillo, 1999). A survey published in 1999 indicated that a substantial majority of animal science departments offered at least one course on "contemporary issues" covering multiple societal issues to animal science undergraduates at the upper-division level (Swanson, 1999) . There is potential not only for improvement in the number of such courses offered, but also to incorporate discussions about social concerns and sustainability into more traditional courses, expose students to these topics earlier during their degree program, and extend this education to students outside of the animal science major.

To meet national and global training needs in animal agriculture during a time of financial constraint, creative approaches will be necessary. One obvious example is increasing the availability of online course offerings not just for undergraduate and graduate students, but to provide opportunities for continuing education (APLU, 2014b). Multiuniversity collaborations are another vehicle for expanding academic education possibilities. The Midwest Poultry Consortium (MPC, 2014) is an example of multiuniversity–industry collaboration designed to fill the gap created by the decrease in academic departments of poultry science in the United States. Students in the 13 participating states can receive 18 units of credits in poultry science at the University of Wisconsin during the summer session that transfer to their home institutions. Course topics span basic avian biology and health to poultry enterprise management, and cover the breeder, hatchery, and processing/product segments. Industry partners participate in lecturing and provide paid internships for the students. There is also the potential for additional industry and foundation funding to be directed toward strengthening academic programs in the animal sciences. Tyson Foods, for example, recently donated $1 million to the USPOULTRY Foundation to endow a fund to recruit students to study poultry science, with contributions to the fund totaling more than $8 million by July 2012, while the Harold Ford Foundation provided nearly $200,000 in grants to universities with poultry science departments to encourage students to enter the poultry industry (Shane, 2014).

The development of innovative programs to train both graduate students and postdoctoral scholars in emerging areas is also critical to future faculty development and research capacity in the animal sciences. An excellent example of an existing USDA NIFA effort to build such research capacity is the Food and Agricultural Sciences National Needs Graduate and Postgraduate Fellowship Grants program. The current fiscal year request for proposals (FY 2014) is funded for $2.8 million to address the following "Targeted Expertise Shortage Areas": (1) animal and plant production, (2) forest resources, (3) agricultural educators and communicators, (4) agricultural management and economics, (5) food science and human nutrition, (6) sciences for agricultural biosecurity, and (7) training in integrative biosciences for sustainable food and agricultural systems. The program has funded 195 projects since 2005, supporting the training of master's and doctoral students, as well as postdoctoral scholars. Additionally, the program has a focus on recruiting fellows from traditionally underrepresented groups in the agricultural sciences. The National Research Council report on AFRI (NRC, 2014) found that there has been a sharp decline in the number of researchers trained as part of the Food and Agricultural Science Enhancement Grants, which are intended, among other areas of emphasis, to train predoctoral and postdoctoral fellows. The committee strongly supports the National Needs Fellowship program and its current emphasis on training individuals in agriculture communication and integrative sustainable food and agricultural systems projects, and recommends continued support and strengthening of this program in order to ensure that future human capital needs in the animal sciences are met. As the NRC (2014) committee noted: "If talented young investigators in agriculture decide to look for higher funding rates outside USDA, they could alter their focus away from agricultural research; some researchers have indicated that this is already happening."

Priorities for Infrastructure

Undergraduate enrollments in animal sciences are robust and increasing. However, a significant shift in demographics has occurred (more enrollees from urban backgrounds), and objectives are veterinary or nonagriculturally related careers. The traditional animal science curriculum structure may not be well matched to meet the interests and needs of these students or structured to improve science literacy and provide a broad knowledge about agriculture. Curriculum reformation should be designed to stimulate students to pursue a career focusing on animal agriculture or animal sciences research as well as train students to meet current employment needs within the food industries.

The number of master's and doctoral degrees conferred in animal sciences appears to be decreasing. Although the committee could not ascertain the reasons for this decline, such a condition could significantly affect future faculty capacity, particularly given the projected pattern of faculty retirements. Providing adequate support for graduate and postgraduate research is a key element of meeting future hiring needs. The USDA NIFA Food and Agricultural Sciences National Needs Graduate and Post-Graduate Fellowship program is an excellent example of a program addressing the need to build research capacity in emerging areas. One priority for infrastructure in this area includes:

- Funding for the USDA NIFA Food and Agricultural Sciences National Needs Graduate and Post-Graduate Fellowship Program should be increased, with periodic evaluation of the program to ensure that it is continuing to adequately address emerging research needs in animal science while developing the next generation of researchers.

5-4 Special Infrastructure Needs

Research, teaching, and outreach in the animal sciences rely heavily on having high-quality research facilities, which include animal housing, handling, and processing facilities. Declining funding has significantly affected these special facilities. In 2012, ARS developed a plan for capital investment in their facilities at the direction of the Secretary of the USDA. Their review indicated that there had been a 25-year period of deferred maintenance and provision of only partial funding for new facility construction, with the exception of the funding directed toward the biocontainment facility in Ames, Iowa. Their analysis of the 122 ARS facilities nationwide showed that 39 of the facilities housed high-priority research, but were in poor condition and needed significant renovation. Funds were also needed for construction of three new facilities to house high-priority ARS research that was currently being carried out in cooperator-owned facilities. Using industry standards for annual investment needs as a function of the capitalization value of facilities, ARS estimated that nearly $150 million in capital investments would be needed on a regular and recurring basis simply to upgrade and maintain their existing facilities.

Although there does not appear to be comparable information available for animal science departments at land-grant universities, anecdotal evidence suggests that they are facing similar challenges with aging and substandard animal facilities that do not reflect current industry housing standards. This makes it difficult to adequately train undergraduate and graduate students for industry employment, and challenging to conduct applied research that can contribute to current animal food production and sustainability needs. It can also create problems for animal science departments in meeting the regulatory requirements of the Animal Welfare Act and the Public Health Service Policy and the certification standards of the Association for the Assessment and Accreditation for Laboratory Animal Care with regard to the care and use of animals in research and teaching. This in turn can constrain the ability of animal scientists to apply for more biomedically oriented research funding. Inadequate facilities also negatively affect the ability of animal science departments to maintain herds and flocks of animals that can be used for research and provide essential hands-on training of students, including veterinary students, and that help to maintain genetic diversity of livestock and poultry populations (Ireland et al., 2008).

In some cases the agricultural industries have contributed funding toward new or upgraded facilities at land-grant institutions. For example, the poultry industry has played a significant role in improving the physical infrastructure for poultry science research and teaching at University of Arkansas, Auburn University, North Carolina State University, and Michigan State University. There is a need for systematic assessment of the current state and capital needs of the physical infrastructure of animal science departments nationally, as well as the development of funding strategies to address deferred maintenance issues and new facilities construction needs.

5-5 Capacity Building to Increase Diversity

It has long been recognized that increasing diversity is a critical part of capacity building for agriculture. The Second Morrill Act of 1890 attempted to extend agricultural education to African Americans, which resulted in some states establishing separate land-grant institutions (1890's institutions, or Historically Black Colleges and Universities). In 1994, Native American institutions were also given land-grant status (NSF, 2000). Despite ongoing discussions about the importance of and barriers to diversity within agriculture in general and animal sciences departments in particular (Beck and Swanson, 2003), it is clear that there are still diversity challenges. Animal science undergraduate students (82 percent), Ph.D. students (80 percent), and faculty (89 percent), and to a lesser extent master's students (74 percent) are overwhelmingly White/Caucasian. This reflects the pattern in U.S. commercial agriculture, where nearly 96 percent of primary operators are White, although the percentage of non-Caucasian primary operators is increasing (USDA, 2012). Note, however, that underrepresented minorities also make up a relatively small percentage of students in STEM fields in general, in 2010 receiving approximately 18 percent of undergraduate degrees, 13 percent of master's degrees, and 7 percent of doctoral degrees (NSF, 2012).

There is also a significant gender gap at the faculty level in the animal sciences. As indicated above, the majority of undergraduate students majoring in animal science are female, as are approximately half of the graduate students. In contrast, only 18 percent of animal science faculty members are female (FAEIS, 2012), which represents a considerably lower percentage than females holding research or teaching appointments at 4-year institutions in the biological sciences or STEM fields overall (NSF, 2012). The percentage of female animal science faculty is higher at the assistant (28 in 2010) than at the associate (21) or full professor (9) ranks, however, suggesting that there may be an increased hiring trend (FAEIS, 2012). Given that the proportion of female animal science doctoral students is similar to the proportions in biological sciences, agricultural sciences, and all university majors, the reason for the gender imbalance in the faculty is unclear.

Gender differences matter for agricultural production in farming systems all over the world where the ownership and management of farms and natural resources by men and women are defined by culturally specific gender roles (World Bank, 2013). Gender differences are also obvious in the staffing and conduct of agricultural research as most agricultural scientists and extension agents are male. Although progress has been made in developing extension systems that are more gender sensitive, unless the sources of new crop, fish, poultry, and livestock varieties and agricultural technologies take women's different needs into account, the products that are being disseminated by extension systems may not meet women's needs and preferences. Therefore, a gender-responsive agricultural research, development, and extension system needs to address women as well as men as both the clients and actors in agricultural research.

> Gender relations are culture and context specific. Men's and women's roles in food and agricultural systems and their involvement in agricultural research depend on the region in which they live. Because gender and cultural issues are inseparable, involving women as well as men in agricultural research issues should take into account existing gender roles and how these can be transformed through education and capacity building (Meinzen-Dick et al., 2011).

One might argue that changing agricultural research, development, and extension systems from being male dominated to gender equitable is a matter of political correctness or ideology. The committee believes that paying attention to gender is not a matter of ideology but rather a matter of developmental effectiveness; incorporating gender issues more widely and systematically in agricultural research, development, and extension systems will contribute significantly to meeting the food needs of the future population or ensuring that productivity translates into the improved welfare of the poor. Meinzen-Dick et al. (2011) note that whereas the fields of health, nutrition, and education have long acknowledged that explicitly addressing gender issues is one of the most effective, efficient and empowering ways to boost development and address poverty, the field of agricultural research has lagged. In the realm of national and international agricultural research, women continue to be underrepresented and underserved, and their contributions are not fully tapped.

It is time to catch up to remedy these gender and other diversity imbalances that negatively affect the ability to conduct and disseminate animal science research that is relevant to improvements in animal agriculture nationally and globally.

5-6 Partnerships for Research, Outreach, and Teaching to Leverage Resources

Land-grant universities should play a critical role in informing public policy related to animal agriculture and food systems. Recent research (Pillay, 2010; Cloete et al., 2011) has demonstrated that cooperation and consensus are key factors in policy making and implementation. Pillay (2010) evaluated three approaches for linking university activity to public policy in South Korea, North Carolina, and Finland. In South Korea, the hand of government is clearly visible in all components of the education system, including oversight of the private sector. Historically, an important network has been developed between the relevant government ministries, the public research institutions, and the large private-sector companies with respect to research and development (R&D). Increasingly universities are becoming an important fourth component of this group as they develop their R&D capacity. Important linkages are developing directly between industry and universities, particularly through initiatives such as the Industry-Academia Collaboration. In addition, a set of networks is being developed between universities, industry, and regional governments as part of initiatives such as the Regional Innovation Committee and the New University for Regional Innovation. In summary, in South Korea

> there has been a dramatic change in the nature of the higher education networks from one historically dominated by central government to one in which the private business sector and regional governments are

starting to play an increasingly important role. Such initiatives are beginning to address both the role of universities in R&D and also the challenge of regional equity in the quality of higher education institutions (Pillay, 2010).

The Pillay (2010) study also shows that other approaches with less government control can be effective. In North Carolina, there are established relationships between higher education systems on the one hand, and government, the private business sector, and civil society broadly on the other to promote economic, social, and environmental development. None of these relationships have been legislated, but instead have come about through a common commitment to the development of the state and region. Pillay (2010) also notes that in Finland, the system is characterized by a high degree of consensus building and cooperation between stakeholders in the higher education system including higher education institutions, government, public funding agencies, and the private sector. This cooperation has been a key factor in stimulating efficiency and effectiveness in the distribution of resources and the development of appropriate education and research outcomes. Moreover, it has facilitated an effective regional development strategy with universities and polytechnics spread throughout the country.

There is a need for a renewed emphasis on investing in higher education in animal sciences and on developing these kinds of partnerships. Investment in human capital formation to promote economic growth, the development of the knowledge economy, and regional and local development is vital to ensure a vibrant research enterprise that contributes to a reliable food security system in general and food animal production. While some of the benefits to higher education are not easily quantifiable, they are indeed real and important for reliable, tangible, and efficient research in animal sciences. Partnerships can contribute to resource mobilization, civil society engagement, and operationalizing capacity-building efforts (NRC, 2009). They can also facilitate access to different levels of expertise, aiding to the diffusion of knowledge and learning (Andonova and Levy, 2003; NRC, 2009) and provide an "avenue for local and regional action, even in the presence of deadlocks or inaction at higher levels" (Andonova and Levy, 2003). In the United States, partnerships between the public and private sectors and academia play an important role in furthering research and leveraging funds in the field of animal agriculture.

Because the private sector is funding and performing a larger share of the overall food and agricultural R&D, these partnerships become even more important during times of stagnant public funding (King et al., 2012). Approximately 70 percent of private funding was targeted to farm production and 30 percent to food manufacturing (Fuglie and Heisey, 2007). Private companies fund nearly all food-processing R&D and perform a growing share of production-oriented R&D for agriculture (e.g., feed additives, feed formulation, feed processing, growth promotants, probiotics, and prebiotics) and animal health (e.g., antibiotics, vaccines, parasiticides, and dewormers). The increase in funding by the private sector reflects the increasing prevalence of public–private partnerships and other policies to enhance private returns on R&D, and the private companies' incentive to capitalize on new opportunities for innovation from prior public investments in the agricultural sciences. King et al. (2012) note that

> most of the growing share of private-sector agricultural R&D supports researchers working in private companies who focus on topics and issues with the highest expected private returns. Relative to the public sector, this leads to a smaller portfolio of research topics and a greater emphasis on short-term research. Public-sector funders and performers of R&D play a largely complementary role by emphasizing social returns in the selection of research topics and valuing rapid and widespread disclosure of new knowledge.

In essence, publicly funded research is needed to ensure that the research agenda in animal agriculture reflects the broad interests of the general public. Fuglie and Toole (2014) note that in developing public–private partnerships, "one of the critical issues is whether public agricultural research complements and thereby stimulates additional private agricultural R&D investments."

There is legal precedent for the public sector to develop partnerships with private-sector entities (Toole, 2013). For example, the Bayh-Dole Act simplified processes for universities and other performers of federally sponsored research to patent and license their research results, which facilitates the transfer of technologies that require additional development by private-sector partners. Additionally, federal laboratories can enter into partnerships with for-profit companies, not-for-profit organizations, state or local governments, and other federal agencies using flexible contracts (e.g., Cooperative Research and Development Agreements [CRADAs]) created by the Federal Technology Transfer Act. CRADAs can be used to transfer technologies or advance research while protecting proprietary information and intellectual property rights. USDA's CRADA program has "been operating longer and has had more agreements per appropriated dollar than similar programs operated by any other Federal agency" (King et al., 2012).

Some select partnerships addressing agricultural research between federal agencies, the public and private sector, as well as with academia are described below. For example, federal agencies frequently partner with land-grant universities to conduct animal research. Several examples include partnerships initiated by the USDA NIFA:

- The National Animal Genome Research Program (NAGRP) which attempts to identify DNA sequences or quantitative trait loci associated with disease resistance or susceptibility and production traits in livestock and poultry species. Key partners in these efforts include Auburn, Texas A&M, Kentucky, Michigan State, Utah State, and Iowa State universities (USDA NIFA, 2009).
- NIFA partners with the Cooperative Extension Service and researchers at universities and other institutions via its Multistate Research Program in order to facilitate research on issues important to the State Agricultural Experiment Stations. Activities include Multistate Research Projects, which facilitate integrated multistate research; 500 Series Projects, which coordinate formal or informal research directed toward acute crises or opportunities; and National Research Support Projects, which focus on enabling technologies, support activities, and sharing facilities needed to accomplish high-priority research.
- NIFA partners with multiple states on implementation and strategies for the National Beef Cattle Evaluation, which provides a focal point for discussion and exchange of information for the many disconnected research activities—biological, statistical, computational, economic—that support National Cattle Evaluation (USDA NIFA, 2009).

Federal agency partnerships have also been forged to address animal agriculture issues. Related to aquaculture, the Interagency Working Group on Ocean Acidification (IWGOA) was chartered in October 2009 and includes representatives from the National Oceanic and Atmospheric

Administration (NOAA), National Science Foundation (NSF), Bureau of Ocean Energy Management, Regulation, and Enforcement, Department of State, Environmental Protection Agency (EPA), National Aeronautics and Space Administration (NASA), U.S. Fish and Wildlife Service, U.S. Geological Survey, and the U.S. Navy. NOAA chairs the group, with vice chairs from NSF and NASA. The agencies represented in the IWGOA have mandates for research and/or management of resources likely to be impacted by ocean acidification. The group meets regularly to coordinate ocean acidification research activities across the federal government (IWGOA, 2014)

Another newly formed foundation, the Foundation for Food and Agriculture Research (FFAR), provides an example of USDA partnering with the private and NGO sectors to foster agricultural research and technology transfer. Created as part of the 2014 Farm Bill (Agricultural Act of 2014, P.L. No. 113-79) with $200 million in initial funding, FFAR can pursue fundraising from individuals, corporations, charitable foundations, and other sources. It can then combine these funds to benefit research that complements existing USDA scientific research plans. FFAR could also coordinate public–private interaction around technology transfer and the translation of scientific discoveries into useful applications. The governance structure for the foundation consists of seven representatives selected from lists of candidates provided by industry and eight from a list of candidates provided by the National Academy of Sciences.

In addition to foundation structures, academic institutions do a fair amount of coordinating funding from the private sector to support agricultural research. For example, the California State University Agricultural Research Initiative program is one of several examples of an academic institution partnering with the private sector, where annually more than $4 million in state and federal funds are matched by mostly private companies to stimulate agricultural research for the public benefit (Benson et al., 2013). Other examples of associations that foster partnerships in animal agriculture include the Federation of Animal Science Societies (FASS) and the Innovation Center for U.S. Dairy. FASS was formed in 1998 to provide a forum for scientific societies to discuss common issues and to coordinate strategies and plans of action to meet public needs and benefit animal agriculture. Members of the Federation include the American Dairy Science Association, the American Society of Animal Science, and the Poultry Science Association. A key contribution of FASS was the publication of the 2012

Farm Animal Integrated Research document, which was based on collaboration among scientists, educators, producers, industry, health professionals, and government representatives, who identified key priorities and strategies for the future of the field. FASS continues to convene and facilitate the dissemination of scientific and technical information through publications and scientific meetings and to promote cooperation among all scientific societies that advance and support animal agriculture (FASS, 2014).

The Innovation Center for U.S. Dairy is another partnership designed "to increase demand for dairy products and ingredients globally, by working with and through industry, academic, government and commercial partners to drive pre-competitive, technical research in nutrition, products and sustainability" (Innovation Center for U.S. Dairy, 2014). Partners include Dairy Management, the National Dairy Council, and the U.S. Dairy Export Council, among others. The Innovation Center for U.S. Dairy also partners with other organizations, such as the World Wildlife Fund. This partnership was developed to mutually advance conservation goals and improve the economic, social, and environmental sustainability of the dairy industry (Innovation Center for U.S. Dairy, 2014).

Despite the productivity of the partnerships cited above, partnerships in general often face similar challenges including the possibility of resource constraints, governance issues, and conflicts about mission and vision. Benson et al. (2013) notes that "public investments in agricultural research and development could be reinvigorated as land grant universities have an opportunity to seek new and innovative partnerships with the private sector to support animal research." Additionally, there are models where "joint ventures with the allied animal industries could leverage strengths of academic institutions with private-sector capital. To accomplish increased funding, partnerships with the allied animal industries must be structured so industry partners receive relevant science while academic scientists remain independent and unbiased." This latter point is important in terms of communicating the results of scientific research to the public and other nonagricultural stakeholders since the credibility of academic scientists is eroded if they are not perceived as "honest brokers" of information (Croney et al., 2012).

Several activities can contribute to the development of successful partnerships, including engaging in problem definition and priority-setting activities, engaging stakeholders upfront, identifying a facilitating agent for brokering and managing the partnership, understanding partner motivations at the outset, and engaging end users in preliminary discussions (NRC, 2009). Also, more attention should be paid to monitoring, evaluating, and communicating results (NRC, 2009).

Priorities for Infrastructure

Partnerships can be an effective tool for leveraging resources for animal science research, teaching, and outreach. Public–private partnerships are playing an increasingly important role with the stagnant level of public funding. Fostering effective partnerships will also be a key element in the successful development of technology and the assurance of its effective transfer. One priority for infrastructure for this area includes:

- Additional partnerships are needed to address animal agriculture research, teaching, and outreach to leverage dollar support. Ongoing engagement of partnerships among federal agencies (e.g., USDA, EPA, and NSF) and those that link animal health and public health, and public–private endeavors needs to be pursued.

5-7 Information Gaps

Although the committee was tasked with assessing human and physical resource needs, there were many gaps in information that meant that only a broad overview of those needs could be provided. Gaps included (1) information about potential shifts in disciplinary/species emphasis and the balances between applied and basic science approaches in animal science departments, and how those might affect curriculum and future hiring needs; (2) current need and projected hiring trends for animal science graduates in the food industries; (3) land-grant physical infrastructure needs and the projected costs of meeting those needs; (4) factors affecting hiring, retention, and diversity in animal science departments; (5) the extent of diversity within extension; (6) factors affecting the decline in the number of doctoral graduates and whether or not the current numbers of students and their expertise are sufficient to meet projected academic and employment demands; and (7) the extent to which public–private partnerships or private funds are utilized to support research, teaching, physical infrastructure, and outreach activities. These and other gaps related to capacity building led the committee to recommend that a strategic planning process be undertaken to identify capacity needs within the animal sciences and develop a roadmap for addressing those needs.

REFERENCES

Alston, A. J., W. Cromartie, C. W. English, and D. Wakefield. 2009. Employer perceptions of graduates of the United States land grant university system's workforce preparation. Online Journal of Workforce Education and Development III(4). Online. Available at http://opensiuc.lib.siu.edu/cgi/viewcontent.cgi?article=1064&context=ojwed Accessed August 18, 2014.

Andonova, L. B., and M. A. Levy. 2003. Franchising global governance: Making sense of the Johannesburg type two partnerships. Pp. 19-31 in Yearbook of International Cooperation on Environment and Development 2003/2004, O. S. Stokke and O. B. Thommessen, eds. London: Earthscan.

Anthony, R. 2014. Integrating ethical considerations into animal science research. Presentation at the Second Meeting on Considerations for the Future of Animal Science Research, May 12, Washington, DC.

APLU (Association of Public and Land-Grant Universities). 2014a. African Higher Education: Opportunities for Transformative Change for Sustainable Development. Online. Available at http://www.aplu.org/document.doc?id=5100. Accessed August 18, 2014.

APLU. 2014b. APLU Sloan National Commission on Online Learning. Online. Available at http://www.aplu.org/page.aspx?pid=311. Accessed August 19, 2014.

Bawden, R. 1992. Systems approaches to agricultural development: The Hawkesbury experience. Agricultural Systems 40(1-3):153-176.

Beck, M., and J. Swanson. 2003. Value-added animal agriculture: Inclusion of race and gender in the professional formula. Journal of Animal Science 81(11):2895-2904.

Benson, M. E., J. M. Alston, and B. L. Golden. 2013. From Innovate 2012: Research in animal agriculture—A high return and a globally valuable investment in our future. Animal Frontiers 3(4):98-101.

Britt, J. H., E. D. Aberle, K. L. Esbenshade, and J. R. Males. 2008. Invited review: Animal science departments of the future. Journal of Animal Science 86(11):3235-3244.

Buchanan, D. S. 2008. ASAS centennial paper: Animal science teaching: A century of excellence. Journal of Animal Science 86(12):3640-3646.

Cloete, N., T. Bailey, and P. Maassen. 2011. Universities and Economic Development in Africa: Pact, Academic Core and Coordination, Synthesis Report. Wynberg, South Africa: Centre for Higher Education Transformation.

Croney, C., M. Apley, J. L. Capper, J. A. Mench, and S. Priest. 2012. The ethical food movement: What does it mean for the role of science and scientists in current debates about animal agriculture? Journal of Animal Science 90(5):1570-1582.

Ellis, S. 2012. NSF Perspective: Constraints and Priorities for Funding Animal Science Research. Available at http://www.asas.org/docs/meetings/steveellispresentation.pdf?sfvrsn=0. Accessed November 3, 2014.

FAEIS (Food and Agricultural Education Information System, U.S. Department of Agriculture). 2012. FAEIS News 16. Available at http://www.faeis.ahnrit.vt.edu/newsletter/april_12/faeis_newsletter_april_2012.pdf. Accessed August 18, 2014.

FAEIS. 2014. Available at http://www.faeis.ahnrit.vt.edu. Accessed August 22, 2014.

FA-RM (Free Agriculture-Restore Markets). 2014. A Short History of the USDA. Online. Available at http://fa-rm.org/usda-history.html. Accessed August 18, 2014.

FASS (Federation of Animal Science Societies). 2014. Testimony Before the Senate Appropriations Committee Hearing: Driving Innovation through Federal Investments. Online. Available at http://www.fass.org/docs/FASS_Senate_Approps_testimony_20140429.pdf. Accessed November 3, 2014.

Fraser, D. 2001. The "new perception" of animal agriculture: Legless cows, featherless chickens, and a need for genuine analysis. Journal of Animal Science 79(3):634-641.

Fraser, D. 2008. Toward a global perspective on farm animal welfare. Applied Animal Behaviour Science 113(4):330-339.

FASEB (Federation of American Societies for Experimental Biology). 2007. Statement for the Record: House Subcommittee on Agriculture, Rural Development, FDA & Related Agencies. Online. Available at http://www.faseb.org/portals/2/pdfs/opa/ag_appropstestimony3.22.07.pdf. Accessed November 3, 2014.

Fuglie, K. O., and P. W. Heisey. 2007. Economic Returns to Public Agricultural Research. Economic Brief 10. USDA ERS (U.S. Department of Agriculture, Economic Research Service).Online. Available at http://www.ers.usda.gov/media/195594/eb10_1_.pdf. Accessed October 2, 2014.

Fuglie, K. O., and A. A. Toole. 2014. The evolving institutional structure of public and private agricultural research. American Journal of Agricultural Economics 96(3):862-883.

Huffman, W. E., and R. E. Evenson. 2006. Do formula or competitive grant funds have greater impacts on state agricultural productivity? American Journal of Agricultural Economics 88(4):783-798.

Innovation Center for U.S. Dairy. 2014. About the Innovation Center: Mission and Vision. Online. Available at http://www.usdairy.com/about-us/about-the-innovation-center/mission-and-vision. Accessed August 14, 2014.

Ireland, J. J., R. M. Roberts, G. H. Palmer, D. E. Bauman, and F. W. Bazer. 2008. A commentary on domestic animals as dual-purpose models that benefit agricultural and biomedical research. Journal of Animal Science 86(10):2797-2805.

IWGOA (Interagency Working Group on Ocean Acidification). 2014. Strategic Plan for Federal Research and Monitoring of Ocean Acidification. Online. Available at http://www.whitehouse.gov/sites/default/files/microsites/ostp/NSTC/iwg-oa_strategic_plan_march_2014.pdf. Accessed August 22, 2014.

King, J., A. Toole, and K. Fuglie. 2012. The Complementary Roles of the Public and Private Sectors in U.S. Agricultural Research and Development. Economic Brief 19. Online. U.S. Department of Agriculture Economic Research Service. Available at http://www.ers.usda.gov/media/913804/eb19.pdf. Accessed October 2, 2014.

Knapp, J. 2014. Summary of animal science degree data. Prepared for the Committee on Considerations for the Future of Animal Science Research, August 1.

Meinzen-Dick, R., A. Quisumbing, J. Behrman, P. Biermayr-Jenzano, V. Wilde, M. Noordeloos, C. Ragasa, and N. Beintema. 2011. Engendering Agricultural Research, Development, and Extension. Washington, DC: International Food Policy Research Institute.

MPC (Midwest Poultry Consortium). 2014. Center of Excellence: Scholarship/Internship Program. Online. Available at http://www.mwpoultry.org/COEhome.html. Accessed August 19, 2014.

NIH (National Institutes of Health). 2014. Dual Purpose with Dual Benefit: Research in Biomedicine and Agriculture Using Agriculturally Important Domestic Animal Species. Online. Available at: http://grants.nih.gov/grants/guide/pa-files/PAR-13-204.html. Accessed September 11, 2014.

NRC (National Research Council). 1972. Report of the Committee on Research Advisory to the U.S. Department of Agriculture. Washington, DC: The National Academies Press.

NRC. 1989. Investing in Research: A Proposal to Strengthen the Agricultural, Food, and Environmental System. Washington, DC: The National Academies Press.

NRC. 2000. National Research Initiative: A Vital Competitive Grants Program in Food, Fiber, and Natural-Resources Research. Washington, DC: The National Academies Press.

NRC. 2005. Critical Needs for Research in Veterinary Science. Washington, DC: The National Academies Press.

NRC. 2009. Enhancing the Effectiveness of Sustainability Partnerships: Summary of a Workshop. Washington, DC: The National Academies Press.

NRC. 2014. Spurring Innovation in Food and Agriculture: A Review of the USDA Agriculture and Food Research Initiative Program. Washington, DC: The National Academies Press.

NSF (National Science Foundation). 2000. Women, Minorities, and Persons with Disabilities in Science and Engineering. (Rep. No. NSF 00-327). Online. Available at http://www.nacada.ksu.edu/Resources/Academic-Advising-Today/View-Articles/Advising-Native-Americans-in-Higher-Education.aspx#sthash.wSkBsH7O.dpuf. Accessed November 3, 2014.

NSF. 2012. Science and Engineering Indicators 2012. Online. Available at http://www.nsf.gov/statistics/seind12/c2/c2s2.htm. October 17, 2014.

NSF. 2014. WebCASPAR: Integrated Science and Engineering Resources Data System. Online. Available at https://ncsesdata.nsf.gov/webcaspar. Accessed August 22, 2014.

Pardue S. L. 1997. Educational opportunities and challenges in poultry science: Impact of resource allocation and industry needs. Poultry Science 76(7):938-943.

Parr, D. M., C. J. Trexler, N. R. Khanna, and B. T. Battisti. 2007. Designing sustainable agriculture education: Academics' suggestions for an undergraduate curriculum at a land grant university. Agriculture and Human Values 24(4):523-533.

Pillay, P. 2010. Linking Higher Education and Economic Development: Implications for Africa from Three Successful Systems. Cape Town: Centre for Higher Education Transformation.

Roberts, R. M., G. W. Smith, F. W. Baze, J. Cibelli, G. E. Seidel, Jr., D. E. Bauman, L. P. Reynolds, and J. J. Ireland. 2009. Farm animal research in crisis. Science 324(5926):468-469.

Schillo, K. K. 1999. An appropriate role for ethics in teaching contemporary issues. Journal of Animal Science 77(S2):154-162.

Shane, S. M. 2014. Chick-cite.com. Poultry Industry News, Comments and More. Online. Available at http://chick-cite.com. Accessed August 19, 2014.

Swanson, J. C. 1999. What are animal science departments doing to address contemporary issues? Journal of Animal Science 77(2):354-360.

Thaxton, Y.V., C. L. Balzli., and J. D. Tankson. 2003. Relationship of broiler flock numbers to litter microflora. Journal of Applied Poultry Research 12(1):81-84.

Toole, A. 2013. Public-Private Partnerships Create Opportunities to Enhance Agricultural Research Systems. U.S. Department of Agriculture Economic Research Service. Online. Available at http://www.ers.usda.gov/amber-waves/2013-november/public-private-partnerships-create-opportunities-to-enhance-the-agricultural-research-system.aspx. Accessed September 4, 2014.

USDA (U.S. Department of Agriculture). 2012. Census of Agriculture. Online. Available at http://www.agcensus.usda.gov/Publications/2012. Accessed September 4, 2014.

USDA. 2014a. Agricultural Research Service (ARS) Summary of Animal Science Investment. From Mark Boggess, April 11.

USDA. 2014b. National Institute of Food and Agriculture (NIFA) Summary of Animal Science Investment. From Mark Boggess, April 11.

USDA ARS (U.S. Department of Agriculture Agricultural Research Service). 2012. The USDA Agricultural Research Service Capital Investment Strategy. Online. Available at http://www.ars.usda.gov/sp2UserFiles/Subsite/ARSLegisAffrs/USDA_ARS_Capital_Investment_Strategy_FINAL_eeo.pdf. Accessed August 19, 2014.

USDA NIFA (U.S. Department of Agriculture National Institute of Food and Agriculture). 2009. National Animal Genome Research Program (NAGRP). Online. Available at http://www.csrees.usda.gov/nea/animals/in_focus/an_breeding_if_nagrp.html. Accessed August 22, 2014.

USDA NIFA . 2014a. Current Research Information System (CRIS). Reports. Online. Available at http://cris.nifa.usda.gov/reports.html. Accessed on December 14, 2014.

USDA NIFA. 2014b. Extension. Online. Available at http://www.csrees.usda.gov/qlinks/extension.html. Accessed August 18, 2014.

Wang, S. L. 2014. Cooperative Extension System: Trends and Economic Impacts on U.S. Agriculture. Choices 20. Online. Available at http://www.choicesmagazine.org/choices-magazine/submitted-articles/cooperative-extension-system-trends-and-economic-impacts-on-us-agriculture. Accessed October 2, 2014.

World Bank. 2013. Agricultural Extension. Online. Available at http://web.worldbank.org/WBSITE/EXTERNAL/TOPICS/EXTGENDER/0,,contentMDK:20208259~pagePK:210058~piPK:210062~theSitePK:336868,00.html. Accessed September 16, 2014.

6

Recommendations

Animal production and the science that informs it are confronted by an emerging and globally complex set of conditions in the 21st century that generate new challenges for sustainable animal production, which in turn requires rethinking about the overall nature of animal science. These challenges include, but are not limited to, growing demand for animal products by an increasingly affluent, global population approaching 10 billion people; the globalization of food systems that cross continents with consequences for individual country and regional concerns about food security; the intensification of production systems in the context of societal and environmental impacts; the development and maintenance of sustainable animal production systems in the face of global environmental change; and the multidecadal stagnation in research funding for animal production. As described throughout this report, a new roadmap for animal science research is required. The findings and recommendations described below will help to inform this new roadmap.

The breadth of the committee's task led to many recommendations being developed. The committee twice deliberated on prioritization of these recommendations. Early in the process the committee chose a limited number of broad and high-level overarching recommendations, which were then refined in subsequent meetings and are described immediately below. At its last meeting, the committee chose its highest priorities from among all of the possible recommendations. These recommendations appear after the overarching recommendations and are specific to what the committee identified as key areas in animal agriculture in both the United States and globally. In addition to its recommendations, the committee identified complementary priorities for research, research support, and infrastructure, which can be found in Chapters 3-5.

Ideally, NRC committee recommendations should include an action statement specifying the specific agency or organization that should follow up. This works well if there is an individual sponsor with a single short-term task; however, the breadth of the tasks and the multiplicity of overlapping national and international public and private organizations involved in sponsoring or performing animal research limited the committee's ability to specify action pathways. Sorting out responsibilities for moving ahead is part of the reason that the committee has recommended the development of a U.S. Animal Science Strategic Plan under the leadership of the U.S. Department of Agriculture (USDA).

Overarching Recommendations

Two central issues have guided National Research Council and other reports regarding the setting of research agendas for animal agriculture in recent years: productivity and sustainability. The committee built on these reports and emphasized the importance of research to sustainably and efficiently increase animal agricultural productivity. The committee's deliberations resulted in the following overarching recommendations:

- To achieve food security, research efforts should be improved through funding efforts that instill integration rather than independence of the individual components of the entire food chain. Success can only be achieved through strong, overarching, and inter- and transdisciplinary research collaborations involving both the public and private sectors. Animal science research should move toward a systems approach that emphasizes efficiency and quality of production to meet food security needs. The recently created Foundation for Food and Agricultural Research (FFAR) needs to incorporate holistic approaches to animal productivity and sustainability (Chapter 5).
- Continuing the research emphasis on improving animal productivity is necessary; however, concomitant research on the economic, environmental, and social sustainability nexuses of animal production systems should also be enhanced. Both public and private funding agencies should incorporate inter- and transdisciplinary approaches for research on animal productivity and sustainability (Chapters 3 and 5).

- There is a need to revitalize research infrastructure (human and physical resources), for example, through a series of strategic planning approaches, developing effective partnerships, and enhancing efficiency. In the United States, the committee recommends that USDA and the newly created FFAR spearhead the formation of a coalition to develop a U.S. Animal Science Strategic Plan or Roadmap for capacity building and infrastructure from 2014 to 2050. The coalition should be broad based and include representation from relevant federal agencies; colleges and universities that are involved in research, teaching, and outreach activities with food animals; NGOs; the private sector; and other relevant stakeholders. Areas of focus should include assessment of resource needs (human and physical infrastructure) to support the current and emerging animal science research enterprise; strategies to increase support for research, outreach, and instructional needs via formula funding, competitive funding, and public-private partnerships; curriculum development and delivery; evaluation of factors affecting hiring, retention, and diversity in the animal sciences; and mechanisms for research, priority setting to meet emerging, local, regional, national, and global needs (Chapter 5).
- Socioeconomic/cultural research is essential to guide and inform animal scientists and decision makers on appropriately useful and applicable animal science research as well as communication and engagement strategies to deal with these extensive challenges. Engagement of social scientists and researchers from other relevant disciplines should be a prerequisite as appropriate for integrated animal science research projects, such as National Institute of Food and Agriculture (NIFA) Coordinated Agricultural Project grants, to secure funding and approval of such projects (Chapters 3 and 5).
- For research in sustainable intensification of animal agriculture to meet the challenge of future animal protein needs, it is necessary to effectively close the existing broad communication gap between the public, researchers, and the food industries. This will require research to better understand the knowledge, opinions, and values of the public and food system stakeholders, as well as the development of effective and mutually respectful communication strategies that foster ongoing stakeholder engagement. A coalition representing universities, federal agencies, industry, and the public should be formed to focus on communications research with the goals of enhancing engagement, knowledge dissemination, stakeholder

participation, and informed decision making. Communications programs within agriculture schools, or in collaboration with other university components, such as schools of public health, could conduct this type of research (Chapters 3 and 5).
- The United States should expand its involvement in research that assists in the development of internationally harmonized standards, guidelines, and regulations related to both the trade in animal products and protection of the consumers of those products (Chapter 4).

Many of the recommendations and priorities discussed in each of the chapters are based on a central theme of the need for strategic planning to meet the challenges of the increased animal agricultural demand that is projected through 2050. These recommendations and priorities include planning for research in the United States and in developing countries and reconsideration of education and training in animal agriculture in the United States, particularly at the university level. These strategic planning activities should be guided by the need for systems approaches that integrate the many scientific disciplines and governmental and nongovernmental stakeholders involved in achieving the goal of food security based on sustainable animal agriculture.

Recommendations for U.S. Animal Agriculture

The committee developed several recommendations that are of high priority for reinvigorating the field of animal agriculture in the United States.

Public Funding

In view of the anticipated continuing increased demand for animal protein, growth in U.S. research related to animal agricultural productivity is imperative. Animal protein products contribute over $43 billion annually to the U.S. agricultural trade balance. Animal agriculture accounts for 60 to 70 percent of the total agricultural economy. In the past two decades, public funding, including formula funding and USDA Agricultural Research Service/National Institute of Food and Agriculture funding, of animal science research has been stagnant in terms of real dollars and has declined in relation to the research inflation rate. A 50 percent decline in the rate of increase in U.S. agricultural productivity is predicted if overall agricultural funding increases in normative dollars continue at the current rate, which is less than the expected rate of inflation of research costs. If funding does meet the rate of research cost inflation, however, a 73 percent increase in overall agricultural productivity between now and 2050 is projected and a 1 percent increase in inflation-adjusted spending is projected to lead to an 83 percent increase.

Despite documenting the clear economic and scientific value of animal science research in the United States, funding to support the infrastructure and capacity is evidently insufficient to meet the needs for animal food; U.S.-based research will be needed to address sustainability issues and to help developing countries sustainably increase their own animal protein production and/or needs. Additionally, animal science research and practices in the United States are often adopted, to the extent possible, within developing countries. Thus, increases in U.S. funding will favorably impact animal production enterprises in developing countries.

With the lack of increase in public funding of animal science research, private/industry support has increased. The focus of industry funding is more toward applied areas that can be commercialized in the short term. Many of these applications are built on concepts developed from publicly funded basic research. With the increased animal protein demands, especially poultry, more publicly funded basic research is needed.

> **RECOMMENDATION 3-1: To meet current and future animal protein demand, and to sustain corresponding infrastructure and capacity, public support for animal science research (especially basic research) should be restored to at least past levels of real dollars and maintained at a rate that meets or exceeds the annual rate of research inflation. This is especially critical for those species (i.e., poultry) for which the consumer demand is projected to significantly increase by 2050 and for those species with the greatest opportunity for reducing the environmental impact of animal agriculture (Section 3-1 in Chapter 3).**

Productivity and Production Efficiency

Regarding productivity and production efficiency, the committee finds that increasing production efficiency while reducing the environmental footprint and cost per unit of animal protein product is essential to achieving a sustainable, affordable, and secure animal protein supply. Technological improvements have led to system/structural changes in animal production industries whereby more efficient food production and less regional, national, and global environmental impact have been realized.

> **RECOMMENDATION 3-2: Support of technology development and adoption should continue by both public and private sectors. Three criteria of sustainability—(1) reducing the environmental footprint, (2) reducing the financial cost per unit of animal protein produced, and (3) enhancing societal determinants of sustainable global animal agriculture acceptability—should be used to guide funding decisions about animal agricultural research and technological development to increase production efficiency (Section 3-2 in Chapter 3).**

Breeding and Genetic Technologies

Further development and adoption of breeding technologies and genetics, which have been the major contributors to past increases in animal productivity, efficiency, product quality, environmental, and economic advancements, are needed to meet future demand.

> **RECOMMENDATION 3-3: Research should be conducted to understand societal concerns regarding the adoption of these technologies and the most effective methods to respectfully engage and communicate with the public (Section 3-3 in Chapter 3).**

Nutritional Requirements

The committee notes that understanding the nutritional requirements of the genetically or ontogenetically changing animal is crucial for optimal productivity, efficiency, and health. Research devoted to an understanding of amino acid, energy, fiber, mineral, and vitamin nutrition has led to technological innovations such as production of individual amino acids to help provide a diet that more closely resembles the animal's requirements, resulting in improved efficiency, animal health, and environmental gains, as well as lower costs; however, much more can be realized with additional knowledge gained from research.

> **RECOMMENDATION 3-4: Research should continue to develop a better understanding of nutrient metabolism and utilization in the animal and the effects of those nutrients on gene expression. A systems-based holistic approach needs to be utilized that involves ingredient preparation, understanding of ingredient digestion, nutrient metabolism and utilization through the body, hormonal controls, and regulators of nutrient utilization. Of particular importance is basic and applied research in keeping the knowledge of nutrient requirements of animals current (Section 3-4 in Chapter 3).**

Feed Technology

Potential waste products from the production of human food, biofuel, or industrial production streams can and are being converted to economical, high-value animal protein products. Alternative feed ingredients are important in completely or partially replacing high-value and unsustainable ingredients, particularly fish meal and fish oil, or ingredients that may otherwise compete directly with human consumption.

> **RECOMMENDATION 3-6.1: Research should continue to identify alternative feed ingredients that are inedible to humans and will notably reduce the cost of animal protein production while improving the environmental footprint. These investigations should include assessment of the possible impact of changes in the protein product on the health of the animal and the eventual human consumer, as well as the environment (Section 3-6.1 in Chapter 3).**

Animal Health

The subtherapeutic use of medically important antibiotics in animal production is being phased out and may be eliminated in the United States. This potential elimination of subtherapeutic use of medically important antibiotics presents a major challenge.

> **RECOMMENDATION 3-7: There is a need to explore alternatives to the use of medically important subtherapeutic antibiotics while providing the same or greater benefits in improved feed efficiency, disease**

prevention, and overall animal health (Section 3-7 in Chapter 3).

Animal Welfare

Rising concern about animal welfare is a force shaping the future direction of animal agricultural production. Animal welfare research, underemphasized in the United States compared to Europe, has become a high-priority topic. Research capacity in the United States is not commensurate with respect to the level of stakeholder interest in this topic.

> **RECOMMENDATION 3-8: There is a need to build capacity and direct funding toward the high-priority animal welfare research areas identified by the committee. This research should be focused on current and emerging housing systems, management, and production practices for food animals in the United States. FFAR, USDA-AFRI, and USDA-ARS should carry out an animal welfare research prioritization process that incorporates relevant stakeholders and focuses on identifying key commodity-specific, system-specific, and basic research needs, as well as mechanisms for building capacity for this area of research (Section 3-8 in Chapter 3).**

Climate Change

Although there is uncertainty regarding the degree and geographical variability, climate change will nonetheless impact animal agriculture in diverse ways, from affecting feed quality and quantity to causing environmental stress in agricultural animals. Animal agriculture affects and is affected by these changes, in some cases significantly, and must adapt to them in order to provide the quantity and affordability of animal protein expected by society. This adaptation, in turn, has important implications for sustainable production. The committee finds that adaptive strategies will be a critical component of promoting the resilience of U.S. animal agriculture in confronting climate change and variability.

> **RECOMMENDATION 3-11.2: Research needs to be devoted to the development of geographically appropriate climate change adaptive strategies and their effect on**

greenhouse gas (GHG) emissions and pollutants involving biogeochemical cycling, such as that of carbon and nitrogen, from animal agriculture because adaptation and mitigation are often interrelated and should not be independently considered. Additional empirical research quantifying GHG emissions sources from animal agriculture should be conducted to fill current knowledge gaps, improve the accuracy of emissions inventories, and be useful for improving and developing mathematical models predicting GHG emissions from animal agriculture (Section 3-11.2 in Chapter 3).

Socioeconomic Considerations

Although socioeconomic research is critical to the successful adoption of new technologies in animal agriculture, insufficient attention has been directed to such research. Few animal science departments in the United States have social sciences or bioethics faculty in their departments who can carry out this kind of research.

RECOMMENDATION 3-12: Socioeconomic and animal science research should be integrated so that researchers, administrators, and decision makers can be guided and informed in conducting and funding effective, efficient, and productive research and technology transfer (Section 3-12 in Chapter 3).

Communications

The committee recognizes a broad communication gap related to animal agricultural research and objectives between the animal science community and the consumer. This gap must be bridged if animal protein needs of 2050 are to be fulfilled.

RECOMMENDATION 3-13: There is a need to establish a strong focus on communications research as related to animal science research and animal agriculture, with the goals of enhancing knowledge dissemination, respectful stakeholder participation and engagement, and informed decision making (Section 3-13 in Chapter 3).

Recommendations for Global Animal Agriculture

Overall, the committee strongly supports an increase in funding of global animal research both by governments and the private sector. The committee also identified several recommendations directed toward global animal agriculture.

Infrastructural Issues

The committee notes that per capita consumption of animal protein will be increasing more quickly in developing countries than in developed countries through 2050. Animal science research priorities have been proposed by stakeholders in high-income countries, with primarily U.S. Agency for International Development, World Bank, Food and Agriculture Organization, Consultative Group on International Agricultural Research, and nongovernmental organizations individually providing direction for developing countries. A program such as the Comprehensive Africa Agriculture Development Programme (CAADP) demonstrates progress toward building better planning in agricultural development in developing countries, through the composite inclusion of social, environmental, and economic pillars of sustainability.

In addition, for at least the last two decades, governments worldwide have been reducing their funding for infrastructure development and training for animal sciences research. Countries and international funding agencies should be encouraged to adapt an integrated agriculture research system to be part of a comprehensive and holistic approach to agriculture production. A system such as CAADP can be adapted for this purpose.

> **RECOMMENDATION 4-1:** To sustainably meet increasing demands for animal protein in developing countries, stakeholders at the national level should be involved in establishing animal science research priorities (Section 4-1 in Chapter 4).

Technology Adoption

The committee finds that proven technologies and innovations that are improving food security, economics, and environmental sustainability in high-income countries are not being utilized by all developed or developing countries because in some cases they may not be logistically transferrable or in other ways unable to cross political boundaries. A key barrier to technological adoption is the lack of extension to smallholder farmers about how to utilize the novel technologies for sustainable and improved production as well as to articulate smallholder concerns and needs to the research community. Research objectives to meet the challenge of global food security and sustainability should focus on the transfer of existing knowledge and technology (adoption and, importantly, adaptation where needed) to nations and populations in need, a process that may benefit from improved technologies that meet the needs of multiple, local producers. Emphasis should be placed on extension of knowledge to women in developing nations.

> **RECOMMENDATION 4-5.2:** Research devoted to understanding and overcoming the barriers to technology adoption in developed and developing countries needs to be conducted. Focus should be on the educational and communication role of local extension and advisory personnel toward successful adoption of the technology, with particular emphasis on the training of women (Section 4-5.2 in Chapter 4).

Animal Health

Zoonotic diseases account for 70 percent of emerging infectious diseases. The cost of the six major outbreaks that have occurred between 1997 and 2009 was $80 billion. During the last two decades, the greatest challenge facing animal health has been the lack of resources available to combat several emerging and reemerging infectious diseases. The current level of animal production in many developing countries cannot increase and be sustained without research into the incidence and epidemiology of disease and effective training to manage disease outbreaks, including technically reliable disease investigation and case findings. Infrastructure is lacking in developing countries to combat animal and zoonotic diseases, specifically a lack of disease specialists and diagnostic laboratory facilities that would include focus on the etiology of diseases. There is a lack of critical knowledge about zoonoses presence, prevalence, drivers, and impact. Recent advances in technology offer opportunities for improving the understanding of zoonoses epidemiology and control.

RECOMMENDATION 4-7.1: Research, education (e.g., training in biosecurity), and appropriate infrastructures should be enhanced in developing countries to alleviate the problems of animal diseases and zoonoses that result in enormous losses to animal health, animal producer livelihoods, national and regional economies, and human health (Section 4-7.1 in Chapter 4).

In addition to the recommendations presented in this chapter, the committee identified complementary priorities for research, research support, and infrastructure that can be found in Chapters 3 through 5.

A

Committee on Considerations for the Future of Animal Science Research

Biographical Information

BERNARD GOLDSTEIN (IOM) is emeritus professor of environmental and occupational health and former dean of the University of Pittsburgh Graduate School of Public Health. He is a physician, board certified in internal medicine, hematology and toxicology. Dr. Goldstein is author of over 150 publications in the peer-reviewed literature, as well as numerous reviews related to environmental health. He is an elected member of the Institute of Medicine (IOM) and of the American Society for Clinical Investigation. His experience includes service as assistant administrator for research and development of the U.S. Environmental Protection Agency, 1983-1985. In 2001, he came to the University of Pittsburgh from New Jersey where he had been the founding director of the Environmental and Occupational Health Sciences Institute, a joint program of Rutgers University and Robert Wood Johnson Medical School. He has chaired more than a dozen National Research Council and IOM committees, primarily related to environmental health issues. He has been president of the Society for Risk Analysis and has chaired the National Institutes of Health Toxicology Study Section, U.S. Environmental Protection Agency's Clean Air Scientific Advisory Committee, the National Board of Public Health Examiners, and the Research Committee of the Health Effects Institute. Dr. Goldstein received his medical degree from New York University and undergraduate degree from the University of Wisconsin.

LOUIS R. D'ABRAMO is William L. Giles Distinguished Professor of Wildlife, Fisheries and Aquaculture in the Department of Wildlife, Fisheries and Aquaculture and dean of the Graduate School Emeritus and associate vice president for Academic Affairs Emeritus at Mississippi State University. His primary research interests are the aquaculture of freshwater and marine organisms with the prevailing goal of the development of sustainable commercial production practices based upon the wise use of natural resources and environmental stewardship. He is past president of the World Aquaculture Society, the largest aquaculture society in the world, and past president of the National Shellfisheries Association. In 2003, Dr. D'Abramo was awarded the highest honor bestowed upon a member of the World Aquaculture Society, the Exemplary Service Medal, for his work in promoting aquaculture research and advancing the knowledge of sustainable aquaculture practices throughout the world. In 2010, he received the Distinguished Life-Time Achievement Award from the U.S. Aquaculture Society for contributions of broad impact to the development of the U.S. aquaculture industry. In 2007, he also received the Meritorious Award from the National Shellfisheries Association for outstanding leadership and dedicated service. Dr. D'Abramo earned M.Phil. and Ph.D. degrees in ecology and evolutionary biology from Yale University.

GARY F. HARTNELL is a senior fellow of the Monsanto Company, St. Louis, Missouri, where he has been employed since 1983. Dr. Hartnell is an expert on the nutritional requirements of food animals and in developing and executing research strategies involving poultry, livestock, and aquaculture, and in the evaluation of genetically modified crops and their co-products, feed ingredients, and additives for regulatory, industry, and consumer acceptance. Dr. Hartnell has been active in every society of importance in the animal nutrition community, including the American Society of Animal Science, American Dairy Science Association (past president), Poultry Science Association, World Aquaculture Society, American Registry of Professional Animal Scientists, Dairy Shrine, the Academy of Science of St. Louis, Sigma Xi, and Federation of Animal Science Societies (past president). He recently co-chaired an International Life Science Institute committee that developed guidelines for conducting livestock feeding studies using biotechnology-derived crops and their byproducts, and he serves on the Board on Agriculture and Natural Resources within the National Research Council. He has authored or co-authored over 104 abstracts, 73 scientific journal articles, 9 books/book chapters, and 121 popular press, symposia, and conference articles. Dr. Hartnell received his Ph.D. from the University of Wisconsin in 1977.

JOY MENCH is a professor and vice chair in the Department of Animal Science and the director of the Center for Animal Welfare at the

University of California (UC), Davis. Dr. Mench conducts research on the behavior and welfare of animals, especially poultry and laboratory animals. She has published more than 120 papers, book chapters, and books on these topics, and also given many invited presentations to national and international audiences. At UC Davis she teaches courses on animal welfare, professional ethics, and the ethics of animal use. Dr. Mench has served on numerous committees and boards related to farm and laboratory animal welfare, including for the Association for the Accreditation of Laboratory Animal Care (AAALAC), the United Egg Producers, the National Chicken Council, McDonald's, Safeway, Sysco, Certified Humane, American Humane Certified, the Food Marketing Institute, the National Council of Chain Restaurants, the World Animal Health Organization (OIE), the U.N. Food and Agriculture Organization, and the European Union. She was president of the International Society for Applied Ethology, and the recipient of the Poultry Science Association Poultry Welfare Research Award in 2004 and the UC Davis Distinguished Public Scholarly Service Award in 2007. Dr. Mench received her Ph.D. in ethology (animal behavior) from the University of Sussex in England in 1983.

SARA PLACE is an assistant professor of Sustainable Beef Cattle Systems in the Department of Animal Science at Oklahoma State University. Her research program focuses on the intersection of management and production practices that optimize animal well-being, nutrient-use efficiency, and the business sustainability of agricultural operations. Prior to Oklahoma State, she worked with the Innovation Center for U.S. Dairy and Winrock International as a livestock production consultant. Dr. Place received her Ph.D. in June 2012 from University of California, Davis in animal biology where her work focused on measurement and mitigation of greenhouse gas emissions from cattle. She earned a B.S. in animal science from Cornell University in 2008 and an A.A.S. degree in agriculture business from Morrisville State College in 2006.

MO SALMAN currently participates in the 2013-2014 Jefferson Science Fellows Program as the senior scientific advisor to the African Bureau under Public Diplomacy and Public Affairs Office in the U.S. Department of State. Dr. Salman is a professor of veterinary epidemiology in the Department of Clinical Sciences and College of Veterinary Medicine and Biomedical Sciences at Colorado State University. He is founder and director of Colorado State University's

Animal Population Health Institute. Dr. Salman's research emphasis is in veterinary epidemiology with interests in analytical veterinary epidemiology, methodology for national and international animal disease surveillance systems, observational and clinical studies on animal populations, and epidemiology of infectious diseases. He has been the principal investigator on several research projects which include the Program for Economically Important Infectious Animal Diseases, enhancement of the technical capability of the National Animal Health and Food Safety Services System in the Republics of Georgia, Armenia, Albania, Kyrgyzstan, and Iraq, among other countries; simulation modeling for foot and mouth disease; training in field investigation for highly pathogenic avian influenza; and the refinement of risk assessment methods for infectious animal diseases that have impact on trade and public health issues. He is a diplomate of the American College of Veterinary Preventive Medicine and a fellow of the American College of Epidemiology. Dr. Salman received his veterinary degree from the University of Baghdad–Iraq, and his both MPVM and Ph.D. from University of California, Davis.

DENNIS H. TREACY is executive vice president and chief sustainability officer, Smithfield Foods, Inc. Mr. Treacy oversees and directs many areas within the company, including government affairs, corporate communications, sustainability initiatives, and the legal department. Mr. Treacy also serves as the executive director of the Smithfield-Luter Foundation, the philanthropic wing of Smithfield Foods that funds education and growth opportunities in communities across America. Additionally, Mr. Treacy serves or has served on dozens of state and national boards and commissions. Prior to joining Smithfield Foods in 2002, Mr. Treacy was director of the Virginia Department of Environmental Quality. Mr. Treacy also served as assistant attorney general in the natural resources section of the Virginia attorney general's office. He is a 2010 Distinguished Environmental Law Graduate from Lewis & Clark Law School in Portland, Oregon, where he graduated in 1983. He completed his undergraduate degree in Forestry and Wildlife at Virginia Tech in 1978, and currently serves on its Board of Visitors.

B. L. TURNER II (NAS) is a Gilbert F. White Professor of Environment and Society in the School of Geographical Sciences and Urban Planning and School of Sustainability at Arizona State University. His research focuses on the study of human–environment relationships. Dr. Turner examines these relationships in the use of land and resources by the ancient Maya civilization in the Yucatan peninsula region, the intensification of land use among contemporary smallholders in the tropics, and land-use and land-cover changes as part of global environmental change. He has contributed journal articles to *Science, Annual Review of Environment and Resources, Annals of the Association of American Geographers*, and many other publications. He is a member of the National Academy of Sciences, and the American Academy of Arts and Sciences. Dr. Turner has served in several editorial positions, including the Editorial Board for *Environmental Science & Policy, Regional Environmental Science*, and Human-Environment Interactions: A Book Series, and is associate editor of the *Proceedings of the National Academy of Sciences*. Dr. Turner received B.A. and M.A. degrees in geography from the University of Austin at Texas and a Ph.D. in geography from the University of Wisconsin at Madison.

GARY W. WILLIAMS is professor of agricultural economics and co-director of the Agribusiness, Food, and Consumer Economics Research Center (AFCERC) at Texas A&M University. He is the AFCERC chief operations officer responsible for managing the research program of the center and leads AFCERC research and outreach projects relating to commodity and agribusiness markets and policy and international trade and policy. He is also an associate member of the Masters of International Affairs faculty in the Bush School of Government and International Affairs and senior scientist with the Borlaug Institute of International Affairs. His areas of teaching and research emphases include commodity promotion programs, international agricultural trade and development, agricultural policy, and marketing and price analysis. Prior to joining the faculty at Texas A&M University, he gained experience as a professor and assistant coordinator of the Meat Export Research Center at Iowa State University, senior economist at Chase Econometrics, agricultural economist for the U.S. Department of Agriculture (USDA), and special assistant to the U.S. Deputy Under Secretary of Agriculture for International Affairs and Commodity Programs at USDA. He is well known for his research on U.S. and world oilseed and oilseed product markets and the U.S. livestock industry including issues related to sheep and lamb markets and the effects of concentration in the beef packing industry. Dr. William recently served as chair of a National Academy of Sciences Committee on the Status and Economic Performance of the U.S. Sheep and Lamb Industry. Dr. Williams holds a Ph.D. and an M.S. degree in agricultural economics from Purdue University (1978 and 1981) and a B.S. in economics from Brigham Young University (1974).

FELICIA WU is a John A. Hannah Distinguished Professor in the Department of Food Science and Human Nutrition and the Department of Agricultural, Food, and Resource Economics at Michigan State University. Previously, she was an associate professor of environmental and occupational health at University of Pittsburgh. Dr. Wu's research interests lie at the intersection of global health, agriculture, and trade. Using the tools of mathematical modeling, health economics, and quantitative risk assessment, she examines how agricultural systems affect health in different parts of the world. For her research on the impact of aflatoxin regulations on global liver cancer, Dr. Wu was awarded a National Institutes of Health EUREKA Award. She is a member of the World Health Organization (WHO) Foodborne Disease Burden Epidemiology Reference Group, as well as the expert roster of the Joint FAO/WHO Expert Committee on Food Additives of the United Nations. She received the 2007 Chauncey Starr Award of the Society for Risk Analysis, given annually to a risk scientist age 40 or under; and serves as the health risk area editor for the journal Risk Analysis. Dr. Wu received her A.B. and S.M. degrees from Harvard University and her Ph.D. from Carnegie Mellon University.

B

Statement of Task

Addressing the economic, social and environmental sustainability challenge of global food security requires an adequate, nutritious food supply produced and distributed cost effectively while improving efficiency across the entire food production system. Recognizing this challenge and the increasing global demand for animal products, an ad hoc committee will conduct a study and prepare a report that will identify critical areas of research and development (R&D), technologies, and resource needs for research in the field of animal agriculture, both nationally and internationally. Specifically, the report will identify the most important needs for future research in this area, including:

1. Assessing global demand for products of animal origin in 2050 within the framework of ensuring global food security;
2. Evaluating how climate change and limited natural resources may impact the ability to meet future global demand for animal products in sustainable production systems, including typical conventional, alternative and evolving animal production systems in the U.S. and internationally;
3. Identifying factors that may impact the ability of the U.S. to meet demand for animal products, including the need for trained human capital, product safety and quality, and effective communication and adoption of new knowledge, information and technologies;
4. Identifying the needs for human capital development, technology transfer and information systems for emerging and evolving animal production systems in developing countries, including the resources needed to develop and disseminate this knowledge and these technologies; and

5. Describing the evolution of sustainable animal production systems relevant to production and production efficiency metrics in the U.S. and in developing countries.

The report will also address the role of governments, nongovernmental organizations, and the private sector in developing partnerships and leveraging these resources to further the field of animal agriculture.

C

Glossary

Adaptive management: Management practices that promote a system's ability to take advantage of opportunities or cope with problems occurring in the environment (FAO, 2009).

Adaptive strategies: Ways in which individuals, households, and communities have changed their mix of productive activities and modified their community rules and institutions over the long term, in response to economic or environmental shocks or stresses, in order to meet their livelihood needs. Adaptive strategies are a mix of traditional livelihood systems, modified by locally or externally induced innovations, and coping strategies that have become permanent (CASL, 2014).

Agrarian society: A culture or community in which agriculture is the primary means of subsistence; an economy that relies heavily on agriculture (Agrarian Civilizations, 2014).

Agricultural intensification: Any practice that increases productivity per unit land area at some cost in labor or capital inputs. One important dimension of agricultural intensification is the length of fallow period (i.e., letting land lie uncultivated for a period) and whether the management approach uses ecological or technological means (FAO, 2009).

Agricultural product/product of agricultural origin: Any product or commodity, raw or processed, that is marketed for human consumption (excluding water, salt, and additives) or animal feed (FAO, 2001).

Agricultural productivity: The output produced by a given level of input(s) in the agricultural sector of a given economy. More formally, it can be defined as "the ratio of the value of total farm outputs to the value of total inputs used in farm production" (Liverpool-Tasie et al., 2011).

Agricultural sustainability: Agricultural sustainability is defined by four generally agreed-upon goals:

- Satisfy human food, feed, and fiber needs, and contribute to biofuel needs.
- Enhance environmental quality and the resource base.
- Sustain the economic viability of agriculture.
- Enhance the quality of life for farmers, farm workers, and society as a whole.

Sustainability is best evaluated not as a particular end state, but rather as a process that moves farming systems along a trajectory toward greater sustainability on each of the four goals (NRC, 2010).

Animal agriculture: Agricultural activities for livestock, poultry, and aquaculture in total (see definitions of livestock, poultry, and aquaculture in this appendix).

Animal husbandry: Controlled cultivation, management, and production of domestic animals, including improvement of the qualities considered desirable by humans by means of breeding. Animals are bred and raised for utility (e.g., food, fur), sport, pleasure, and research (Merriam-Webster, 2014).

Animal products: A product made from animal material.

Animal protein: Dietary components derived from meat, fish, and animal products such as milk.

Animal sciences: Refers to all disciplines currently contributing to animal food production systems. These disciplines are generally housed in departments focused on conventional animal sciences, animal husbandry, food sciences, dairy husbandry, poultry husbandry, veterinary science, veterinary medicine, and agricultural economics.

Animal welfare: How an animal is coping with the conditions in which it lives. An animal is in a good state of welfare if (as indicated by scientific evidence) it is healthy, comfortable, well nourished, safe, able to express innate behaviour, and if it is not suffering from unpleasant states such as pain, fear, and distress. Good animal welfare requires disease prevention and veterinary treatment, appropriate shelter, management, nutrition, humane handling, and human slaughter/killing. Animal welfare refers to the state of the animal; the treatment that an animal receives is covered by other terms such as animal care, animal husbandry, and humane treatment (OIE, 2014).

Anthropogenic: Of, relating to, or resulting from the influence of human beings on nature (Merriam-Webster, 2014).

Antibiotic: A metabolic product of one microorganism or a chemical that in low concentrations is detrimental to activities of specific other microorganisms. Examples include penicillin, tetracycline, and streptomycin. Not effective against viruses. A drug that kills microorganisms that cause mastitis or other infectious diseases (FAO, 2009).

Antimicrobial: Destroying or inhibiting the growth of microorganisms and especially pathogenic microorganisms (Merriam-Webster, 2014).

Antimicrobial resistance (AMR): Resistance of a microorganism to an antimicrobial drug that was originally effective for treatment of infections caused by it (WHO, 2014).

Aquaculture: Also known as fish or shellfish farming, aquaculture refers to the breeding, rearing, and harvesting of plants and animals in all types of water environments including ponds, rivers, lakes, and the ocean. Researchers and aquaculture producers are "farming" all kinds of freshwater and marine species of fish, shellfish, and plants. Aquaculture produces food fish, sport fish, bait fish, ornamental fish, crustaceans, mollusks, algae, sea vegetables, and fish eggs (NOAA, 2014).

Beef: Meat from cattle (bovine species) other than calves. Meat from calves is called veal (EPA, 2012).

Biofloc technology: A system that has a self-nutrification process within culture pond water with zero water exchange (Yoram, 2012).

Bioinformatics: Collection, classification, storage, and analysis of biochemical and biological information using computers especially as applied to molecular genetics and genomics (Merriam-Webster, 2014).

Biosecurity: Security from exposure to harmful biological agents; also: measures taken to ensure this security (Merriam-Webster, 2014).

Capacity building: Strengthening groups, organizations, and networks to increase their ability to contribute to the elimination of poverty (FAO, 2014a).

Carbon footprint: The amount of greenhouse gases and specifically carbon dioxide emitted by something (e.g., a person's activities or a product's manufacture and transport) during a given period (Merriam-Webster, 2014).

Climate change: A change in the state of the climate that can be identified (e.g., by using statistical tests) by changes in the mean and/or the variability of its properties and that persists for an extended period, typically decades or longer (OECD, 2014). Climate change may be due to natural internal processes or external forcing, or to persistent anthropogenic changes in the composition of the atmosphere or in land use.

Dairy cow: A bovine from which milk production is intended for human consumption, or is kept for raising replacement dairy heifers (EPA, 2012).

Domesticated fowl: Poultry. A bird of one of the breeds developed from the jungle fowl (Gallus gallus) including some specialized for meat production and others for egg laying, for fighting, or purely for ornament or show (Merriam-Webster, 2014).

Ecosystem: A system in which the interaction between different organisms and their environment generates a cyclic interchange of materials and energy (OECD, 2014).

Environmental footprint: The effect that a person, company, activity, etc., has on the environment, for example, the amount of natural resources that they use and the amount of harmful gases that they produce (Cambridge Dictionaries, 2014).

Epidemiology: A branch of medical science that deals with the incidence, distribution, and control of disease in a population; the sum of the factors controlling the presence or absence of a disease or pathogen (Merriam-Webster, 2014).

Epigenetics: The study of heritable changes in gene function that do not involve changes in DNA sequence (Merriam-Webster, 2014).

Ethology: The scientific and objective study of animal behavior, especially under natural conditions (Merriam-Webster, 2014).

Feed conversion ratio (FCR): A measure of feed efficiency that is used for all livestock production (New and Wijkström, 2002).

Food animal: An animal used in the production of food for humans. Includes, in common usage, the species and breeds that also supply fiber and hides for human use (FAO, 2014a).

Food security: When all people at all times have access to sufficient, safe, nutritious food to maintain a healthy and active life (FAO, 1996).

Functional genomics: A branch of genomics that uses various techniques (as RNA interference and mass spectrometry) to analyze the function of genes and the proteins they produce (Merriam-Webster, 2014).

Freshwater aquaculture: Produces species that are native to rivers, lakes, and streams. U.S. freshwater aquaculture is dominated by catfish but also produces trout, tilapia, and bass. Freshwater aquaculture takes place primarily in ponds and in on-land, manmade systems such as recirculating aquaculture systems (NOAA, 2014).

Genome: One haploid set of chromosomes with the genes they contain (Merriam-Webster, 2014).

Genomics: A branch of biotechnology concerned with applying the techniques of genetics and molecular biology to the genetic mapping and DNA sequencing of sets of genes or the complete genomes of selected organisms, with organizing the results in databases, and with applications of the data (as in medicine or biology) (Merriam-Webster, 2014).

Greenhouse gases (GHGs): Carbon dioxide, nitrous oxide, methane, ozone, and chlorofluorocarbons occurring naturally and resulting from human (production and consumption) activities, and contributing to the greenhouse effect (global warming) (OECD, 2014).

Growth promoters: Synthetic substances that are included to the feed in order to maximize growth of animals; when applied to a plant, they promote, inhibit, or otherwise modify the growth of a plant (FAO, 2009).

Growth promotants: Among the many sophisticated tools used by feedlots and other producers to raise more rapidly, using less feed, while maintaining high standards of animal health, carcass quality, and food safety. Growth promotants include ionophores, growth implants, and beta-agonists (Beef Cattle Research Council, 2013).

Holistic approach: Looks at the whole picture. The totality of something is much greater than the sum of its component parts and they cannot be understood by the isolated examination of their parts (Encyclo.co.uk, 2014).

Human capital: Productive wealth embodied in labor, skills, and knowledge (OECD, 2014).

Industrial agriculture: A form of modern farming that refers to the industrialized production of livestock, poultry, fish, and crops. The methods of industrial agriculture are technoscientific, economic, and political. They include innovation in agricultural machinery and farming methods, genetic technology, techniques for achieving economies of scale in production, the creation of new markets for consumption, the application of patent protection to genetic information, and global trade (FAO, 2009).

Life-cycle assessment: Life-cycle assessment is an objective process to evaluate the environmental burdens associated with a product, process, or activity by identifying energy and materials used and wastes released to the environment. LCA addresses the environmental aspects and potential impacts throughout a product's life cycle from raw material acquisition through production, use, and end-of-life treatment (FAO, 2009).

Livestock: Includes cattle, sheep, horses, goats, and other domestic animals ordinarily raised or used on the farm. Turkeys or domesticated fowl are considered poultry and not livestock within the meaning of this exemption (29 CFR § 780.328).

Marine aquaculture: Refers to the culturing of species that live in the ocean. U.S. marine aquaculture primarily produces oysters, clams, mussels, shrimp, and salmon as well as lesser amounts of cod, moi, yellowtail, barramundi, seabass, and seabream. Marine aquaculture can take place in the ocean (i.e., in cages, on the seafloor, or suspended in the water column) or in on-land, manmade systems such as ponds or tanks. Recirculating aquaculture systems that reduce, reuse, and recycle water and waste can support some marine species (NOAA, 2014).

Meat: Tissue of the animal body that are used for food (EPA, 2012).

Meat products: Meat that has been subjected to a treatment irreversibly modifying its organoleptic and physicochemical characteristics (OIE, 2014).

Medically important antibiotics: Those belonging to seven classes, specific entities of which are also used in human medicine (Animal Health Institute, 2014).

Micronutrient: Organic compound (e.g., vitamin) essential in minute amounts to the growth and health of an animal (Merriam-Webster, 2014).

Mycotoxin: Toxic substance of fungal origin (e.g., aflatoxin) that proliferates on crops at specific levels of moisture, temperature, and oxygen in air (FAO, 2014).

NAHMS: The National Animal Health Monitoring System (NAHMS) Program Unit conducts national studies on the health and health management of U.S. domestic livestock and poultry populations (USDA, 2014).

Natural system: A biological classification based upon morphological and anatomical relationships and affinities considered in the light of phylogeny and embryology; specifically: a system in botany other than the artificial or sexual system established by Linnaeus (Merriam-Webster, 2014)

Nutrient cycle: A repeated pathway of a particular nutrient or element from the environment through one or more organisms and back to the environment. Examples include the carbon cycle, the nitrogen cycle, and the phosphorus cycle (OECD, 2014).

Organic agriculture: A holistic production management system that promotes and enhances agroecosystem health, including biodiversity, biological cycles, and soil biological activity. It emphasizes the use of management practices in preference to the use of off-farm inputs, taking into account that regional conditions require locally adapted systems. This is accomplished by using, where possible, cultural, biological, and mechanical methods, as opposed to using synthetic materials, to fulfill any specific function within the system (FAO, 2009).

Pastoralism: Livestock raising; social organization based on livestock raising as the primary economic activity (Merriam-Webster, 2014).

Phytosanitary: Of, relating to, or being measured for the control of plant diseases especially in agricultural crops (Merriam-Webster, 2014).

Production efficiency: The most efficient means of producing a given good (FAO, 2014a).

Proteomics: A branch of biotechnology concerned with applying the techniques of molecular biology, biochemistry, and genetics to analyzing the structure, function, and interactions of the proteins produced by the genes of a particular cell, tissue, or organism, with organizing the information in databases, and with applications of the data (Merriam-Webster, 2014).

Poultry: All domesticated birds, including backyard poultry, used for the production of meat or eggs for consumption, for the production of other commercial products, for restocking supplies of game, or for breeding these categories of birds, as well as fighting cocks used for any purpose.

Birds that are kept in captivity for any reason other than reasons referred to in the preceding paragraph, including those that are kept for shows, races, exhibitions, competitions, or for breeding or selling these categories of birds as well as pet birds, are not considered to be poultry (OIE, 2014).

Smallholder farmers: Those marginal and submarginal farm households that own or/and cultivate less than 2.0 hectares of land (Singh et al., 2002).

Sociocultural: Of, relating to, or involving a combination of social and cultural factors (Merriam-Webster, 2014).

Socioeconomic: Of, relating to, or involving a combination of social and economic factors (Merriam-Webster, 2014).

Subsistence: The minimum necessary to support life (Merriam-Webster, 2014).

Subtherapeutic: Not producing a therapeutic effect (Merriam-Webster, 2014).

Sustainability: To create and maintain conditions, under which humans and nature can exist in productive harmony, that permit fulfilling the social, economic, and other requirements of present and future generations (National Environmental Policy Act of 1969; Executive Order 13514 [2009]; NRC 2010).

Sustainable aquaculture: A dynamic concept and the sustainability of an aquaculture system will vary with species, location, societal norms, and the state of knowledge and technology. Several certification programs have made progress in defining key characteristics of sustainable aquaculture. Some essential practices include environment practices, community practices, and sustainable business and farm management practices (Word Bank, 2014).

Sustainable intensification: Maximization of primary production per unit area without compromising the ability of the system to sustain its productive capacity. This entails management practices that optimize nutrient and energy flows and use local resources, including horizontal combinations (e.g., multiple cropping systems or polycultures), vertical combinations (e.g., agroforestry), spatial integration (e.g., crop-livestock or crop-fish systems), and temporal combinations (rotations) (FAO, 2009).

Systems approach: The consideration of different interacting parts of a distinct entity (i.e., system). In a food system, this involves the integration of all biophysical and sociopolitical variables involved in the performance of the system (FAO, 2009).

Therapeutic: Of or relating to the treatment of disease or disorders by remedial agents or methods (Merriam-Webster, 2014).

Traditional agriculture: An indigenous form of farming, resulting from the coevolution of local social and environmental systems, that exhibits a high level of ecological rationale expressed through the intensive use of local knowledge and natural resources, including the management of agrobiodiversity in the form of diversified agricultural systems (FAO, 2009).

Tropospheric ozone (O_3): A major air and climate pollutant. It causes warming and is a highly reactive oxidant, harmful to crop production and human health. O_3 is known as a "secondary" pollutant because it is not emitted directly, but instead forms when precursor gases react in the presence of sunlight (UNEP, 2014).

Water footprint: An indicator of water use that looks at both direct and indirect water use. The water footprint of a product (good or service) is the volume of freshwater used to produce the product, summed over the various steps of the production chain (FAO, 2014b).

Wicked problem: A term from the 1970s social planning literature. Such a problem has the essential characteristic that it is not solvable; it can only be managed. The combination of Rittel and Webber (1973) with Conklin (2006) provides a lengthy list of relevant criteria that characterize wicked problems. Four of these criteria are adopted here as an efficient set to define the concept:

- No definitive formulation of the problem exists.
- Its solution is not true of false, but rather better or worse.
- Stakeholders have radically different frames of reference concerning the problem.
- The underlying cause-and-effect relationships related to the problem are complex, systemic, and either unknown or highly uncertain (Peterson, 2013).

Zoonotic: Any disease or infection that is naturally transmissible from animals to humans (OIE, 2014).

REFERENCES

Agrarian Civilizations. 2014. What is an Agrarian Society? Online. Available at http://agrariansocieties.weebly.com/what-is-an-agrarian-society.html. Accessed December 12, 2014.

Animal Health Institute. 2014. Animal Health Industry Supports FDA Antibiotic Judicious Use Guidelines. Online. Available at http://www.whitehouse.gov/sites/default/files/microsites/ostp/PCAST/carnevale_richard.pdf. Accessed September 15, 2014.

Beef Cattle Research Council. 2013. Explaining growth promotants used in feedlot cattle. Online. Available at http://www.cattlenetwork.com/cattle-news/Explaining-growth-promotants-used-in-feedlot-cattle-186847742.html. Accessed September 18, 2014.

Cambridge Dictionaries. 2014. Online. Available at http://dictionary.cambridge.org. Accessed September 19, 2014.

CASL (Community Adaptation and Sustainable Livelihoods). 2014. Adaptive Strategies. Online. Available at http://www.iisd.org/casl/intro+defs/def-adaptivestrategies.htm. Accessed December 12, 2014.

Conklin, J. E. 2006. Dialog Mapping: Building Shared Understanding of Wicked Problems. Napa, CA: CogNexus Institute.

Encyclo.co.uk. 2014. Online. Available at http://www.encyclo.co.uk/define/Holistic%20approach. Accessed September 19, 2014.

EPA (U.S. Environmental Protection Agency). 2012. Glossary. Online. Available at http://www.epa.gov/oecaagct/ag101/glossary.html. Accessed September 15, 2014.

FAO (Food and Agriculture Organization of the United Nations). 1996. Rome Declaration on World Food Security. World Food Summit, November 13-17, 1996, Rome, Italy. Available at http://www.fao.org/docrep/003/w3613e/w3613e00.HTM. Accessed August 14, 2014.

FAO. 2001. *Codex Alimentarius*—Organically Produced Foods. Online. Available at http://www.fao.org/docrep/005/y2772e/y2772e04.htm. Accessed September 15, 2014

FAO. 2009. Glossary on Organic Agriculture. Available at http://termportal.fao.org/faooa/oa/pages/pdfFiles/OA-en-es-fr.pdf. Accessed September 15, 2014.

FAO. 2014a. FAOTERM. Online. Available at http://termportal.fao.org/faoterm/main/start.do?lang=en. Accessed September 19, 2014.

FAO. 2014b. FAOWATER. Online. Available at http://termportal.fao.org/faowa/main/start.do. Accessed September 19, 2014.

Liverpool-Tasie, L. S., O. Kuku, and A. Ajibola. 2011. A review of literature on food security, social capital and agricultural productivity in Nigeria. *Nigeria Strategy Support Program (NSSP) Working Paper*. Abuja: International Food Policy Research Institute.

Merriam-Webster. 2014. Dictionary. Online. Available at http://www.merriam-webster.com/dictionary. Accessed September 18, 2014.

New, M. B., and U. N. Wijkström. 2002. Use of Fishmeal and Fish Oil in Aquafeeds: Further Thoughts on the Fishmeal Trap. FAO Fisheries Circular No. 975. Online. Available at http://www.fao.org/3/a-y3781c.pdf. Accessed September 15, 2014.

NOAA (National Oceanic and Atmospheric Administration). 2014. What Is Aquaculture? Online. Available at http://www.nmfs.noaa.gov/aquaculture/what_is_aquaculture.html. Accessed September 15, 2014.

NRC (National Research Council). 2010. Toward Sustainable Agricultural Systems in the 21st Century. Washington, DC: The National Academies Press.

OECD (Organisation for Economic Cooperation and Development). 2014. OECD Glossary of Statistical Terms. Online. Available at http://stats.oecd.org/glossary. Accessed September 18, 2014.

OIE (World Organisation for Animal Health). 2014. Terrestrial Animal Health Code: Glossary. Online. Available at http://www.oie.int/index.php?id=169&L=0&htmfile=glossaire.htm#terme_bien_etre_animal. Accessed September 15, 2014.

Peterson, H. C. 2013. Sustainability: A wicked problem. Pp. 1-9 in Sustainable Animal Agriculture, E. Kebreab, ed. Wallingford, UK: CAB International.

Rittel, H. W. J., and M. M. Webber. 1973. Dilemmas in a General Theory of Planning. Policy Sciences 4:155-169.

Singh, R. B., P. Kumar, and T. Woodhead. 2002. Smallholder Farmers in India: Food Security and Agricultural Policy. Bangkok, Thailand: FAO Regional Office for Asia and the Pacific. Online. Available at http://www.fao.org/docrep/005/ac484e/ac484e04.htm. Accessed December 12, 2014.

UNEP (United Nations Environmental Programme). 2014. Definitions. Climate and Clean Air Coalition. Online. Available at http://www.unep.org/ccac/Short-LivedClimatePollutants/Definitions/tabid/130285/Default.aspx. Accessed September 19, 2014.

USDA (U.S. Department of Agriculture). 2014. National Animal Health Monitoring System (NAHMS). Online. Available at http://www.aphis.usda.gov/wps/portal/banner/help?1dmy&urile=wcm%3Apath%3A/APHIS_Content_Library/SA_Our_Focus/SA_Animal_Health/SA_Monitoring_And_Surveillance/SA_NAHMS. Accessed September 15, 2014.

WHO (World Health Organization). 2014. Antimicrobial resistance. Online. Available at http://www.who.int/mediacentre/factsheets/fs194/en. Accessed September 19, 2014.

World Bank. 2014. Sustainable Aquaculture. Online. Available at http://www.worldbank.org/en/topic/environment/brief/sustainable-aquaculture. Accessed February 9, 2015.

Yoram, A. 2012. Biofloc Technology: A Practical Handbook, 2nd ed. Baton Rouge, LA: World Aquaculture Society.

D

Key Strategies Involving Animal Agriculture Being Focused on by USDA Research, Education, and Economics (REE)

The USDA's Research, Education, and Economics component works to direct scientific knowledge related to agriculture through promoting research and education. The work of REE influences the funding available for animal science research, and the following are areas that were identified as future research focuses:

1. Invest in the research, development, and extension of new varieties and germplasm, practices, and systems of interest (both domestically and in developing countries) to safely and sustainably increase animal and crop production and its nutritional value. Improve feed and forage use efficiency in animals and identify alternative feed and forage options for animal systems that do not compete for human food and energy needs. Develop and populate a framework for understanding the sustainability (productivity, economic, and environmental) outcomes of agriculture/food/forestry practices and systems.

2. Invest in research, development, and outreach of new varieties and technologies to mitigate animal/plant diseases and increase productivity, sustainability, and product quality. Establish more sustainable systems that enhance crop and animal health.

3. Generate new fundamental knowledge through research in genomic sciences and the applications of systems approaches required to enhance the sustainability of agriculture while increasing productivity. Preserve, characterize, and deploy genetic diversity to ensure economic and environmental sustainability and to maintain American agriculture leadership in a global, biobased economy. Conduct biotechnology risk and benefits assessment research that accurately and scientifically inform regulators, product development, and consumer

acceptance, and provide information to FAS relevant to trade issues.
4. Explain the processes driving the direct and indirect effects of climate variability on natural and managed ecosystems, including feedbacks to the climate system.
5. Develop knowledge and tools to enable adaptation of agriculture, forestry, and grasslands to climate variability and to improve the resilience of natural and managed ecosystems and vulnerable populations.
6. Develop knowledge and tools to enhance the contribution of agriculture, forestry, grasslands, and other land management practices to mitigate atmospheric greenhouse gas (GHG) emissions.
7. Develop and provide the best available science and technology to inform decision-making and improve practices on water conservation, use, and quality.
8. Explain the determinants of socioeconomically viable and environmentally sound livestock, forage, and forest production systems
9. Provide research that helps to understand and define the microbial populations (pathogens and normal flora) in foods and surrounding environments.
10. Provide research to understand the biology and behavior of foodborne pathogens.
11. Develop technologies for the detection and characterization of food supply contamination from microbial pathogens, toxins, chemicals, and biologics.
12. Develop intervention and control strategies for foodborne contaminants along the food production continuum.
13. Provide research strategies, models, and data that identify and characterize effective management strategies and incentives for food safety improvement and the costs and benefits of improved safety for public health and industry viability.
14. Recruit, cultivate, and develop the next generation of scientists and leaders with a highly skilled workforce for food, agriculture, natural resources, forestry, and environmental systems, and life sciences to out-educate our global competitors.
15. Provide effective research, education, and extensions that inform public and private decision making in support of rural and community development.

E

USDA ARS Proposed FY 2015 Priorities

USDA ARS has proposed in their 2015 budget to spend $1.1 billion of which 36 percent will be directed toward crop research, 18 percent to food safety and nutrition, 18 percent to environmental stewardship, 16 percent to livestock, 8 percent to product quality/value added, and 4 percent to other. Major focuses will include climate change, genetic improvement and translational breeding, livestock production, feed safety, and livestock protection.

- *Food Safety ($110 million requested)*—The goal is to yield science-based knowledge on the safe production, storage, processing, and handling of plant and animal products, and on the detection and control of toxin-producing and/or pathogenic bacteria, fungi, parasites, chemical contaminants, and plant toxins.
- *Livestock Protection ($87 million requested)*—Goal of the animal health program is to protect and ensure safety of the U.S. food supply through improved disease detection, prevention, control, and treatment.
- *Livestock Production ($83 million requested)*—Goal is to (1) safeguard and utilize animal genetic resources, associated genetic and genomic databases, and bioinformatic tools; (2) develop a basic understanding of the physiology of livestock and poultry; and (3) develop information, tools, and technologies that can be used to improve animal production systems. The research will be heavily focused on development and application of genomic technologies to increase the efficiency and product quality of beef, dairy, swine, poultry, aquaculture, and sheep systems.
- *Climate change ($44 million requested)*—Goal is to better understand the effects of climate change and develop adaptive strategies and technologies to address its impacts.
- *Genetic Improvement and Translational Breeding ($25.9 million requested)*—Goal is to strengthen U.S. agricultural productivity and resilience by developing new breeds, lines, and strains with better

climate adaptation, drought tolerance, disease resistance, nutritional value, enhanced production efficiencies, and reduced environmental impacts. Translational breeding will be advanced through application of genomic knowledge to breeding programs (classical and genomic-enabled), expansion of access to genetic resources, knowledge, and tools.

F

Animal Health Priorities from a 2011 USDA NIFA Workshop

The USDA's National Institute of Food and Agriculture (NIFA) is a leader in supporting food and agriculture research, overseeing federal support and funding for research and extension programs. USDA NIFA plays a substantial role in the future of animal science research, through both priority identification and the provision of competitive research grants, and facilitates communication and collaboration between stakeholders related to animal agriculture.

BY INDUSTRY

High priority areas of animal issues developed by stakeholders participating in the 2011 ARS/NIFA workshop include:

Beef Industry
1. Top Priority: Elucidate the effects and interplay of host, indigenous microbial communities and the production environment on components of production efficiency, sustainability, and product value (including healthfulness and safety). A holistic approach for research was prioritized focused on systems optimization to improve the following: forage and feed efficiencies; product quality, product safety, and healthfulness; environmental sustainability; economic sustainability; and animal well-being. Ultimate goal is to improve global food security by improving sustainable beef production and production efficiencies.
2. Improve efficiency of nutrient utilization including forages.
3. Mitigate antimicrobial usage and better understand their effects.
4. Enhance the healthfulness of beef products.

5. Measurement of, and best management practices for, animal well-being across the production system.

> "A major challenge over the next 30 years will be to generate or discover the significant biological production and production efficiency increases and to generate improved economic efficiencies to meet the global demands for meat production and to ensure profitability for producers."

Dairy Industry

1. Top Priority: Reproduction and reproductive efficiency as one of the largest drivers of profitability.
2. Translational genomics of dairy cattle which impact all phenotypes and production traits associated with dairy cattle.
3. Improved dairy nutrition and nutrient utilization as a primary factor for impacting production costs—feed costs currently representing approximately 70 percent of the cost of dairy production.
4. Animal well-being, related to dairy production and management and consumer acceptance of dairy products and production practices.
5. Improving performance for the transition cow.
6. Improved heifer development as related to lifetime production and transition cow challenges.

Goat Industry

1. Top Priority: Comparative physiology and genomics to accelerate technological and genetic progress.
2. Conduct studies to develop economic analysis of the goat industry, promote goat production, and foster domestic and international market development.
3. Develop improved meat and milk products to enhance local food security and provide products that are higher quality and more nutrient dense.
4. Develop more comprehensive sustainable production systems.
5. Development of sustainable, low-input, forge-based production systems that optimize costs and resources including approaches to address genetic, reproductive and production traits with the goal of maximizing profit.

APPENDIX F

6. Improvement of goat performance on high concentrate rations in confined feeding operations with focus on growth, nutrient efficiency, and environmental sustainability.
7. Develop research programs focused on improved production traits including health, immunity, growth, efficiency, reproduction, parasite resistance, product quality, and animal well-being.
8. Focus on systems to promote reproductive efficiency on a year-round basis.

Swine Industry

1. Top Priority: Improved production efficiencies for growing swine post-weaning in response to increasing feed and grain costs.
2. Improved production efficiencies in the breeding herd to increase (optimize) sow production, longevity, and lifetime performance.
3. Development of systems-based research models to optimize swine production for maximum profitability and competitiveness.
4. Proactive management issues related to swine production that may affect swine industry competitiveness and profitability.
5. Improved quality and demand for fresh and processed pork products.
6. Adaptation of animals and production systems in response to climate change temperature, humidity, and other environmental changes.
7. Development of production models that optimize economic and environmental sustainability of the pork industry.

Poultry—Layer Industry

1. Top Priority: Develop expanded research in poultry layer well-being, including alternative housing systems so informed decisions can be made by producers facing pressure from society to modify existing housing.
2. Enhance feed and nutritional efficiencies of broiler breeders and develop alternative feed ingredient options to decrease the cost of feed and production.
3. Enhance environmental sustainability and reduce the environmental footprint of the poultry layer industry.

4. Develop comprehensive programs in functional genomics to enable the prediction of a phenotype from a genotype.
5. Develop research in poultry reproduction to improve hatchability, fecundity; address management issues associated with feed restriction and satiety; and improve the genetics of egg lay and fertility.
6. Improve the ability to preserve poultry germplasm through development of improved techniques and technologies—improve the ability to regenerate a specific genetic line from preserved germplasm and improve the ability to maintain current production lines.

Poultry—Broiler Industry

1. Top Priority: Improve preharvest food safety to provide wholesome poultry products for end users with maximum attainable shelf life.
2. Enhance feed and nutritional efficiencies of broilers and develop alternative ingredient options to decrease cost of production.
3. Enhance broiler production efficiency to improve the domestic and international competitiveness.
4. Improve the understanding of broiler physiology to optimize bird well-being, productivity, and efficiency.

BY RESEARCH DISCIPLINE

Genetics, Genomics, and Genetic Technologies

1. Top Priority: Revise the USDA Blueprint for Animal Genetics to include direction for expanding the bioinformatic and quantitative genomic capacities for research to facilitate manipulation and analysis of large datasets.
2. Focus research programs to cost-effectively increase the genome sequencing of individual animals and organisms. Promote the development of the next-generation genomic and genetic technologies and increase genetic progress through improved breeding and selection programs,
3. Expand and coordinate research programs in phenomics, including development of comprehensive phenotype databases required for genetic characterization.

4. Maintain basic research in genetic modification and genetic engineering of food animals.
   5. Maintain research and programming to secure, preserve, and collate animal genetic resources to ensure future access to genetic variability for food animal industries.

See reference for recommendations for extension and educational programs.

Physiology: Reproduction

   1. Optimize fertility in livestock as a primary driver of productivity, production efficiency, and profitability for the food and animal industries.
   2. Assess the effect of genetic, epigenetic, and environmental influences and relationships on reproduction (genotype to phenotype).
   3. Develop innovative reproductive biotechnologies to improve reproductive function and efficiency in food animals.

Physiology: Lactation

   1. Top Priority: Optimize or improve lactation performance and lactation efficiency.
   2. Improve understanding of mammary development and growth.

Growth Biology and Nutritional Efficiency

   1. Top Priority: Develop means to potentiate nutrient utilization to reduce the relative cost of feed and forages in food animal production systems.
   2. Characterize and manipulate the microbiome of the gastrointestinal tract to elucidate the role of the microbiome for growth, nutrient utilization, immune function, and greenhouse gas emissions.
   3. Optimize current population genetic and production systems to maximize food animal product synthesis and quality. Specific areas of research include nutrition and nutrient partitioning; metabolic modifiers; fundamental biology of productive tissues, including muscle adipose, mammary gland, liver wool, etc; and adaptability, including thermal stress in response to climate change and another environmental factors.

4. Developmental programming to assess how prenatal and postnatal environment affects lifetime productivity, longevity, product quality, and composition and related traits.

Animal Well-Being, Stress, and Production

1. Top Priority: Quantify the relationship between animal well-being, production, and economic factors including genetics/genomics, behavior, housing, health, nutrition, management, production level, profitability, production efficiencies, and food safety. Specifically develop objective criteria to assess animal comfort and care within specific production environments.
2. Evaluate the effect of current management practices and procedures on animal well-being, stress, and productivity. Develop and validate cost-effective alternative management practices and procedures that improve animal well-being.
3. Identify alternatives to subtherapeutic levels of antibiotics in livestock production.
4. Develop strategies to enhance recovery of livestock from stressful events and disease challenges.

Meat Quality and Muscle Biology

1. Top priority: Enhance food animal meat product quality to increase consumer demand for food animal products.
2. Assess and enhance the nutritional value and human health benefits of meat products.
3. Increase saleable product yield from food animals, including increased lean deposition and increased yield of high-value meat cuts.
4. Enhance meat product value through development and investigation of novel technologies and their application to the meat Industry.

Forage and Forage Utilization

1. Top priority: Enhance basic rumen ecology through improved understanding of the following priorities: (1) enhanced nitrogen utilization in the rumen and post-ruminally; (2) elucidate the role and function of the rumen microbiome in forage digestion; and (3) elucidate the role and function of fatty acids in health and production efficiencies of ruminant food animal species.

APPENDIX F 393

2. Develop comprehensive, systems-based, applied or translational sustainable forage systems to increase production efficiencies of ruminants.
3. Improve forage utilization efficiency—on farm.
4. Improve vegetative quality and nutrient availability of forages in an integrated plant and animal system.

Environmental Aspects of Sustainability: Climate Change, Greenhouse Gases, Manure Management, Water, and Air Quality

1. Top priority: Define impacts of pollutants and mitigation practices through use of comparative life-cycle assessment (and other tools) for various production systems.
2. Characterize, quantify, and mitigate air and water pollutants, including greenhouse gases, from livestock and poultry operations.

Alternatives to Antimicrobials for Production

1. Top Priority: Conduct comprehensive risk/benefit analysis of the use of antimicrobials for specific application in food animal production.
2. Identify and discover new technologies, compounds, or agents to promote animal health and well-being in the absence of conventional antimicrobials.
3. Physiology of animals and bacteria: Better understand how commonly used subtherapeutic antibiotics work in the context of food safety, animal well-being, health, and profitability.

Food Animal Production and Energy Evolution

1. Top Priority: Ensure economic and environmental sustainability of livestock enterprises by characterizing energy partitioning and evaluating opportunities for energy alternatives and more efficient use.
2. Close energy loop in animal production and increase efficiency to minimize energy impacts on profitability.
3. Develop technologies and methods needed to improve efficiency in food animal production.

Current and Promising Technologies in Genomics and Bioinformatics

1. Top Priority: Develop animal agriculture specific tools for biological information, including software to manage large

complex datasets, across multiple species. Deploy relevant applications in order to improve data integration and visualization from disparate sources to support evidence-based functional genomics.
2. Improved sampling and data collection procedures and protocols for metagenomics.
3. Train and develop human capital to serve as next-generation scientists in genetics, biology, and bioinformatics.

Develop low-plex, low-cost, cost-effective genotyping technologies.

Food Animal Products: Nutritional Value, Healthfulness, Emerging Consumer Trends

1. Top Priority: Enhance the healthfulness of animal products and identify factors that control variation in nutritive value of animal-source food products.
2. Identify ramifications of changing nutritive value of animal-source products on human health.

G

Results of a USDA ARS- and NIFA-Sponsored Workshop on Animal Health

In 2010, the USDA Agricultural Research Service (ARS) held a joint animal health research planning workshop with the National Institute of Food and Agriculture (NIFA). ARS plays a key role in identifying and addressing research needs within the animal science discipline. Stakeholders from this workshop developed the following prioritized list of animal health priorities for future research (Table G-1).[1]

TABLE G-1 Animal Health Priorities for Future Research

Species	Ranked 1	Ranked 2	Ranked 3	Ranked 4	Ranked 5	Ranked 6
Beef cattle	Bovine respiratory disease complex including BVD	-Mycobacterial diseases (TB and Johnes) -Vectorborne diseases -Infectious reproductive disease			Minimize impact of emerging infectious diseases	Animal well-being

---

[1] Additional information can be found at http://www.ars.usda.gov/SP2UserFiles/Program/103/ARSNIFAWorkshop/PrioritiesCondensedFinal.pdf. Accessed September 2, 2014.

Species	Ranked 1	Ranked 2	Ranked 3	Ranked 4	Ranked 5	Ranked 6
Poultry breeder/layers	Housing systems' influence on health/welfare	Salmonella enteriditis	Tumor viruses	Colibacillosis	Mycoplasma gallisepticum	Infectious laryngotracheitis
Poultry/broilers/meat	Functional genomics for disease resistance	GI disease/integrity/host microbial interactions	Disease affecting world trade	Respiratory disease complex	Vaccines and their limitations	Tumor viruses (Marek's, ALV, REV)
Dairy	Lameness	Johnes	Tuberculosis	Mastitis	Transition cow	Infertility Nutrition and metabolic disorders
Equine	Emerging & reemerging diseases	Noninfectious diseases of economic importance	Reproductive and developmental health	Equine genomics	Foreign diseases & zoonoses	
Goat	Gastrointestinal parasites (worms and protozoa)	Species-specific approvals for necessary pharmaceuticals	Control measures for caseous lymphadenitis	Mastitis control and treatment	Q Fever	Eradicate scrapie
Swine	PRRS elimination	Emerging and zoonotic diseases	Optimize health of growing pig	Peripartu-rient production efficiency	Healthy pig production with restricted antimicrobial access	Foreign animal diseases

Species	Ranked 1	Ranked 2	Ranked 3	Ranked 4	Ranked 5	Ranked 6
Sheep	Research on bighorn/domestic sheep compatibility	Eradicate scrapie	Control and prevention of ovine progressive pneumonia	Prevent malignant catarrhal fever in bison and cattle	Genetic/genomic solutions to economically significant sheep diseases	Improved diagnostics for ovine Johnes, Q Fever, and Brucella ovis
Specialty species	Tuberculosis rapid diagnostic tools	Prevent sheep-associated malignant catarrhal fever in specialty farmed species	Epizootic hemorrhage disease/bluetongue	Bacterial pneumonia—Pasturella/Fusobacteria	Parasite control	Tools and resources
Turkey	Clostrial dermatitis (turkey cellulitis)	Preharvest food safety	Influenza in turkey breeders	Enhanced gut health	Histomoniasis	Understanding the adaptability of pathogens to current treatments

H

Summary of NOAA/USDA Findings on Alternative Feeds for Aquaculture

The National Oceanic and Atmospheric Administration (NOAA) partnered with the USDA to examine the issue of aquaculture feed and to create strategies for the development of alternative feed for use in aquaculture farming. The findings of this effort are as follows:[1]

1. Fish meal and fish oil are not nutritionally required for farmed fish to grow.
2. Farming of fish is a very efficient way to produce animal protein and other human nutritional needs.
3. Feed manufacturers making diets for carnivorous fish and shrimp have already reduced their reliance on fish meal and fish oil.
4. Economics is currently the major driver of using alternate feed ingredients in feed mills.
5. The net environmental effects of the production and use of alternate feeds should be considered.
6. The human health implications of using alternative feeds needs to be better understood and considered.
7. Fish meal and fish oil are minor contributors to the world protein and edible oil supply.
8. Recovery and utilization of fisheries processing waste should be encouraged and increased.
9. Plants produce the vast majority of protein and edible oils in the world, accounting for 94 percent of total protein production and 86 percent of total edible oil production.
10. Algae-based biofuel may present opportunities for feed ingredient production because protein is a byproduct of oil recovery from algae, and marine algae produce the long-chain

---

[1] NOAA/USDA. 2011. The Future of Aquafeeds (Alternative Feeds Initiative). NOAA Technical Memorandum NMFS F/SPO-124. 93 pp

omega-3 fatty acids and certain amino acids important to fish and human health.
11. There will likely be increased demand for and production of ethanol and bioplastics. Byproducts from these industries could make good ingredients for fish diets.
12. As replacements, many alternatives are higher in cost per unit fish gain (biological value) than fish meal and fish oil.
13. Fish have dietary needs and preferences for specific compounds not found in plants, so there is a need for specialized products that supply these compounds and/or add flavor to the diet.
14. Alternative sources of protein and oil are common commodities used in livestock and companion animal feeds and come from novel byproducts from other industries, underutilized resources, or completely novel products.
15. Plants and other alternatives contain some compounds (antinutrients) that are detrimental to fish.
16. Harvest of lower-trophic-level species, such as krill, for fish meal and oil production may be possible, but the environmental benefits afforded to the marine ecosystem from these species should be considered along with the economic and nutritional aspects of their use.
17. The use of bycatch for production of fish meal and fish oil could provide a substantial amount of these products without increasing the current impact from the wild capture fisheries.
18. Demand for long-chain omega-3 fatty acids for both direct human consumption and feed ingredients is likely to increase beyond the amounts available from marine resources.
19. Farmed fish species are being increasingly domesticated and performance is improving through conventional genetic selection and selection for performance on plant-based and/or low-fish meal–based aquafeeds.
20. Scientific information on the nutritional requirements of farmed fish species, and feed ingredients, and the interaction between the fish and the diet, will need to expand greatly to make substantial improvements in feed formulation by commercial aquaculture feed producers.

I

Goals for Priorities Identified by the EU Animal Task Force

The EU Animal Task Force comprises stakeholders from various parts of the livestock production chain and provides guidance on investment into livestock-related initiatives, with many specifically related to sustainability. Priorities identified by this group are as follows:[1]

Resource Efficiency

- *Efficient and robust animals*: Improve resource efficiency of animals by more efficient and robust animals that are more healthy, are more resilient, have an increased well-being, have a lower feed conversion rate.
- *Efficient feed chains*: Create new opportunities to improve the efficiency of feed chains by optimizing the quantity of feed available for the animal, reducing losses, making better use of local resources, and creating new feed chains of alternative feed resources and byproducts of the food chain, thereby reducing wastes.
- *Improving the use of residues in animal production*: More efficient recovery and recycling of food, feed, water, and animal waste, including P losses and N emission reduction. Reduced energy cost across animal farming.
- *Precision livestock farming*: Develop and implement future options for innovation in livestock systems that will make Europe's livestock systems more efficient and sustainable. Achieve integration of knowledge between biological, veterinarian, social, economic,

---

[1] Animal Task Force. 2013. Research & Innovation for a Sustainable Livestock Sector in Europe. An Animal Task Force White Paper: Available at http://www.animaltaskforce.eu/Portals/0/ATF/documents%20for%20scare/ATF%20white%20paper%20Research%20priorities%20for%20a%20sustainable%20livestock%20sector%20in%20Europe.pdf. Accessed June 16, 2014.

engineering, and ICT scientists. Combine research and development with product and service development by matching academic and industrial communities.

Responsible Livestock Farming Systems

- *Assessing EU animal production*: Develop an integrated European approach for the assessment of current systems and their efficiency that will lay out the future options for improving and redesigning animal production systems that contribute to social, environmental, and economic gains. Gain a better understanding of the contribution of animal production systems to community sustainability, or fragility, in Europe, in terms of other values than food products and to include this in the development and evaluation of production systems, for example, the possibility to use livestock farms in "green care," for recreation, for education, or development of rural areas and coastal zones. This requires the development of multicriteria assessment of livestock systems and food chains.
- *Improving protein and energy autonomy of the animal production sector in Europe*: Understand and find opportunities for improving protein and energy autonomy of livestock in Europe by the development of viable systems for optimizing the use of current and emerging resources such as co-products from nonfood industry (biofuels).
- *Productive grassland-based system*: Develop an integrated approach for grassland management that is cost-effective, environmentally sound and manageable, is essential in the context of the development of large-scale dairy enterprises with highly productive, healthy animals that benefit from high welfare standards, and in the context of remote rural areas that need to be grazed for landscape maintenance for recreation.
- Climate-smart animal production: Develop climate-smart, low-emitting, productive, resilient, and robust animal production systems.

Healthy Livestock and People

- *Prevention, control, and eradication*: Develop integrated approaches to disease control and explore the combined impacts (and tradeoffs) of combinations of individual approaches. Develop both the necessary elements of individual control systems (e.g., management procedures/biosecurity; vaccines; disease-resistant genotypes; feeding systems; etc.) and the cost-effective approaches required to

combine these elements into integrated systems for disease control. Create operating networks and paradigms and apply them to the appraisal of risks to European livestock (at national, transnational, and local levels) and human health of endemic, exotic, and emerging diseases.
- *The microbiome, animal and human health*: Enhance the understanding of the interactions in the gut between digestion products of animal feeds, residing microbes, and host immune cells (host genetics) with a view to identify routes for the implementation of this knowledge in the management of improved immune competence in livestock species and reduced health risk to humans.
- *Nutritional quality of animal products*: Improve the nutritional value and health-promoting properties of food of animal origin in sustainable production system (e.g., the fatty acid profile of animal products, the amount of essential trace elements such as iodine or selenium and also critical nutrients such as calcium, zinc, or folate) by understanding the interaction between nutritional composition of feed and genomics. Assess the impact of new feed resources on the nutritional value of animal products.
- *Feed and food safety*: Provide tools and practical guidelines to ensure supplies of food and feed that are microbiologically and toxicologically safe.

Knowledge Exchange Toward Innovation

- *Knowledge exchange with farmers and industry toward innovation*: Ensure that new technologies are developed in a context that improves the uptake of research results into practice, and allows for (a) a positive impact on farm incomes and (b) the exploration of new business models within systems of production and consumption.
- *Improving systems for the implementation of "omics" tools*: The "omics" approach has the overall goal to improve knowledge of the genetic, genomic, and transcriptomic control of traits in order to assist in breeding decisions and in herd management. The primary goal is to link the changes achieved in breeding with the expression of genes, quantification of proteins and pathways, and the metabolic outcomes of these changes. This will also provide us with tools to better understand and use the potential "omics" in creating a more sustainable livestock sector. Omics technologies give also new means to ensure animal well-being, health, and fertility by

understanding the interactions between the nutrient composition of the feed and animal genomics.
- *Ensuring animal welfare*: Evolve techniques and new concepts to improve the implementation of animal welfare in farming practices and to provide a level playing field regarding monitoring of animal welfare, in combination with other sustainability aspects (profitability, environmental load, etc.) in a global perspective. Deliver innovation of production systems to ensure intrinsic animal welfare in combination with other sustainability requirements. Enable the achievement of high standards of animal welfare across local production and societal circumstances.

Opportunities and Needs in "Excellent Science"

- *Diet-host-microbiome interactions*: Gain an understanding of the symbiotic functions of key members of the microbiota, their metabolism and ecology. Predictive understanding of the factors that affect interactions between the gut microbial community (and lung microbial immunity) and host function, and the means to exploit this understanding for practical benefit.
- *Long-term consequences of environmental effects in early life*: Better understanding of key environmental factors that impact later life; quantify the relative impact of early-life conditions on later health, welfare, and performance; and develop a chain approach to integrate this new knowledge in an integrated chain approach.
- *Enabling the predictive understanding of phenotypic expression*: Create capability across Europe for the development of theory and applications to deliver predictive understanding of phenotypic expression in mammals and birds.
- *Immune regulation at mucosae*: Ensure a preventive animal health care allowing a responsible use of medicines and antihiotics.

J

USDA NIFA Investment in Animal Science by Species (Subject of Interest)

TABLE J-1 Animal Science Investment by Subject of Interest

	2003	2004	2005	2006	2007	2008	2009	2010	2011	2012
Egg-type chicken	1.9	2.7	2.1	1.7	3.0	2.1	2.9	2.6	4.7	3.2
Meat-type chicken	3.4	4.6	4.5	4.2	6.1	5.7	4.4	5.1	5.7	4.1
Turkey	1.7	1.7	2.6	1.1	2.3	1.8	1.6	1.9	0.5	1.4
Duck and goose	0.0	0.0	0.0	0.0	0.0	0.2	0.2	0.1	0.0	0.0
Ratites	0.0	0.0	0.0	0.0	0.0	0.0	0.0	0.0	0.0	0.0
Poultry meat	1.4	0.7	1.7	1.0	0.9	1.3	1.5	1.1	0.5	0.7
Eggs	0.1	0.1	0.2	0.7	0.2	0.3	0.1	0.2	0.4	0.4
Other poultry products	0.1	0.1	0.3	0.2	0.1	0.1	0.1	0.1	0.0	0.1
Poultry, general	3.2	3.7	4.1	3.7	1.8	3.5	2.5	3.7	3.1	2.8
Beef cattle	13.2	8.8	12.5	13.3	11.8	10.9	12.5	9.6	13.6	14.2
Meat, beef cattle	2.1	1.5	1.1	2.7	1.4	1.6	1.3	1.7	1.6	1.9
Other beef products	0.3	0.0	0.0	0.1	0.0	0.2	0.1	0.2	0.1	0.2

	2003	2004	2005	2006	2007	2008	2009	2010	2011	2012
Beef cattle, general	2.7	5.8	2.5	2.6	4.1	4.5	4.5	4.1	2.4	3.3
Dairy cattle	12.1	12.8	12.1	15.2	14.1	13.2	14.4	15.0	15.8	13.8
Butter	0.1	0.1	0.1	0.0	0.0	0.0	0.0	0.0	0.0	0.0
Cheese	0.6	0.3	0.7	0.3	0.2	0.3	0.2	0.9	0.2	0.3
Meat, dairy cattle	0.0	0.0	0.0	0.0	0.2	0.3	0.1	0.0	0.2	0.1
Milk	1.8	1.3	3.3	1.2	1.0	1.2	1.0	0.8	0.9	0.9
Ice cream	0.0	0.0	0.0	0.0	0.0	0.1	0.2	0.1	0.1	0.1
Other dairy products	0.5	0.3	0.4	0.6	0.7	0.3	0.3	0.5	0.1	3.7
Dairy cattle, general	2.7	6.1	2.3	2.8	3.2	3.7	2.4	4.2	1.4	4.3
Swine	7.8	9.4	6.4	4.5	8.0	5.8	5.2	4.9	10.7	8.2
Meat, swine	0.8	1.6	1.0	1.2	1.2	1.0	0.9	1.0	1.0	1.2
Other swine products	0.0	0.0	0.0	0.1	0.1	0.1	0.1	0.9	0.1	0.2
Swine, general	2.8	3.3	4.1	6.6	2.4	4.2	4.4	1.5	4.2	2.5
Sheep	3.0	2.5	5.5	3.4	3.0	2.4	5.0	3.5	3.1	5.2
Meat, sheep	0.2	0.2	0.7	0.3	0.5	0.3	0.5	0.4	0.2	0.6
Wool fiber	0.1	0.2	0.0	0.3	0.0	0.1	0.1	0.1	0.2	0.0
Sheep and wool, general	0.8	0.7	1.1	0.6	0.5	0.6	0.4	0.5	0.1	0.1
Catfish	3.6	2.4	1.8	2.2	2.0	2.1	3.3	2.3	1.4	1.5
Trout	2.4	2.1	2.1	2.0	3.0	1.5	1.8	1.1	1.3	1.4
Salmon	1.5	2.6	0.8	0.4	0.8	0.8	0.0	0.4	0.3	0.4

	2003	2004	2005	2006	2007	2008	2009	2010	2011	2012
Striped bass	1.0	0.6	1.1	0.7	0.3	0.4	0.3	0.3	0.2	0.2
Tilapia	0.9	0.9	0.8	0.9	0.9	0.8	0.4	1.2	0.6	0.5
Baitfish	0.4	0.6	1.1	0.8	0.6	0.3	0.5	0.7	0.6	0.5
Ornamental finfish	0.4	0.2	0.4	0.3	0.2	0.5	0.3	0.4	0.2	0.1
Other cultured finfish	3.2	2.9	3.0	4.2	2.8	3.8	3.7	3.2	1.6	2.8
Crawfish	0.7	0.3	0.2	0.2	0.0	0.1	0.1	0.1	0.1	0.2
Marine shrimp	4.5	3.0	4.1	4.4	1.2	3.3	2.9	3.1	0.5	0.2
Freshwater shrimp	0.5	0.4	0.6	0.2	0.1	0.1	0.1	0.2	0.2	0.2
Oysters	1.5	1.2	0.8	2.4	0.7	0.8	1.5	0.9	0.5	0.4
Clams and mussels	0.3	0.4	0.1	0.5	0.1	0.4	0.3	0.9	0.3	0.3
Ornamental shellfish	0.0	0.3	0.0	0.1	0.0	0.0	0.5	0.0	0.0	0.0
Other cultured shellfish	0.2	0.2	0.2	0.1	0.4	0.0	0.1	0.1	0.1	0.1
Cultured aquatic animals, general	1.2	1.5	1.2	1.0	1.5	1.3	1.0	1.3	0.7	1.5
Horses, ponies, and mules	2.3	2.4	1.9	2.4	3.4	2.5	2.6	2.6	1.6	4.0
Goats, meat, and mohair	4.8	4.6	4.6	3.4	3.8	2.4	4.2	6.4	5.5	4.0
Other animals, general	1.6	0.6	0.5	0.9	0.7	1.3	0.5	1.9	1.5	0.9
Cross-commodity research	2.2	2.2	2.9	2.2	6.9	7.1	5.3	5.6	4.6	4.2
Animal research,	3.2	2.1	2.5	2.2	3.6	4.8	3.7	2.6	7.1	3.3

	2003	2004	2005	2006	2007	2008	2009	2010	2011	2012
general Total:	100.0	100.0	100.0	100.0	100.0	100.0	100.0	100.0	100.0	100.0

USDA NIFA Investment in Animal Science by Discipline (Knowledge Area)

K

Animal Science Investment by Knowledge Area

TABLE K-1 Animal Science Investment by Knowledge Area (%)

	2003	2004	2005	2006	2007	2008	2009	2010	2011	2012
Reproductive performance	14.3	11.8	13..1	11.8	14.0	12.7	12.4	10.0	13...1	11.9
Nutrient utilization	12...1	10.6	11.2	11.7	11.2	11.3	11.5	13.1	14.0	15.2
Genetic improvement	8.1	6.8	8.4	8.1	5.0	9.2	8.7	8.6	6.3	7.8
Animal genome	5.0	11.5	9.6	10.5	12.6	8.2	9.5	9.0	9.7	7.1
Animal physiological	9.5	6.0	6.7	6.0	6.1	5.3	5.9	5.0	5.6	6.3
Environmental stress	1.2	1.6	1.7	1.4	2.3	1.1	1.2	1.5	1.1	1.4
Animal management systems	12.0	9.8	10.7	11.0	10.4	9.8	9.0	11.1	8.9	8.8
Improved animal products	1.8	2.2	2.5	4.2	2.5	2.5	2.4	2.6	5.5	6.4
Animal diseases	18.0	23.5	18.0	17.3	20.5	25.0	22.0	20.6	18.1	18.0
External parasites and pests	1.4	1.4	1.3	1.2	0.7	0.7	1.0	0.7	0.4	0.6
Internal parasites	1.4	1.1	1.7	1.0	1.3	0.5	0.4	1.1	1.3	1.4
Toxic chemicals	0.5	0.3	0.3	0.5	1.3	0.3	0.2	1.3	0.2	0.4

	2003	2004	2005	2006	2007	2008	2009	2010	2011	2012
Animal welfare	1.3	2.0	2.5	1.9	2.3	3.8	5.2	3.3	6.5	3.2
Structures, facilities	0.1	0.0	0.1	0.2	0.3	0.4	0.3	0.1	0.1	0.1
Engineering systems	0.8	0.7	0.2	1.5	0.6	0.3	0.3	0.4	0.3	0.2
Waste disposal, recycling	0.8	0.8	1.0	1.6	0.7	0.8	0.9	0.9	0.7	0.6
Instrumentation	0.3	0.5	0.3	0.6	0.5	0.4	0.5	0.2	0.1	0.1
Drainage and irrigation	0.0	0.0	0.0	0.0	0.0	0.0	0.0	0.0	0.0	0.0
Food processing	3.5	2.5	3.0	3.3	1.8	1.4	2.2	2.0	1.7	2.1
Food products	1.5	1.7	1.8	1.1	1.3	1.9	1.7	1.5	1.7	1.7
Food quality maintenance	0.9	1.6	0.8	0.7	0.7	0.4	0.5	0.3	0.6	0.4
Food service	0.0	0.0	00	00	0.0	0.0	0.0	0.0	0.0	0.0
Non-food products	1.2	0.5	0.8	0.5	0.2	0.3	0.8	0.8	0.4	0.3
Quality maintenance, non-food	0.0	0.0	0.0	0.0	0.0	0.0	0.0	0.0	0.0	0.0
Economics of agriculture	1.2	1.0	0.8	0.8	1.6	1.1	1.0	2.0	1.4	1.2
Business management	0.1	0.0	0.1	0.1	0.1	0.0	0.1	0.0	0.0	0.1
Market economics	1.1	0.4	0.4	0.7	0.5	0.6	0.5	1.0	0.4	0.7
Marketing and distribution	0.7	0.8	0.6	0.9	0.6	0.4	0.3	1.0	0.2	0.3
Natural resource and environment	0.1	0.1	0.3	0.1	0.2	0.1	0.1	0.2	0.2	0.2
International trade	0.1	0.0	0.0	0.1	0.1	0.1	0.1	0.1	0.1	0.1
Consumer economics	0.1	0.1	0.1	0.1	0.1	0.0	0.1	0.2	0.2	0.3

APPENDIX K

	2003	2004	2005	2006	2007	2008	2009	2010	2011	2012
Community resource planning	0.1	0.0	0.0	0.0	0.1	0.0	0.1	0.0	0.1	0.1
Economic theory	0.0	0.0	0.1	0.0	0.0	0.0	0.0	0.1	0.1	0.1
Domestic policy analysis	0.3	0.3	0.3	0.3	0.1	0.3	0.3	0.6	0.1	0.2
Foreign policy and program	0.5	0.4	0.4	0.4	0.0	0.4	0.3	0.3	0.0	0.0
Zoonotic diseases and parasites	0.1	0.0	0.9	0.4	0.2	0.5	0.3	0.3	1.1	2.6
Total:	100.0	100.0	100.0	100.0	100.0	100.0	100.0	100.0	100.0	100.0

L

USDA ARS Animal/Agriculture Research FY 2010–FY 2014

United States Department of Agriculture
Agricultural Research Service

TABLE L-1 Animal/Aquaculture Research FY 2010-FY 2014

Program	Description	FY 2010	FY 2011	FY 2012	FY 2013	FY 2014	Total	Appropriation	Animal Research
101	Food Animal Production	$50,882,000	$49,437,000	$45,712,000	$41,793,000	$51,225,000	$239,049,000	4	20
103	Animal Health	70,823,000	66,510,000	65,205,000	60,554,000	66,794,000	329,886,000	6	27
	Veterinary, Medical and Urban	12,810,000	11,241,000	11,240,000	10,458,000	11,702,000	57,451,000		
104	Entomology	35,670,000	31,213,000	28,839,000	26,951,000	29,616,000	152,289,000	1	5
106	Aquaculture	4,320,000	3,869,000	3,357,000	3,126,000	3,126,000	17,798,000	3	13
107	Human Nutrition	0	0	0	0	0	0	0	1
108	Food Safety	52,458,000	52,344,000	54,787,000	49,953,000	58,627,000	268,169,000	5	22
203	Air Quality	262,000	261,000	0	0	0	523,000	0	0

Program	Description	FY 2010	FY 2011	FY 2012	FY 2013	FY 2014	Total	Appropriation	Animal Research
204	Global Change	307,000	306,000	0	0	0	613,000	0	0
206	Manure and Byproduct Utilization Water Availability and Water	9,411,000	9,205,000	0	0	0	18,616,000	0	2
211	Management Climate Change,	80,000	241,000	0	0	0	321,000	0	0
212	Soils, and Emissions	525,000	0	261,000 1,154,000	241,000 1,089,000	391,000 1,089,000	1,418,000 4,041,000	0	0
213	Biorefining Agricultural and	355,000	354,000	0	0	0	25,620,000	0	0
214	Industrial Byproducts Pasture, Forage and	0	0	8,888,000	8,236,000	8,496,000	41,822,000	1	2
215	Rangeland Systems Agricultural System Competitiveness and	9,001,000	8,820,000	7,750,000	7,418,000	8,833,000	1,688,000	1	4
216	Sustainability Plant Genetic Resources, Genomics and Genetic	426,000	377,000	311,000	287,000	287,000	1,012,000	0	0
301	Improvement Crop Protection and	204,000	204,000	203,000	188,000	213,000	647,000	0	0
304	Quarantine Quality and Utilization of	324,000	323,000	0	0	0	45,127,000	0	0
306	Agricultural Products Bioenergy and	10,143,000	9,515,000	8,855,000	8,307,000	8,307,000	1,602,000	1	4
307	Energy Alternatives	802,000	800,000	0	0	0	1,207,692,000	0	0
	Total Research in Support of Animals	258,803,000	245,020,000	236,562,000	218,601,000	248,706,000		22	100

Program	Description	FY 2010	FY 2011	FY 2012	FY 2013	FY 2014	Total	Appro-priation	Animal Research
	ARS Total Appropriation	$1,179,639,000	$1,133,230,000	$1,094,647,000	$1,016,948,000	$1,122,482,000	$5,546,946,000		